AF679249

Iron and Steel in Ancient Greece

Iron and Steel in Ancient Greece

Artefacts, Technology and Social Change in Aegean Thrace from Classical to Roman Times

Maria Kostoglou

BAR International Series 1883
2008

Published in 2016 by
BAR Publishing, Oxford

BAR International Series 1883

Iron and Steel in Ancient Greece

ISBN 978 1 4073 0239 3

COVER IMAGE *Attic red-figured cylix depicting Thracian peltast on the* tondo *(Robinson 1931,* Corpus Vasorum Antiquorum *2, plate III)*

BAR Publishing is the trading name of British Archaeological Reports (Oxford) Ltd. British Archaeological Reports was first incorporated in 1974 to publish the BAR Series, International and British. In 1992 Hadrian Books Ltd became part of the BAR group. This volume was originally published by John and Erica Hedges Ltd. in conjunction with British Archaeological Reports (Oxford) Ltd / Hadrian Books Ltd, the Series principal publisher, in 2008. This present volume is published by BAR Publishing, 2016.

Printed in England

BAR
PUBLISHING

BAR titles are available from:

BAR Publishing
122 Banbury Rd, Oxford, OX2 7BP, UK
EMAIL info@barpublishing.com
PHONE +44 (0)1865 310431
FAX +44 (0)1865 316916
www.barpublishing.com

Preliminaries

Abstract

Science based archaeology has had a profound effect over the past twenty years on the study of technology, the process which results in material production and requires knowledge of the physical properties of the material, chosen techniques of manufacture and successful application of them. The realisation that science alone is insufficient to find answers to complicated human phenomena came soon after, along with the application of interdisciplinary approaches to archaeology. On one hand, ignorance of the implications and limitations of archaeological methods confuses scientists. On the other hand, archaeologists remain sceptical about the application of scientific techniques in archaeology, arguing for more social oriented input into science-based archaeology. Archaeologists have not attempted consistently to ask questions about the social role of the material. They often take the role of the material as given because of its natural properties or its function. They do not attempt to explain why one given aspect of material culture and not another is used to express certain social relations. Since the 1970s anthropologists, ethnologists and sociologists have emphasised how techniques can be understood as cultural choices that depend on the social, economic and ideological setting of technological acts. Therefore, technologies can only be understood if the analyses consider the wider socio-cultural context in which those technologies are embedded. This work examines the iron metallurgy of Aegean Thrace as a culturally generated activity and on both a local and regional scale. The methodology used consists of a holistic approach, which combines:

1. Archaeological evidence coming from recent excavations in the area.
2. Analytical data coming from the analyses of industrial waste (bloom, slag) and iron artefacts (weapons, agricultural and domestic tools) with optical (metallographic microscope) and chemical analyses (SEM-EDAX micro analyser) and
3. Documentary and iconographic evidence.

The aims for this work are:

1. To establish the level of effectiveness in iron technology as it was practised in Aegean Thrace from Classical to Roman times.
2. To understand the indigenous involvement in iron production.

3. To interpret the social context of iron technology. Do changes in technology (in the form of innovation or invention) cause changes in society and vice versa? What happens when one of these factors stays the same?

The long-term aim of this project is to determine whether there are aspects of technological, economic, artistic or symbolic uses of iron, which are reflected, in culturally distinct groups living within the same area (in this case, Thracian versus Greek or Roman). Finally, an attempt is made to provide a theoretical and methodological model for the archaeological and archaeometallurgical study of interaction between ancient technology and society.

Acknowledgements

Different people and institutions contributed to the successful completion of this work. I am deeply intended to Prof. A.B.Knapp for his supervision, for useful and always inspiring discussion on archaeometallurgy and archaeology, and for his assistance in difficult times. I would like also to thank Dr. E. Photos-Jones and Dr. R. E. Jones, University of Glasgow, for discussing various issues of my analytical work. Dr. P. van Dommelen and Dr. M. Given for providing me with a better and broader understanding of Mediterranean archaeology. Prof. A. Hendry, Department of Mechanical Engineering, University of Strathclyde, for kindly providing access to lab facilities and introducing me to the study of metallography. The lab technicians, Mr. J. Kelly, Metallographic Suite, and Mr. D. MacMahon, lab of Metrology, University of Strythclade, and Mrs Lesley Farrell, Department of Archaeology, Glasgow, for teaching me how to take good care of the instruments and tolerate or solve practical problems in the laboratories.

Dr. Y. Bassiakos, Lab of Archaeometry, National Centre for Scientific Research " Demokritos", Athens, for sharing with me his knowledge on the mining areas of North Greece and taking me to some of them. I would like also to thank the geologists of the Institute of Mineralogical Exploitation (IGME) in Xanthi for allowing me to use their library and copy unpublished reports.

I would like also to thank Mr. D. Triantafyllos, 19th Ephoreia of Prehistoric and Classical Antiquities, (19th EPKA) Komotini and the Central Archaeological Committee of the Ministry of Culture for permitting the study of the material; and the archaeologists of 19th EPKA P. Tsatsopoulou and K. Kallintzi, as well as the staff and personnel of the archaeological Museums in Komotini, Kavala, and Avdera for their assistance and friendship during the months of recording and sampling the material under study.

This thesis could not be possible without the financial support of various institutions. I am grateful to the 'A.S.Onassis' Public Benefit Foundation for honouring me with a three year Ph. D. scholarship and for additionally assisting the cost of my field work visit to Bulgaria; the Carnegie Trust for the Universities of Scotland for funding my field work in Thrace in 1999 and the study trip to Princeton University in 2000 to contact library research; and the Historical

Metallurgy Society- London, for financial support to attend seminars and deliver papers to conferences abroad.

Nothing of this would be possible without the love of my family. My graditute will be always with my mother Eugenia for teaching me the value of hard work and with my father Ioannis for teaching me to believe in dreams. I would like also to thank Kostas for giving me the courage to work for my dreams and for much more; my friends from Greece, especially Anastasia, Dina and Zoe; my friends in Glasgow for introducing me to the pub culture and making the endless rain days sunnier; and my friends in Manchester for tolerating me during the last stages of this work.

Abbreviations

A.A.A.	Archaeologika Analekta ex Athinon.
AD	Archaeologiko Deltio.
AD	Ante Domine.
A.E.	Archaeologiki Ephemeris.
A.E.M.Th.	To Archaeologiko Ergo sti Macedonia kai Thraki. Thessalonica.
Al	Aluminium.
AVD	Avdera
Au	Gold.
BC	Before Christ.
BCC	Body Centred Cubic
C	Carbon.
c.	Circa.
^{o}C	Degrees Celcius.
Ca	Calcium.
CVA	Corpus Vasorum Antiquorum.
Cu	Copper.
E	East.
E.A.E.	To Ergo tis Archaeologikis Etaireias.
EPKA	Ephoreia Proistorikon kai Klassikon Archaeotiton.
Ed (s)	Editor (s).
et al.	Et al.
etc	Etcetera.
F	Ferrite.
FCC	Face Centred Cubic.
Fe	Iron.
Fig.	Figure.
gr	Gram.
Hv	Vickers Hardness
i.e.	For example.
IGME:	Institouto Geologikon kai Metaleutikon Ereunon, epexigimatiko teuchos tou metallogenetikou charti tis Elladas, Athens 1983.
K	Potassium.
KL	Kalyva-Kastro.
kl	Kilogram.
km	Kilometre.
km^2	Square kilometre.
M	Martensite.
m	Meter.
m^2	Square meter.
MES	Messemvria-Zone.
Mg	Magnesium
Mn	Manganese.
N	North
Na	Sodium.

Ni	Nickel.
O	Oxygen.
P+	Pearlite.
P	Phosphorus.
p.	Page.
Pb	Lead.
PBG	Mixed sulphurous minerals.
P.A.E.	Praktika tis en Athinai Archaeologikis Etaireias.
S	South
SEM	Scanning Electron Microscopy.
Si	Silicon.
TTT	Time-Temperature-Transformation diagram.
vol. Thrace	edited by the general Secretariat of the Region of East Macedonia and Thrace, Thessalonica 1994.
X	Magnification (times).
W	West.
wt %	Weight per cent.
Zn	Zinc

List of Figures in the text

List of tables in the text.

Chapter I. Introduction.

Chapter one provides a description of the landscape and geomorphology in order to introduce the reader to the area under study. A short historical background based on the existing literature is also included in an attempt to illustrate the main conventions of the current interpretations and set the background for this research.

I. 1. Setting the scene for the study of Aegean Thrace.

Aegean Thrace called the northeastern part of the Modern Greek state, lies between the rivers Nestos and Evros, the Rodhope Mountains and the Aegean Sea. In antiquity the boundaries and the population of the Aegean Thrace varied. According to Herodotus (V, 3), Thrace in the 5th century BC extended from the Aegean Sea in the south to the Istros (Danube) River in the north and from Macedonia in the west to the Euxine Pontus (Black Sea) in the east (Fig. 1). This territory attributed to the first Macedonian Meris, after the geographical division imposed by the *"Synedrion"* of Amphipolis in 167 BC; it is in this extended territory that Roman administrative policy created the province of Thracia in AD 46. This extended territory is now under the borders of three states, namely, Greece, Bulgaria and Turkey. The treaties of Berlin, Sevres, and Lausanne from1885 to1923 (Vakalopoulos 1994, 231-241) divided "Thrace" into three different states (Fig.1):

a. *Roumelia or Northern Thrace* (now Southern Bulgaria).
b. *Eastern Thrace* (now European Turkey).
c. *Western Thrace* (now the northeaster part of Greece, known as *Aegean Thrace*).

Today different people live in the area: Greeks, Turks, Pomaks, Armenians, and Greek immigrants from the former Soviet Union.

Archaeology in the Balkans, as indeed elsewhere, is nationalistic and played a vital role in the *ethnogenesis* of the countries. Archaeology in Bulgaria, having a central role under the State emphasised the greatness of the Thracians who became the famous ancestors of the Bulgarian nation. In an attempt to establish historical unity and continuity a new subject called 'Thracology' has been created to promote the picture of the great lost past. Bulgarian archaeologists (Hoddinott 1981, 10-15) proposed more extended boundaries as far as Transylvania in the north, Slovakia in the east, and Moldavia in the west. The same author also used terms such as "Thracian nation" (Hoddinott 1981, 158) implying the existence of a nation in modern terms in antiquity.

Archaeological research in Greek Thrace started systematically ten years ago. After the travellers of the 19th century who visited the area (Leake 1841), the first notable study and collection of sculptures, inscriptions, and coins was created and housed in the Greek Litreary Society of Constantinople (1861). Other societies (such as The Thracian Phi-Educational Society in Raidestos, 1871) and individuals (such as A. Stamoulis, a businessman from Sylemvria) started to collect antiquities from Thrace (Bakirtzis 1994, 151). Antiquarian activities stopped during the Balkan and the World Wars, except the discovery of some antiquities during military construction work. G. Bakalakis (1959) was the first archaeologist in the area, whose work laid the foundations for the study of Greek Thrace. The first excavation began at Avdera by D. Lazaridis (1971) in 1950 that remained in the area until 1966. After 1970 the foundation of the archaeological museum of Komotini and the Ephoreia of Prehistoric and Classical Antiquities (EPKA) of Thrace gave a new dimension to the research of the area. The main purpose was to record and to protect the sites, but not until 1981 did the excavations at Avdera, Maroneia, and a little later at Messemvria-Zone change to a systematic character. After 1987, with the support of the European Union, archaeological research in Thrace expanded. Both excavations and surveys are still in progress.

Greek archaeology in Aegean Thrace had to justify the Hellenic character of an area, which was the last one to be added to the independent Greek state in the early 1920s, has been under Bulgarian domination during the Balkan and the world wars, and which had a mixed population. Excavation was concentrated in the coastal Greek colonies. The short coastline was colonised intensively in 8th and 6th century BC by people from Athens, the Aegean Islands (Thassos, Chios, Samothrace), and Asia Minor. These colonies flourished from the 5th to the 3rd cent BC. All these people are considered as Greeks as opposed to the indigenous population of the area who in the ancient Greek and Latin literature are known under different tribal names and under the collective name the Thracians. The same conventional names for the people (Greeks, Romans, and Thracians) will be used in this work in order to avoid confusion but the differences in their meaning will be clarified according to their archaeological or historical context. The term Thracian is also used in its geographical meaning. Arguments based upon architecture, pottery, sculpture, jewellery and the ancient literature came to claim that the indigenous population of the Aegean Thrace has been either assimilated or been pushed to the interior, to the Rodhope mountains and far beyond leaving the Thracian coastline to the Greek colonists. This process according to the existing literature took place in the following stages:

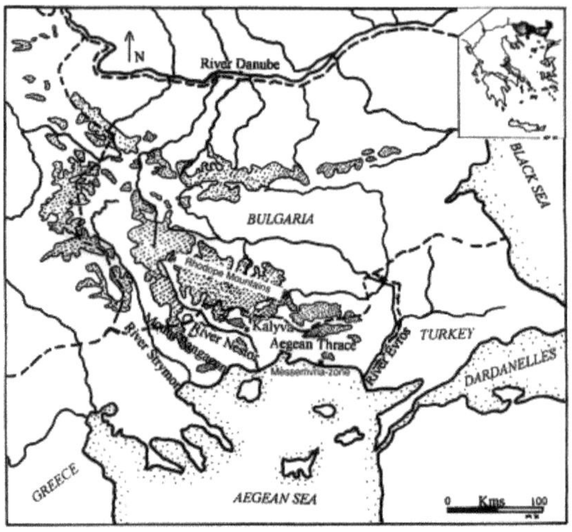

Fig.1. Location map showing the territory of ancient Thrace and the area of Aegean Thrace within (schematic drawing by Maria Kostoglou based on Casson 1968).

Conventional historical framework and interpretations.

Dispersion of Thracian tribes (1200-800 BC). Their migration to the territory of ancient Thrace took place towards the end of the Late Bronze Age and the beginning of the early Iron Age. Iron Age settlements, tombs and monuments testify to their presence in the area. The literature refers to this as the "first colonisation" (Sakellariou 1980, Samsaris 1980).

Founding of Greek colonies (800-550 BC). This is known as the *"second colonisation"* and took place in two phases (Triantafyllos 1994, 54-87):

c. 800-650 BC. The metropolis was a small rural town with an aristocratic elite. The large clans of noble landowners owned almost all the fertile land, with the result that within a short period of time, a population of landless peasants was created which sought new territory to cultivate. In this early stage, Klazomenai (in the coast of Asia Minor) founded Avdera next to the river Nestos c.750 BC. It seems that the indigenous people did not allow the new colony to survive for long because towards the end of this phase colonists from another city in Asia Minor, Teos, established the second settlement with the same name. Koukouli-Chryssaanthaki, who excavated the archaic part of the city, bellieves that this was probably a period when the old (Klazomenians) and the new (Teians) colonists lived together (Koukouli-Chryssanthaki 1988, 53-54). The continuity between the two phases of colonisation is based on the recent archaeological finds in the archaic cemetery of Avdera, which was in continuing use from 8th to 6th century BC. To the same period bellongs the founding of Maroneia by Chios (one of the Aegean islands). No evidence for this early phase of the history of the city was uncovered during the recent excavations.

The colonising process proved difficult. Greeks met the Thracian tribes, which occupied and defended their territories. The conventional history of Avdera tells of many wars between the colonists and the Thracians (Isaac 1986, 291).

c. 650-550 BC. The metropoles realised the profit from their colonies and began to settle "commercial" colonies (emporia). These are more mercantile in character, and they were founded in areas of key communication and near sea routes. To this phase bellong the Samothracian colonies between Maroneia and the Evros delta, amongst which the best known from recent excavations is Messemvria-Zone. Many colonies of the first phase, such as Avdera and Maroneia, developed into trading centres, although not discarding completely their agricultural character.

The Persian campaign in Thrace. Documentary evidence certifies that the Thracians were friendly to Darius, 491 BC, and to Xerxes, 480 BC (Isaac 1986, 89-90).

Founding of the Odryssian State, after 480 BC. This state was founded by King Teres after the Persian wars in 480-460 BC. Teres was said to be the one who managed to unite the many Thracian tribes under his rule, but it was his son Sitalkes (455-424 BC) who succeeded in spreading the frontiers of the kingdom from the River Strymon to the Euxine and from the Aegean to Danube. He organised an army and cavalry, laid many roads for the development of trade and favoured the settlement of Greeks and helped the Hellenisation of Thrace. Along the coastline many Greek colonies, such as Avdera, had to pay heavy taxes to the Dristan State (Isaac 1986, 96). The Thracian elite imitated Greek culture (Archibald 1998). The Odryssian State lost its power due to the military campaigns of Philip II of Macedon.

Philip II in Thrace, 346-342 BC. Philip II captured cities in Thrace, amongst them Avdera and Maroneia, but he did not succeed in moving further inland. He developed treaties with the Odryssian State. In his time the Odryssae lost their power and the dispersion of the Greek language and civilisation to the interior of the Balkans began (Hammond 1972, vol. II, 264-267, 554-566).

Alexander the Great and the Thracians. No clear historical evidence exists. However it is known

(Hoddinott 1981, 122) that some of the Thracian tribes like the Agriani and Trivalli, took part in Alexander's campaign. The archaeological evidence from this period, mainly epigraphic and numismatic, suggests that most of the material used in this period and the language are Greek.

Lysimachos in Thrace, 306 BC. After Alexander, Lysimachos declared himself King of Thrace and built a new capital, Lysimachia (in Propontis). From this time onwards Thrace had a provincial character.

Roman Period (146 BC - AD 330). The important events of this period are the beginning of the Roman possession of Thrace (146 BC) when Avdera and Maroneia were declared free, and the creation of the Roman Province of "Thracia" by Emperor Claudius in AD 46 (Avramea 1994, 135). The ancient and modern geographical, ethnological, and political subdivisions of the area under study, are reflected on the recent literature. This established historical interpretation is monolithic and approaches acculturation as a one-way process, ignoring interaction between people and ideas. Because scholars (Isaac 1986; Hoddinot 1981 and 1975; Triantafyllos 1994) have studied only well defined modern geographical entities or well-known populations such as the Greeks or Romans, no comprehensive treatment of Ancient Thrace currently exists. Moreover, comparisons between the different parts of Thrace appear difficult because of the difference between the chronological systems used in Greece, Turkey and Bulgaria. For the same reasons, this work deals with the part of ancient Thrace today bellonging to the Greek State and comprising the prefectures of Xanthi, Rodhope, and Evros. Despite the fact that they are part of the same geographical and historical region, the prefectures east of the river Strymon (Kavala, Drama, Serres) and the islands of Thassos and Samothrace, are excluded from this work because of the administration policy for study permits.

I.2. Landscape and geomorphology.

Western Thrace occupies the north-east part of Greece, and, as mentioned previously in this chapter, its frontiers are the Rodhope mountains in the North, the Aegean Sea in the South, the River Nestos in the West, and the River Evros in the East; its surface covers 8578km^2. From the northern frontier the Thracian terrain comprises three parallel zones (Fig. 2) of increasing altitude (Delopoulos 1994, 351):

- Lowland (altitude until 200m) 4139km^2 or 48.3% of the total surface (green on Fig. 2).
- Upland (altitude 200-600m) 2156km^2 or 25.1% (light brown on Fig. 2).
- Highland (altitude more than 600m) 2.283 km^2 or 26.6%, where the mountainous pine and fir forests are situated (dark brown on Fig. 2).

The climate in the Lowland and Upland zones is characterised by extreme differences in temperature: frost and rain during the winter and spring months, and dry during the summer characterise Lowlands whilst the Highlands receive even a greater amount of rain and snow. The Thracian plains are highly fertile with a rich hydrological system. Most of the plain (62.2%) is situated West of the River Evros.

The distinctive geographical feature of western Thrace is the Rodhope Mountains with the highest points being the Papikion and Ismaros mounts (Fig. 2), which are made up of metamorphic and igneous rocks. Other characteristic rocks are: (a) crystalline schist (Rodhope, Ismaros), (b) Eocene, Oligocene and Neogene Sediments, (c) Holocene deposits in plains of Nestos, Evros, Ardas rivers, (d) Platonites (mainly granites), (e) Tertiary volcanic rocks. The subsoil contains:

(a) Mixed sulphurous mineral (PBG in Fig. 3) of iron pyrites, galena, zinc and mineral of the oxidation zone in the region of Thermes and Kalyva (Fig. 2 and 3).
(b) Sporadic copper (Cu in Fig. 3) and magnetite (indicated as Fe in Fig. 3) at Kimmeria (Fig. 2 and 3).
(c) Iron ores (Fe in Fig. 3) Southeast of Maroneia (Fig. 2 and 3).
(d) Mixed sulphurous (PBG in Fig. 3) ores at Kirke (Fig. 2 and 3).
(e) Manganese ferrous ores (Mn in Fig. 3) in the area of Kalyva and Messemvria-Zone (Fig. 2 and 3).

The evidence for possible exploitation of the above minerals in antiquity will be discussed in Chapter II. 2, iron working in Greece. These sites will also be mentioned often in chapter V during the discussion of the analytical data. As can be seen from maps 2 and 3 the areas with minerals related to iron are:

- Kimmeria and Thermes, north of Xanthi in the hinterland of Avdera.
- Kalyva and the region in the upper valley of the River Nestos.
- Maroneia on the coast and Kirke in the hinterland of Messemvria-Zone.

In Fig. 3 the wider region of each site (Kalyva, Thermes, Kimmeria, Maroneia, and Kirke) is indicated with a rough circle around the name of the site. These regions also include important woodland and water resources. These regions are artificial for the purpose of clarity and they will be used as such during this work. The River Evros rises from Bulgaria and enters the open gulf of

Ainos (East of Evros) into two branches, where it forms a large fertile delta. It stretches 204km into Greek territory and it has two tributaries, the River Ardas in the North, and the River Erythropotamos nearby Dydymoteicho (Fig. 2). The River Nestos and the ancient city of Avdera founded, for the first time in the 8th century BC on the Thracian coast (see chapter IV), define an area where large geomorphologic changes occurred. According to recent geological investigation (Syrides 1996, Psilovikos and Syrides 1992), palaeogeographic and palaeoenvironmental evolution followed 4 stages: Stage I. During the Holocene (i.e. post – Pleistocene/Ice age) the sea coastline was several kilometres north of the present coast, approaching the margins of the hilly terrain. A peninsula composed of metamorphic rocks (gneiss-schist) dominated the area. Ancient Avdera was built up during the 8th century BC on this peninsula, and in the West Coast some port facilities were constructed.

Stage II. The enormous supply of clastic sediments from the neighbouring Nestos Delta in the West caused a gradual sediment deposition and swallowing of the sea, shifting the coastline southwards and forming marshlands and coastal lagoons. The alluvial deposits between today's Nestos route and Avdera, which occurred after Herodotus' times, perhaps explain the absence of surface prehistoric archaeological evidence in this area (there is no report at any Archaeologiko Ergo, Archaeologikon Deltion, Praktika tis en Athinaes Archaeologikis Etaireias, between 1970 and 1996).

Stage III. The gradual filling up of the area continued. Along the present coastline an extensive (12km long) offshore sand bar was formed, parallel to the coast. Communication with the open sea was cut off and marshlands and lagoons started to form again. According to archaeological criteria (Koukouli-Chryssanthaki 1988, 46), this stage can be probably dated to the 3rd to 2nd centuries BC and is related to the decline of the city (see chapter IV). An oblitreated port, as a result of extensive silting, may be the main reason of this decline.

Stage IV. Extensive marshlands dominated the area and the coastline approached the present day coastline. The end of this stage is marked by the intensive human impact with extensive reclamation of the marshlands for cultivation causing rapid change of the morphology.

The present day coastline between the rivers Nestos and Evros is straight with many small lagoons. Because of its special features (Delta of the Rivers Nestos and Evros, Lake Vistonis) it is an important wetland zone (Nature 1987, 1993) protected by European law since 1983 (treaty of Bern, L.1335/1983).

A wetland environment is characterised by the presence of water with a depth less than 50m, usually next to delta or marshlands formations. Their special ecosystem offers an ideal home for many species of flora and fauna (see Appendixes). Some of them are endangered species. Birds (such as ducks, swans, and coots) that specialise in seeking food in wet terrain utilise the flooded summer tundra, marshes and swamps, lake and seashores (Alerstam 1990, 68). Almost 200 different species of birds live in Thracian wetlands, amongst them 36 species of rare eagles (Love 1989, 26-29).

The ancient Thracian coastline however was different. In the case of Avdera we know that the sea extended further north (stage I), before the alluvial deposits of the river Nestos shift the coast southwards forming lagoons and marshlands. Moving due east, the coastline in the area of ancient cities of Maroneia (Fig. 1 and 2) and Messemvria-Zone (Fig. 1 and 2) would have extended further south. This is an assumption based on the archaeological data, because no geomorphologic study of these areas exists. The Department of Antiquities of Thrace (P. Tsatsopoulou) and the Department of Underwater Archaeology (C. Kazazis) began a project for the study of the ancient coastline in Messemvria-Zone in September 1996, but it is still unpublished. The excavated parts of Maroneia and Messemvria-Zone have shown that the southern part of the ancient town walls (in both cities dated in the 4th century BC) is destroyed. Both cities are built next to the sea. Pottery and building stones coming from the present day coastline suggest that a small part of the towns may be now in the sea. Moreover, the sandstone quarries in Messemvria-Zone lying today in the sea bottom, support this theory (see chapter IV).

It has been suggested that the Thracian coast offers no good harbourage (Boardman 1980, 226-243; Carry 1949, 293) because it is straight and strong N-NW winds blow in the area. The only natural bay in this area is in Lake Vistonis, one of the largest Greek lakes. Today the lake is separated from the sea by a narrow strip of land (map 2), which would not have existed in the Byzantine Period (after 4th century AD) when a city called Poroi (Bakirtzis1994, 157) was built there. Poroi means "pass" and the city took its name for its location on the straits leading from the open sea to Lake Vistonis. Although no earlier archaeological evidence exists we have a terminus *ante quem* (4th century AD) to indicate some geomorphologic changes in the area.

Fig. 2. The landscape of Aegean Thrace with sites mentioned in the text (adapted from Avramea 1993)

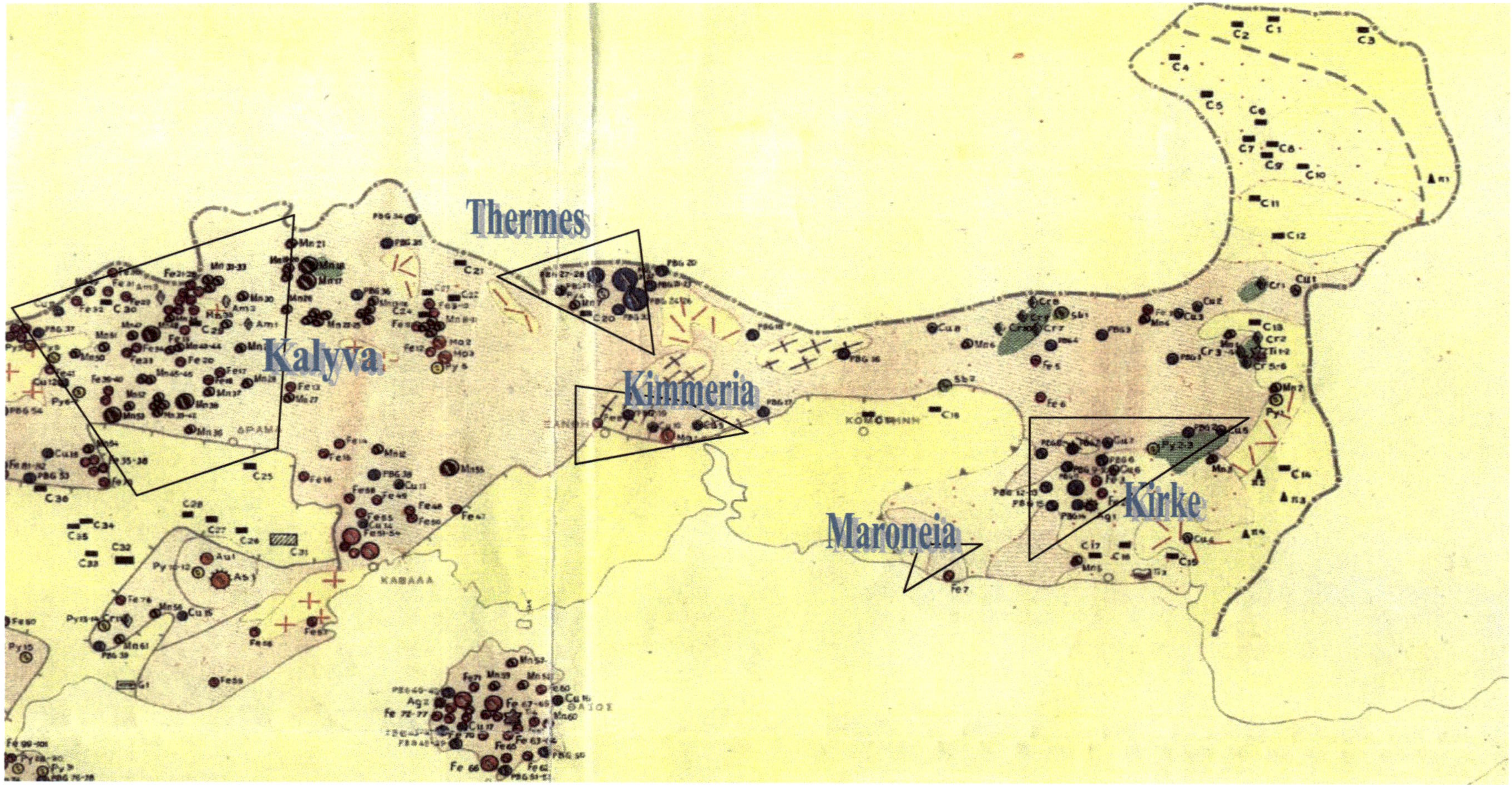

Fig. 3. Metal occurrences in Aegean Thrace (adapted from IGME 1973).

I. 3. Routes of communication.

The land frontier that marks off Balkan lands from Greece is difficult to cross at many points. It has been argued that in ancient times only the coastline fringe came within the purview of the ancient Greek World and much later Roman armies succeeded in opening up the western and northern highlands (Carry 1949, 293-294; Avramea 1994, 138-139). Most of the information about the road network comes from Roman times, but we can assume that the same roads were in use earlier. Romans organised the road system by repairing and extending the already existing one. Roman milestones (*miliaria*) found in several places indicate the distances between the different stations (*stationes*) that in turn verify the use of the roads.

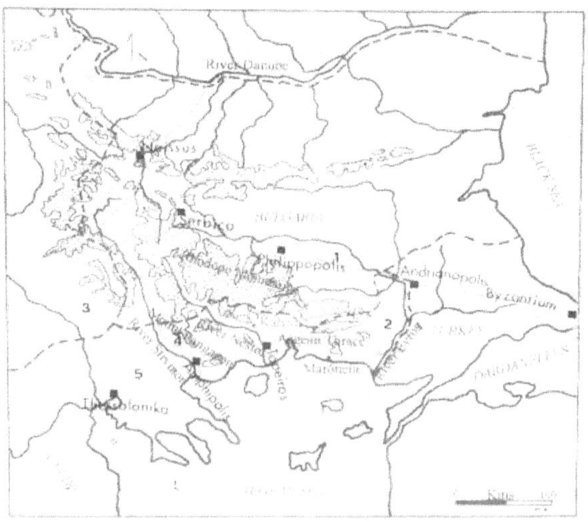

Fig. 4. Routes of communication between the Aegean and the Balkan hinterland (based on Hupchick 1996).

1. Naissus (Nish) - Serbica (Sofia) - Andrianoupolis - Byzantium (*the diagonal).*
2. Naissus (Nish) - Serbica (Sofia) - Philippopolis - Thracian coast (next to the River Evros).
3. Naissus (Nish) - Thessalonica (next to the River Axios).
4. Serbika (Sofia) - Amphipolis (next to the River Strymon).
5. Thessalonica - Amphipolis - Topeiros - Byzantium (Via Egnatia)
6. Road next to the River Nestos.

1. The route from Naissus (present - day Nish), in the upper Morava valley, to Serbica (present-day Sofia) which descends to the basin of the river Evros and the plain of south-east Thrace; from here it is easy to travel as far as Byzantium (Konstantinoupolis, present day Istanbul). This road, called *the diagonal* (Carry 1949, 298) has been the main line of communication between the European mainland, the Mediterranean, and the Near East (Fig.4).

2. The route from Naissus (Nish) to Serbica (Sofia) and via the West part of the River Evros valley to the Thracian coastline plains. Thracian Iron Age settlements found along this route (Fig. 3) suggest that this road has been used as early as the Early Iron Age by Thracian tribes during their dispersion due South. Furthermore, according to Strabo (fragment 48), the River Evros was navigable from its seaport Aenos (in the East, now in Turkey) up to Philipopolis. Despite the absence of archaeological evidence, navigation earlier than Roman times seems probable given the stable geomorphology of the River Evros route.

3. The route from Naissus (Nish) to Thessalonica (next to the River Axios).

2. The route from Serbica (Sofia) to Amphipolis (next to the River Strymon) is part of this chapter because it could explain the early presence of Thracian tribes in the region of Macedonia (Hammond 1972). Silver Thassian coins (Pantos 1980, 391) found in the upper route of the River Strymon valley, support the hypothesis that the same route was one of the most important for the trade between the Aegean Sea and the Balkan hinterland.
3. Via Egnatia, the Roman conversion of the old route leading from the Adriatic to the Propontis, the Bosporus and Asia Minor, was the most important W-E inland route. The route leading from Dyrachion and passing through Thessalonica, Amphipolis and the Thracian city Topeiros is located on the West shores of the River Nestos valley traversed the Thracian plain and terminated at Byzantium. Most of the Roman cities were founded along this road, in contrast to the Greek colonies (Fig. 2) that were located next to the sea to control major trade routes in the Aegean. In these earlier times the W-E communication in lowland Thrace should have been generally easy because no impassable natural limits exist in the area.

4. Finally, another route might have existed next to the river Nestos valley linking the Aegean with the Danube. Casson (1926, 9) notes that the river valley of Nestos could not have afforded passage in antiquity. According to Herodotus (58, 108-109) the Persian army did not pass the river in spring of 480 BC. However, there is a pass east of the river through the hills and the Rodhope Mountains (Isaac 1986, 74). This is an important modern military highway from Xanthi to Bulgaria and

perhaps is the road mentioned by Thucydides leading to Danube from Avdera. Although the sources of information about this road need more examination, it seems that the river Nestos valley offered an important land route to the north with clear economic significance. The archaeological evidence support this hypothesis:

(a) The city-state of Avdera was located next to the river probably to exploit part of this road.

(b) Thassian silver coins found in the upper Nestos valley (Pantos 1980, 391) prove the importance of the road from as earlier as the end of 6th century BC.

(c) Philip II built two forts (Aeriko and Kalyva, Triantafyllos 1994, 92) on both shores of the river (Fig. 2) to protect and control this strategically important route.

(d) The Roman city Topeiros was built on the West bank of the river (Fig. 2) where the present day bridge exists. Occupation in Topeiros continued up to Byzantine period. The historian Procopius in the 6th century AD (Bakirtzis 1994, 158) says that the Slavs in AD 549 "dominated the city of Topeiros completely". It could be assumed that the Slavs followed the road next to the river Nestos as well.

In conclusion, we can say that the main routes at all periods for inland communication in the Balkans are the two parallel arteries:

1. The diagonal, and the Via Egnatia, linking the West and the East, and

2. The North-South oriented routes between them, which follow the river valleys (Evros, Nestos, Strymon, Axios).

The identification of possible routes of communication in the highland zone remains a problem. We do not know how the Thracians established settlements in the Rodhope Mountains. Even today it is difficult to find a passage between the mountains and one can visit the archaeological sites only with the assistance of locals who have a good knowledge of the area. Excavation in the mountainous zone is limited (Triantafyllos 1972, Tsimpidis-Pentazos 1971).

Communication by sea is also difficult because of the geography of the Thracian coastline. The northern end of the Aegean is the least inviting part of that sea. On the other hand, strong N - NW winds such as the *Boreas* (Greek name for the N wind) make it difficult for navigation. It was on this coast that Odysseus first became a victim of wind and wave (Homer, *Odyssey* ix, 67) and that the Persians lost their first invading armada (Herodotus VI, 44, 2-3). Boardman (1980, 226-245) and Carry (1949, 300) talking about Greek colonies on Thrace such as Avdera and Maroneia, have argued that none of these had good harbourage and none were founded for maritime trade. According to new archaeological data, this does not seem to be the case. Maroneia (Triantafyllos 1994, 71) had a harbour and Avdera had two harbours (Triantafyllos 1994, 62). Coins found as far as Egypt and Syria (Koukouli-Chryssanthaki 1988) and imported pottery from Corinth, the Aegean Islands and Asia Minor found in the excavations of the sites, indicate that the economy of these colonies flourished due to their long distance trade activity (Boardman 1980, 226-245). The main maritime roads linked the Thracian coast with the coast of Macedonia, the Aegean Islands, the Mediterranean Sea, the coast of Asia Minor, and the Black Sea, and along with the inland road system form a complex network for the circulation of people, goods, and ideas over the centuries.

Summary: chapter I has introduced the landscape and the geomorphology of Aegean Thrace with particular reference to the areas with metal occurrences. It also presented the conventional interpretation of the historical and archaeological data in order to set the background of the present study. As can be noted the current literature is dictated by (a) environmental determinism (coast versus hinterland, difficult inland crossing and harbourage, direct the relationship between Greeks, Romans and Thracians); (b) cultural historical assumptions based on uncritically used Greek and Latin sources; and (c) colonial biases about the predominance of Greek material culture. Chapter II investigates the historical causes behind these ideas and sets the theoretical and methodological aims of this work.

I. 4. What this study will do and how it will proceed.

The first chapter, as mentioned earlier, introduces the landscape and the geomorphology as well as the conventional historical framework of the area under study. A short description of both the inland and maritime roads linking the Aegean Sea with the Balkan hinterland is also included in order to provide better insight into the routes of communication and interaction between people and cultures.

Chapter II is divided in three sections. Chapter II.1 deals with current theoretical issues related to the study of material culture and cultural identity. Both these issues are crucial in the discussion about technology and the cultural change that follows. Cultural change interpreted as 'Hellenisation' or 'Romanisation' and the assumed technological superiority of some cultures in the present historical/archaeological record are strongly questioned in the light of post-colonial theory. Chapter II.1 concludes with the description of the theoretical framework suggested by the present work on the archaeological and archaeometallurgical study of ancient technology in general and iron metallurgy in particular.

A short presentation of the technological processes involved in iron making follows in chapter II.2 where a discussion of the current state of the study of iron metallurgy is included. The final part of this section, chapter II.3, presents the methodology used for data collection, sampling and analysis of the material used in this research as well as the aims of the study being undertaken.

Chapter III presents the documentary evidence. Greek and Latin sources have unavoidably been used, given the absence of direct Thracian written evidence. A critical approach has been applied. Ancient authors provide interesting information about Thracian socio-political structure, life and customs, describing a society based on prehistoric structures of organisation. References related to the Thracians as warriors or to their involvement in mining and other stages of metal production are particularly discussed. Although the ancient authors have a rather incomplete and sometimes confused geographical and cultural knowledge, because they rarely had visited the area personally, the extensive use of their texts in the archaeological literature makes a critical re-evaluation of these texts more than necessary. The ultimate aim of this chapter is to discuss the issue of Thracian identity versus the Greek or Roman and the Thracian involvement in iron production.

Chapter IV presents the archaeological data from both surveys and excavations in the area. References are made to:

a. Preliminary reports (published in the *Archaeologikon Deltion, Archaeologiki Ephemeris, Praktika tis en Athinaes Archaeologikis Etaireias* from 1970 to 1996).
b. Papers in local conferences (*Archaeologiko Ergo sti Macedonia kai Thraki* 1987 - 1999, *Kavala and its Region* 1988) or international meetings (*2nd Symposium of Thracian Studies*, Alexandroupoli 1992).
c. Excavation reports. These are mainly hand-written excavation diaries and the archaeologists working in the area kindly have provided access to them.

No final publications of the excavated sites exist so far. The volume *"Thrace"* (published by the region of eastern Macedonia and Thrace in 1994) has also been used because it is the only existing publication. It covers the history of the area from the early Iron Age to modern times and, although very descriptive, it presents most of the recent archaeological finds. Nevertheless, excavation reports have been used to extract contextual information related to the finds under study. Chapter IV presents data on the typology, distribution and deposition of the archaeological finds under study. Discussion and conclusions are drawn about (a) the economic and social patterns of iron production as these were practised on local and regional levels; and (b) about the visibility of the cultural identity of the people in the settlement patterns, burials and other archaeological finds. It is argued that the discontinuity in material culture and cultural identity suggested by the theories of 'Hellenisation'/'Romanisation' is a product of culture historical approach imposed on archaeology of the region. It is based on the questionable assumption that some cultures (and the material related to them) are superior. The archaeological record is open to alternative interpretations. The technology of iron can provide some crucial evidence for studying social changes and transformations of identities from Classical to Roman times.

Chapter V presents the analytical data from the study of both the industrial waste and the artefacts by site in order to illustrate the technological processes involved from the mining to the finished products. Technological choices related to the choice of ores, fuel, mining, processing, smelting and manufacturing techniques are discussed in an effort to understand the level of effectiveness in iron technology, in order to investigate their social implications. Technological innovations and/or inventions apparent from the analytical data presented in this chapter form an important part of this research; they help to establish the kind of changes occurring in the iron technology of the area and to establish if they are local or imported developments. After establishing the level of technology in each site, Chapter V concludes with the interpretation of iron technology, in the region from Classical to Roman times, arguing for the existence of continuous local technological traditions.

Chapter VI, interpretation, brings together all the data presented in the previous chapters. The analytical data are compared with the archaeological and textual in order to see if they contradict or complement each other. The technology of iron production at a local (Messemvria-Zone, Avdera, Kalyva-Kastro) and regional (Aegean Thrace) level is discussed in the light of the archaeological context to extract information about the production, distribution and consumption of iron in Aegean Thrace from Classical to Roman periods. Technological continuity or discontinuity is evaluated along with other changes noted in the archaeological record in order to see if changes in technology are part of an entirely new socio-economic structure or not, and if technological choices are culturally oriented. This chapter concludes with the discussion of the existence, meanings and transformations of Thracian identity in antiquity from Classical to Roman times. Vital clues are thus provided about 'who is who' in Aegean Thrace and whose metallurgy we are studying. General conclusions are drawn about the social, economic, artistic and likely cultural meaning of iron in Aegean Thrace from Classical to Roman times.

Chapter II. Theory and methodology.

Chapter II includes three sections. Chapter II.1 (the study of material culture) explores uses and interpretations of material culture, cultural identity and Classical archaeology in the light of current theoretical concerns and post-colonial discourse. These issues are crucial in understanding the biases of current literature, as these were noted at the end of chapter I. They are also crucial in determining the inter-relation between technology and cultural change that is the main question of this thesis. Chapter II.1 concludes with the description of theory and methodology that combines the theoretical concerns of the previous section with the scientific limitations imposed by the nature of iron making described in the next section.

Chapter II.2 (the process of iron making) introduces the scientific background of this study. This section outlines some basic principles and methods of studying iron. Sections 2 and 3 of chapter II are essential for understanding how the theoretical/methodological model of chapter II.2 was constructed and how the analytical data are interpreted in order to determine the type(s) of iron technology practised in Aegean Thrace (chapter V). Finally, the last part, section II.3 presents the aims and methodology employed throughout this thesis.

II. 1i. The study of material culture and cultural identity.

The study of material culture.
Archaeology by its nature deals with material remains and people who no longer exist. The concept of culture provided the connecting link between people and material (artefacts) for most of the archaeological work of this century (Shanks and Tilley 1992, 117). In this scenario, artefacts as well as other types of archaeological evidence, such as settlement patterns or burial practices are first classified according to their style in terms of size, form, and decoration. Then, they are grouped in stylistic groups and largeger families. On a step further, specific styles are connected with specific groups of people (cultural, but more often ethnic). Finally, the distribution of artefacts across a geographic area is explained by the physical distribution of people. This leads to the assumption that similarities in material culture mean close contact between groups, continuity in terms of style means also cultural (and even biological) continuity, and discontinuity can separate clearly epochs, cultures and people.

The term culture has been perceived differently by generations of scholars. Its development was also different in Britain and America where scholars were interested in various approaches (from culture history to functionalism or cultural ecology). In contrast, Continental, Eastern European and the Mediterranean countries had no interest in other approaches but only in archaeological cultures and the distribution of these cultures across a geographical area. History and archaeology, as well as geography, developed in these countries with little or no integration with theories developed in Britain and America (Hodder 1982, 2). Political and linguistic barriers kept the countries of Eastern Europe closed to Western ideas. Here I will refer mainly to British and American scholars to illustrate how culture is perceived and how it has affected archaeology.

For American anthropology culture has been understood as a mental structure of rules and procedures similar to language (Jones 1997, 46-47). Geertz put an emphasis on the symbolic meanings within the society, which are shared by people during well-organized public events (Jones 1997, 65). Thus, from the primordial point of view culture is not simply a mental state but it is created and exists through symbolic rituals, ceremonies etc. Furthermore, social philosophers argued for the importance of practice. Culture it is not in peoples' minds but it is in their everyday life. For Bourdieu's *habitus*, culture is built up by individuals in order to structure proper activities - personal versus family, private versus public, sacred versus profane (Jones 1997, 88-90; Hodder 1986, 73-79). For practice theory culture thus is a historical dynamic and continually evolving phenomenon. This theory allows cultures to develop during time and provide a better explanation of social changes. It can also close the dichotomy between functionalistic approaches where human agency is subordinated to environmental determinism (Jones 1997, 116-117) and structuralistict approaches where "an abstract code of meaning appears to exist before action and divorced from context" (Hodder 1989, 71). In all the previously described theories, culture is expressed and shared by people through material. Material is the connecting link between people and archaeology. Materialization has been defined as 'the transformation of ideas, values, stories, myths and the like into a physical reality that can take the form of ceremonial events, symbolic objects, monuments and writing' (Earle 1997, 155). It is this process which creates and contains culture codified usually as ideology. In this framework, material objects are ideal signifiers of social positions and relationships because they are 'portable, exchangeable, controllable (in terms of their production and circulation), and they can be owned, inherited, and transferred' (Earle 1997, 155). Some scholars are very critical about the use of material culture theories. Some argue that material culture studies take as self-evident that objects were produced and used by specific cultural groups or peoples and therefore reflect almost directly these groups. Others argue that these theories also neglect a symbolic meaning that the material might carry. Hodder defined as symbolic every 'object or situation in which a direct or litreal meaning also designates another indirect, secondary and figurative meaning' (Hodder 1982, 11). Finally, others such as behavioural archaeologists say that in material culture studies artefacts are subordinated to a cultural frame 'acknowledging objects but denying the materiality of human life' (Schiffer 1999, 6). Scholars are still debating the meaning of material culture. However, in the post-modern world, social archaeologists agree that:

"past 'meanings' of material goods were never fixed in the objects themselves but, rather, were 'read' into them from other experiences and expectations, and acted upon: social life is therefore played out against material conditions which are themselves revised or reworked according to an individual's interpretations of some material reality" (Knapp 1996, 150).

Provisional conclusion: Therefore, multiple and often alternative readings of the same material are more than possible since its meaning is not fixed. Although scholars are still debating the meaning of material culture, what unite most theoretical concerns (from Braudels *'long duree'* and Bourdieu's *'habitus'* theories to 'contextual archaeology' (Shanks and Tilley 1987) and to the archaeology of agency (Dobress and Robb 2000) is that, when it comes to the production and consumption of artefacts, people and groups have choices. These choices have socio-cultural significance (Lemmonnier 1993; Herbert 1993; Lechtman and Steinberg 1979) that can better be understood if studied on a micro scale, time and place specific context (Dobress and Robb 2000; Lightfoot 1995). It is therefore argued that the study of any given technology, from the actions and choices of its operators to the use and consumption of its artefacts, provides a solid case study for the long debated relationship between people and materials, society and individuals.

Archaeological cultures.

In archaeology, the term "culture" found its triumph in the culture historical approach, which has provided the dominant framework for the archaeological analyses throughout most of the world during the 20th century. It is beyond the scope of this work to place culture-history into its' placement in the history of archaeology. However, the historical conditions under which it came about are pivotal in understanding the evolution and use of this much-debated concept. Briefly, dissatisfaction with economic crises and growing foreign competition during the Industrial Revolution led to a greater sense of racial grouping; a conservative view of social change became more popular promoting ideas about the evolutionary and biologically determined nature of the society (Trigger 1989, 150-162). Layered onto this came the archaeological 'culture' that by the late 19th and early 20th centuries had began to be linked with the concept of an ethnic group identified by certain styles of material at a specific time in the archaeological record and at a specific geographical location; it was not until Kossinna's settlement archaeology in 1911 that this concept was systematically applied in an archaeological study (Trigger 1989, 163). Kossina's, racially orientated, views of culture groups provided a turning point in a move from evolutionary archaeology to one accounting for geographical variation.

Under culture-history, British archaeological research was formed around Vere Gordon Childe's views of society that were different from Kossinna's (Thomas 1996, 20-25). From the ecological setting of Childe and the role of economic factors of Clarke, to the system function of Binford (Trigger 1999), culture always connects material with people. The main assumption underlying the culture historical approach is that bounded uniform archaeological material is correlated with particular people, ethnic groups, tribes or races (Jones 1997, 63-64). It is thus assumed that "culture" is made up of a set of shared ideas and belliefs, all results of continuous cultural tradition; material homogeneity is regarded as a product of regular contact and interaction while discontinuities reflect physical or cultural changes. Diffusionism is in the core of the argument. There is an advocation that cultural ideas would transfer because the superiority of the high culture was obvious to those societies coming into contact with it.

Cultural identity

In most of the recent studies about cultural identity (Jones 1997; Graves-Brown et al. 1996; Shennan 1989), ethnicity is perceived as the central component of cultural identity. For purposes of clarity, it is necessary to define terms such as ethnic group, ethnic identity, and ethnicity (Jones 1997: XIII) because all of them are often used in this text.

Ethnic group: any group of people who set themselves apart and/or are set apart by others with whom they interact or co-exist on the basis of their perceptions of cultural differentiation and/or common descent.

Ethnic identity: that aspect of a person's self-conceptualisation which results from identification with a broader group in opposition to others on the basis of perceived cultural differentiation and/ or common descent.

Ethnicity: all those social and psychological phenomena associated with a culturally constructed group identity as defined above. The concept of ethnicity, according to Jones (1997, XIII), focuses on the way in which social and cultural processes interact with one another in the identification of ethnic groups. This concept has been much discussed (Jones 1997, 88): from the Primordialist focus on the essential importance of ethnic identity (Frederic Barth) to Instrumental theories (Manchester School, A. Cohen), where greater emphasis is placed on the difference between culture and ethnicity, and to *habitus-praxis* theory (Bourdieu 1977, Bentley 1987) which focuses on a more fluid and dynamic type of ethnic identity. It is therefore clearly understood that ethnicity has been a controversial issue. However, for the purposes of this work an effort will be made to describe some of the factors, which determine an ethnic group. The oldest "definition" of an ethnic group can be found in Herodotus (VII. 144), who summarised all genetic, linguistic and cultural aspects of ethnicity with reference to Greeks of the 5th century BC:

"the kinship of all Greeks in blood and speech and the shrines of gods and the sacrifices that we have in common, and the likeness of our way of life".

According to Renfrew (1996, 130) Herodotus' definition is close to that of the modern anthropologists:
"Ethnos...can be defined as a firm aggregate of people, historically established on a given territory, possessing in common relatively stable particularities of language and culture, and also recognising their unity and differences from other similar formations (self-awareness) and expressing this in a self-appointed name (*ethnonyme*)".
(Dragadze 1980, 162, cited in Renfrew 1996,130)
Ethnicity thus can been seen as based on a number of factors such as (Dragadze 1980, 162, in Renfrew 1996, 130);
1. Shared territory or land;
2. Common decent, "blood" as in the ancient Greek term;
3. A shared language;
4. A community of customs, or culture;
5. A community of belliefs, or religion;
6. A name (*ethnonyme*) to express the identity of the group; it is important that they have given this name to themselves;
7. Self-awareness, self-identity; ethnicity is what the people in question bellieve it to be;
5. A shared history, or myth of origin.

Finally the distinction between ethnic groups and nations should be made clear. It is argued that "nations" are distinct from ethnic groups by virtue of their political independence in the form of a sovereign state. Nevertheless, both ethnic and national identities are subjective constructions rather than "natural" ones (Jones 1997, 6).

The question of ethnic identity in antiquity came to be of crucial importance in a number of recent publications on Classical archaeology. Scholars, such as J. Hall, accept that ethnic groups can be identified in ancient Greece on the basis of the characteristics described by Herodotus (Hall 1997, 19-33), although he accepts that ethnic identity is neither static nor monolithic (Hall 1995, 8). He bellieves that ethnic identity can be more evident in burials where archaeologists can study both the cultural (funerals are a direct expression of a society's belliefs) and physical dimensions of identity. He draws attention to the contribution by physical anthropology and particularly pathology because dietary habits can be connected to ethnic attitudes towards food (Hall 1997, 111).

Since Greek culture has found its expression both in texts and in material, defining Greekness might not be as difficult as trying to define other groups. Ancient groups such as the Thracians were non-litreate. They are mainly known from Greek and Latin sources. To distinguish historic truth from propaganda and rhetoric in ancient literature (chapter III) is of critical importance because it is more likely that these texts reflect Greek and Roman representations of others, rather than the self- perceptions of these other groups (Diaz-Andreu 1996, 49). This adds to the importance of archaeological evidence in the attempt to create a social history of the area and its people despite the fact that a particular view about the pre-dominance of the Hellenic high culture in antiquity is inherent (Whitley 2001, 26-37). Culturally distinct archaeological material can include:

"city-states, with a known autonomous territorial unit, self-governing, issuing their own coinage, having a common spoken language and written script, practicing the same religious, political or military structure" (Cherry 1987, 155)

These characteristics can be found in the traces that the Greek colonies have left on the Thracian coastline. These traces define what could be referred to as the Greek culture in Aegean Thrace. On the other hand, the Thracians, in terms of their political, linguistic or ethnic characteristics, can be classified as another "culture" known mainly because of its difference from the Greek civilisation. Thracian customs, rituals, eating and drinking habits, clothing, language, music (chapter III), are described in ancient Greek literature only to stress the 'barbaric' character of the people and its civilisation. Therefore Greeks and Thracians appear to be two distinctive cultures at least for the first centuries of their coexistence. After the 5th century BC only traces of Greek culture can be seen in the area until the time Romans come to Aegean Thrace. At this stage the researcher should either:

(a) Accept the necessity of cultural and/or biological assimilation of the Thracians by the Greeks or the Romans (conventional history of the area); or
(b) Question the existence of identity in antiquity, its expression in material culture and the meaning of terms such as Greek, Thracian, and Roman during the different periods. In continuity with the apparatus of this research I follow the second option.

One way to refer to the identity of these people is to use the previously defined terms but always being aware of their connotations and modern political implications. This thesis discusses transformations of ethnic identity (based on origin, language etc.) towards a socio-political identity (based on peoples experiences and belliefs) juxtaposing traditional ideas to modern instrumentalist approaches. It should be also made clear that terms such as Greek nation or Thracian nation are constructive entities created to provide the symbolic evidence of a nation's historical continuity and hence its authenticity (Woolf 1996, 15). In this respect, the choice of the Classical Antiquity with the political system of city-state rather than the multi-national Byzantine Empire was more appropriate to the dominant nationalistic ideology of the newborn Greek state influencing the Greek archaeology of almost all 20th century (Hamilakis and Yalouri 1996; Kotsakis 1991).

Cultural change, acculturation and Greek archaeology.

Greek archaeology has been used as a "spectre of history" and has been "intimately involved with a two-century-old project of understanding 'Europeanness'"(Morris 01994, 8). For Ian Morris the crisis in Classical archaeology occurred because of "the gradual disappearance of the social arrangements...which had made Hellenism an important academic discourse. In the contemporary post-modern world of the new Europe and post-colonial international relations, it no longer does. So the Classical disciplines as a whole and Greek archaeology in particular have been left without adequate intellectual justification" (Shanks 1996, 108). In the context of current approaches to archaeology Classical archaeology is marginalized. The problem is more complicated than the juxtaposition between the 'traditional' and 'new' approaches to Classical archaeology. Some scholars tried to trigger the problem by studying the history of the discipline, its practice and its institutionalisation (Shanks 1996; Morris 1994). These studies, although very critical, are more introverted and try to understand the past of the discipline and its development in the specific historical and cultural contexts. Others try to apply well-tested methods of post-modern archaeology, such as regional survey, to the study of Classical or Roman past (Alcock 1996) creating alternative interpretations. Narratives based predominately on the archaeological finds and their interpretation are rare, with few exceptions (Hall 1997; Morris 1994; Shanks 1996, Snodgrass 1987, 1982) who combines archaeology and history in an attempt to create a social history.

However, most of Classical archaeology and its practitioners never escaped from the culture historical frame where stylistic studies and textual evidence dominate the record. The consideration of archaeological cultures drove to the creation of monolithic explanations such as 'Hellenisation' or 'Romanisation'. The above theories regard cultural change as a one-way process leading to the assimilation of weaker cultures (such as those called barbaric, the Thracian in this study) by the stronger ones (such as the 'Hellenic' and later the 'Roman'). Ancient literature also is interpreted along the same lines. Thus, the Classical and Roman past is full of invasions, colonisation, migrations of people with distinctive cultures and archaeological remains, which are expected 'to speak to us in the same clear language as that of historical events,' (Snodgrass 1987, 38).

Post-colonial theory has strongly criticised the biases of traditional approaches. First of all it is acknowledged that representations of Classical and Roman antiquity are strongly influenced by modern colonial conceptions of the Western world, successfully applied in antiquity (Dietler 1998; van Dommelen 1998, 22-24; Hingley 1996, 35; Mattingly 1996, 53; Webster and Cooper 1996).

Secondly, collective terms, such as 'the Romans' or 'Romanisation' are used to describe a specific culture or acculturating process in terms of material and people. In the culture historical framework Romanisation (as indeed Hellenisation) is perceived as acculturation based on the assumption "of a one-to-one correlation between culture and ethnic identity, and the idea that cultural contact and conquest result in the rapid transmission of cultural traits and ideas" (Jones 1997, 34). Terms such as Romanisation imply an artificial uniformity that it is not evident in the archaeological record (Barret 1997); they exclude the variability of the process of acculturation (Hingley 1996, 42) and they are incapable of accommodating local agency (Dietler 1998, 288).

In other words, local people and cultures appear to be passive receivers with no other choice than to accept the new culture. The relationships between the Greco-Roman world and local cultures were better understood under the center-periphery model. The theory of centre-periphery often associated with world systems analysis was first used in politics and geography to explore the relationship between different parts of the world in a capitalist economy (Shanks 1996, 160). In this model the two worlds, namely the Mediterranean civilised city-states and the 'barbaric' periphery were interdependent in terms of supply and demand for raw materials and slaves. The centre-periphery model is not yet accepted as fully explanatory. It imposes strong oppositions such as:

(a) Greco-Roman/civilized versus indigenous/barbaric (Webster 1996); and
(c) Town/consumer versus country/supply (Dietler 1998, 295-299; Champion 1989; Rowlands *et al.* 1987).

The centre-periphery model minimises the variety of human and societal relationships and neglects the different roles that material can play in this relationship (see section II.1 on material culture). Today more archaeologists have become aware of the fact that people (both as individuals and as members of a group) have and make choices that may be seen in the material record (Dobres and Robb 2000).

The third point of criticism derives from the use of colonialism and/or imperialism to explain cultural change. In the first case we should include the Greek colonization of the area, and in the second the Roman annexation. 'Colonialism' implies exploitation by the ruling power through a relative small number of local agents (Bernstein *et al.* 1992, 171). Also, 'colonialism' and 'colonies' generally denote all instances of direct political control of people by a foreign state irrespective of the numbers of settlers (Webster 1996, 5). Thus no \matter how many Greeks or Romans actually came to the region they were enough to change the local culture.

Although aware of the criticism and of the colonial biases carried in terms such as Greeks, Romans and Thracians, in this study I have decided to use the previously mentioned terms. The creation of a new terminology to describe past social relationships, although challenging, is beyond the limits of this work and of the expertise of the author. Furthermore, the non-recallable absence of minds and social relations of the producers and users of material culture (Thomas 1996, 25) would make the utility of new terminology again limited. During this work I use the established terms in a time and place specific context avoiding generalisations and being aware of historical fallacies underlying their use in archaeology. Therefore, the terms Greek and Greeks are used for colonists (assuming that they arrived in large numbers of settlers and after one-two generations settled in the area). The terms Roman and Romans will be used in their imperial meaning, assuming that they arrived in the province of Thracia in small, mainly military units (no settlers, or merchants). The term Macedonians is used to describe the people of Macedonia under Philip II during the brief military occupation of some Thracian forts such as Kalyva-Kastro. Aditionally, the terms, Thracian and Thracians are used either in their geographical meaning or to describe the indigenous people of the region.

Finally, an extra difficulty stems from the fact that more recent studies of 'Romanisation' refer to the Western provinces of the Roman Empire (Randsborg 1991, 166-68; Millet 1990, Woolf 1990; Young 1990). Unfortunately, there is not a single study for the east. The story of ancient Thrace of the Roman province of Thracia and even of a smaller region of it (Aegean Thrace) does not fit either with the West nor the East (chapter IV). How then can we study the impact of the Roman rule in the archaeological record of the region? What kind of processes are involved in the cultural changes noticed in the region? What forms did Roman rule take in the region, what agents employed, in what material was manifested? And why is technology a good medium in facilitating this study?

II. 1ii. The study of technology and Science based archaeology.

The above discussion is concerned with the meaning of material culture; however, material has another aspect, its materiality, which had not been appreciated before the introduction of scientific techniques into archaeology in the early 1960s. More than two decades later, archaeologists and scientists today disagree strongly that science alone cannot answer the difficult questions. Today more complicated techniques are applied and long tables with analytical data are still produced (for a good account see Knapp and Cherry 1994, 15-40). However, the new methodology created more questions than it answered. The realisation that science alone is insufficient to find answers to complicated human phenomena came soon after along with the application of interdisciplinary approaches to archaeology. On the one hand scientists are confused because of ignorance of the implications and limitations of the archaeological methods (Knapp 2000; Killick and Young 1997). On the other hand archaeologists seem to remain sceptical of the application of scientific techniques in archaeology and they argue for a more humanistic input into science-based archaeology (Muhly 1995; Vaugham 1995; Dunnel 1993). The present study hopes to be another step towards that direction.

II. 1iii. The archaeology and anthropology of technology.

Archaeologists dealt with the study of technology in an indirect way to accomplish other objectives such as putting sites in time or space, reconstructing past practices or identifying status differences (Speth 1992, vii). In studying technology and its products they form stylistic groups and they are captured by the form or design of the object. As previously noted (chapter II.1.i) most archaeological studies have been devoted to the artefacts, which have been used by archaeologists of the culture historical approach to explain behavioural phenomena in the past such as movements of people, inter-group interaction or boundary creation and maintenance. However, artefacts consist of only one part of technology. Technology is a social phenomenon and can be studied only in this framework. Archaeologists did not attempt to ask questions about the social role of the material. Often they take the role of the material as given because of its natural properties or its function. They do not attempt to explain why one given aspect of material culture and not another is used to express certain social relations (Lemmonier 1992, 1-4). Since the 1970s mainly anthropologists, ethnologists and sociologists have emphasised how techniques can be understood as cultural choices, which depend on the social, economic and ideological setting of the technological act. Therefore, technologies can only be understood if the analyses consider the wider socio-cultural context in which those technologies are embedded (Lemmonier 1993, 1992, 1986; van der Leevw 1993).

Another aspect of technology, which is particularly relevant to the present work, is the fact that most of the technologies practised by a given society are interrelated; they may share the same actors, the same places, and the same tools. Sometimes a given technique uses as raw materials the results of other techniques. For example, a carpenter, a miner or an artist can use iron tools that are the product of the activity of blacksmiths. In this context, technologies involved in objects made from metal, clay or glass can be connected since they all incorporate a good knowledge of pyrotechnological processes. The pattern that emerges here is that technologies cannot only be regarded as interrelated but as interactive also. Thus, change or innovation in one technology can cause change to others as well. The study of technology should consider if there is any relationship between different technologies and between technologies and other social phenomena, for example building or burial practices.

II. 1iv. Theoretical model: putting ancient technology in a social context.

The elements of the study of technology so far discussed in this chapter are closely related to the interpretation of iron technology in Aegean Thrace, which this thesis aims to provide. My methodology takes into account two major disciplines, namely archaeology and scientific techniques applied to the study of ancient metallurgy. Looking for inspiration in my attempt to put the technology of iron in its socio-cultural context, I also consulted anthropological studies of technology by Lemmonier (1992, 1993), studies of African Iron production (Herbert 1993) and Andean metallurgy (Lechtman 1984).

The study of technology involves first the study of processes. One way to do that is by looking at the sequence of technological acts involved at every step of the transformation of the raw material to the usable product. This series of steps has been described as 'chaine operatoire' (Leroi-Gourhan 1964 and 1965). At each step there is a series of choices involved, in terms of choice of ores, fuel, furnace types, smelting/smithing techniques, tools, products).

In an attempt to elucidate the cultural significance of the technological choices, which constitute the technology under study, the archaeologist must ask why one choice was made instead of another. (S) He also has to consider alternative choices available to the people practicing this technology at a given time.

Questions are asked about both the physical properties of the materials used and their cultural context. In the study of any given technology there are five areas where technological choices can be seen (Sillar and Tite 2000, 4):

1. Raw materials including choice of iron ores, clay and stones for the building of the furnaces, skin for belllows, water or oil for quenching.
2. Tools used to process the raw materials, for example, saws for wood cutting, spades for charcoal, hammers for ore crushing, vessels for ore washing, belllows, moulds, anvils.
3. Energy sources for the transformation of raw material to consumable product, namely, fuel (grass, charcoal, coal) for roasting, smelting or smithing, water or air to drive belllows, people to perform the technological acts.
4. Techniques used to coordinate all the different actions in order to achieve the particular product, for example, mining and preparing the ore for smelting, preparing charcoal, firing the furnace, collecting the bloom, smithing the bloom to a shape appropriate for trading, manufacturing artefacts.
5. The sequence in which all the above acts take place, including the number of individuals involved in each action, the time and the location of acts, the order of the techniques used, the type and the distribution of products and by-products.

Technological choices can be influenced by a number of factors associated both with the physical properties of the materials involved and the cultural context in which these materials are embedded in terms of the social, economic, and ideological background of the people involved. Since archaeologists rarely can identify individual acts within a technological framework, we accept that the previously described areas are related to the technology that a given group of people or a given society practices at a specific period of time or over a longer time span. The following table summarises the theoretical and methodological model used during this research

Chapter II. 2. The process of iron making.

II.2i. Bloomery iron.

Due to the absence of data related to excavated furnaces, belllows, charcoal pits and other related material from the under study region, this work draws conclusions about iron metallurgy in Aegean Thrace by studying only smelting / smithing slag, bloom and artefacts found in safe archaeological contexts in Messemvria-Zone, Avdera and Kalyva-Kastro. Despite the limited evidence, the excavated material has two large advantages: (a) it comes from safe stratigraphic context; and (b) it is well dated. It is generally acknowledged that industrial waste such as slag cannot be dated on the basis of its morphology. One other option is absolute dating (by C_{14}). This is possible only in the case of charcoal preserved in the slag and none of the slag sectioned preserved any. Dating is the most serious problem related to large-scale evidence (mining tunnels, slag heaps) preserved on the landscape of north Greece. Section II.2.iv presents and discusses old and new landscape evidence in north Greece. The mining landscape of north Greece although important in the general discussions about metallurgy in Aegean Thrace (chapter VI), cannot be used for direct comparisons and correlation with the archaeological data because has been not yet studied systematically.

A metallurgist might consider this material be incomplete. It is also the case that more analyses could be carried out in order to determine with higher accuracy the scientific details of the metallurgical processes involved, but this is beyond the scope and time of the present work. Since the aim of this thesis is to determine the social context of iron production, the focus remains on the archaeological data and their interpretation. Because this study is time and space specific, the analytical work focuses on the excavated material that is the only surviving direct evidence. Laboratory generated data are used as a methodological tool sufficient to answer the main technological questions of this study, as stated in the previous section, and to set the background for future research.

As will be demonstrated in chapter V, the analytical data from all three sites suggest that iron was produced locally by the bloomery process. It is therefore necessary to introduce the main principles of the process so the reader can appreciate both the limitations and the advantages of the analytical work undertaken.

Bloomery smelting is the reduction of an iron ore to solid metal with charcoal to make wrought iron. The main iron ores used in antiquity for the production of iron are the following:

a. Hematite, iron oxide (Fe_2O_3): is a major mineral ore of iron, which is also found as an accessory mineral in many rocks.
b. Magnetite (Fe_3O_4): one of the most common ferromagnetic minerals; it is an oxide mineral with a spinel crystal structure.
c. Limonite ($FeO.OH.nH_2O$): Hydrated iron oxide mineral.

d. Laterite: Residual deposit of iron and aluminium hydroxides formed by the weathering of rocks in humid, tropical conditions.

e. Bog Iron: a soft porous limonite found in freshwater environments.

f. Iron sands: magnetite sands.

Charcoal is the primary fuel. Successful smelting demands charcoal made by hardwoods that would burn slowly at high temperatures. According to experimental data for a successful smelting the fuel to ore ration ranges between 10:1 and 20:1 (Rostoker and Bronson 1990, 97). The use of ores is related to the furnace types. Different variants of the bowl and the shaft furnace (Fig. 5) is bellieved to be used in pre-industrial iron technology. In both, the ore is reduced to metal by direct contact to charcoal. The operation is described belllow. The necessary high temperatures can be achieved with the appropriate air supply. One or more tuyères can be used for this purpose and they are usually located above the bottom of the furnace in order to leave enough room for the formation of slag and metal. It has been suggested (Tylecote, 1962) that natural-draft smelting furnaces were used in prehistoric Europe. Most of the recorded examples are 2m high with an average internal diameter at tuyères level of about 1m. There are also smaller examples with 1m tall shafts and 0.60m pits. The whole structure was placed over a tunnel, into which the slag was tapped by knocking out a plug in the base of the pit. The ore and the fuel were placed into the furnace from the top. The type with a possible use in antiquity is the slag-tapping furnace. The furnace consists of a shallow clay basin with a diameter of 0.50m or larger. The ore must be roasted and there is only a single tuyère. The slag was tapped periodically through a side hole into a pit dug against one side, and the iron was obtained in the form of many small pieces or in a single bloom.

Fire-resisted stones make furnace buildings and the interior is lined up with clay. Clay is also used for the construction of tuyère. Both furnaces and tuyère are used only for one smelting. The removal of the bloom destroys the furnace. Tuyères on the other hand cannot be used for more than one or two smeltings since the part inside the furnace usually melts due to high temperatures. Therefore, smelting is an activity with a lot of industrial waste left behind. How much of this unwanted material survives is a matter related to other factors such as recycling and weathering.

The preparation of ore, charcoal and the building of the furnace are processes involving considerable amount of raw materials (stones, clay, wood), labour and cost, and they should be completed before the actual smelting started. Anthropological work in Iran showed that the preparation of charcoal takes five days for two men working: 'it took one day to dig and line the pit in Shiraz, one day to carbonise the wood and two to extinguish the fire, plus an unspecified amount of days to cut the wood and fill the pit. However, they apparently worked a series of pits at once, actually producing 60kg a day, 30kg each' (Horne 1982, 11). There is no direct evidence for ore, charcoal and furnace preparations within the finds under study.

Generally, the process of bloomery smelting includes the following stages:

1.The iron oxide in the ore is reduced to solid, metallic iron by reaction with carbon monoxide gas, released by burning charcoal. The temperature rises to 1100°C -1150°C by forcing a lot of air through the tuyère into the furnace and carbon dioxide is produced ($C + O_2 \rightarrow CO_2$). Carbon from charcoal reacts further with the carbon dioxide produced and releases carbon monoxide ($CO_2 + C \rightarrow 2CO$), which is necessary for the reduction of the metal that follows ($Fe_2O_3 + 3CO \rightarrow 3Fe + 3CO_2$).

2. The solid iron is separated from the earthy material (gangue) that is always present in the ore, by converting some or all of the gangue to liquid slag that could be drained away from the metal. This slag is formed by the reaction between the gangue and part of the iron oxide in the ore ($2FeO + SiO_2 \rightarrow 2FeO. SiO_2$). The lowest possible melting point of slag formed by the reaction of iron oxide with gangue minerals is about 1200°C. During the formation of slag air supply should be slowed down in order to avoid oxidisation of the metal.

3. The metal particles formed by the reduction of the ore were agglomerated into the bloom, which is a mass of iron and slag.

4. The bloom is removed from the smelting furnace and hammered while still soft, to weld the iron particles together and expel enough of the slag to make solid metal. The bloom is a roughly finished metallic product; specifically, it is the spongy mass of iron produced in a bloomery furnace in which the iron is reduced *in situ* and is not molten during reduction. The crude iron bloom must be extensively hammered at red heat to consolidate the iron and remove excess slag and charcoal. The product of this process is called wrought iron. Wrought iron usually contains slag stringers that have been elongated and flattened in the process of working from the bloom.

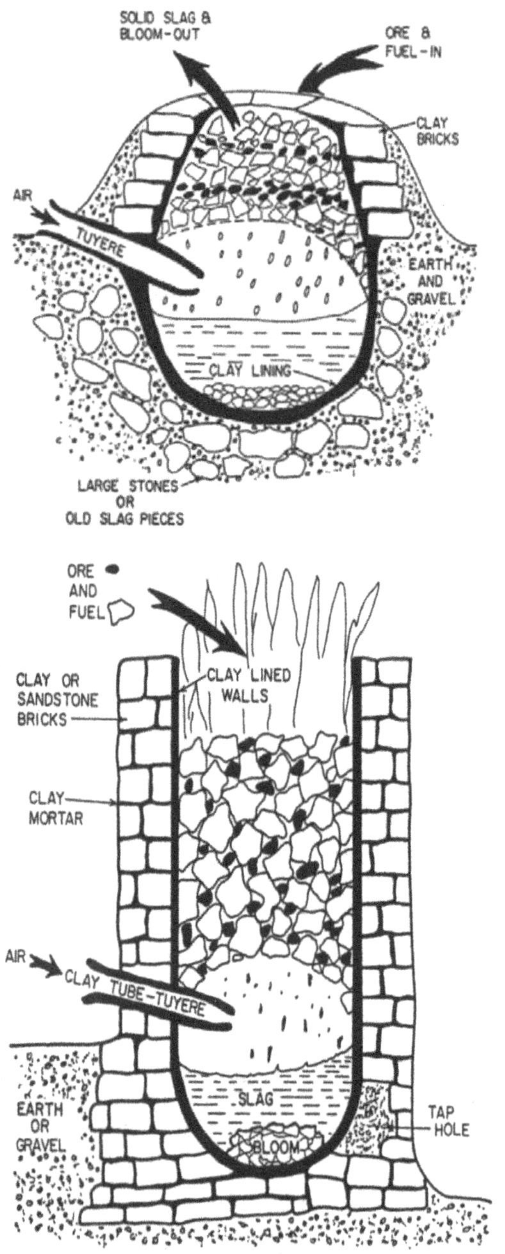

Fig.5. Schematic representation of bowl and shaft furnaces (Rostoker and Bronson 1990, 29).

Bloomeries operated on all the inhabited continents (with the possible exception of Australia and South America), and were in use from the early second millennium BC until the middle of the 20th century (Tylecote 1962). It has been suggested that bloomery iron from natural-draft furnaces was used in prehistoric Europe, Asia, North America and Africa (Killick 1988). Since pre-industrial bloomeries could produce between 4 and 10 litres of slag for each litre of iron (Rostoker and Bronson 1990, 92) slag is the type of industrial waste most likely to survive.

Heating also produces slag, sometimes accidentally, upon organic or inorganic material. Slag would be the by-product of any process involving inorganic material and high temperatures. For example, there is slag produced out of a furnace for firing clay, glass or out of a limekiln. Macroscopic characteristics, such as size, shape, weight, colour, and texture are not enough for the identification of fragments. Nevertheless, porosity and magnetic density are often associated with metallurgical slag.

Fragments of ores, charcoal, and fluxes hardly survive in considerable amount and they can easily escape recognition. Fragments of furnace lining, tuyères, crucibles, hammer stones and crushing plates are rarely found in the archaeological record and most of the time their pure preservation condition does not provide enough information for reconstruction of the process they are related to.

Because of the absence of adequate data, archaeometallurgists turned to experimental smelting in order to learn how the bloomery process works and what material residues are produced. Most of the experiments in bloomery smelting have taken place in Europe (for bibliography see Tylecote and Merkel, 1985). In these experiments, professional archaeometallurgists recorded and measured everything, from temperature to air delivery and gas composition and both raw materials, and smelting products have been analysed by chemical and metallographic techniques. There is a problem, however, with experiments of this kind: professional metallurgists and engineers have an understanding of metallurgy that the prehistoric people did not have. This is why ancient material related to metallurgy is the only direct evidence for smelting. The use of ores is inevitably related to the formation of slag, because slag collects the impurities introduced into the process by the ores, the gangue, the furnace lining, and the charcoal. The purer the ore the less slag will be formed. Some ores have the ideal chemical composition for the formation of slag. These ores are called self-fluxing. If this property is not present no successful separation of the metal will occur. Early metallurgists understood the importance of fluxes in order to produce a good slag, namely a slag of low melting point and viscosity that could take up all the impurities (Bachmann 1982, 9). A considerable amount of metal is also wasted for the formation of liquid slag. Ancient bloomery slag contains up to 70% iron. Their high metallic component leads to their re-smelting in later times.

Morphologically, smelting slags are divided into three main categories: furnace slag, tap slag and cakes of slag. Each of them indicates a different furnace type and a different process of how slag is removed from the metal. Furnace slag is typically formed at the bottom of the furnace and thus it reveals its size and shape; metal is solidified underneath the slag. Tap slag is tapped out of the furnace and it is solidified in pits or small hearths. Cakes of slag are formed when special care for the construction of a pit has been taken. Slag can be produced from smithing as well. Morphologically, fragments of smelting and smithing slag are often indistinguishable. A combination of their macroscopic characteristics with their metallographic and chemical composition (see appendix) can help identification. Slag in metallurgy is considered as an artificial and not natural product. Bachmann (1982, 1) defines slag as 'once molten silicate or silicate mixture, sometimes including oxides, phosphates, sulphides, carbides, pure metals, etc'. In this composition fragments of charcoal, and unreduced ore are also included. Thus the composition of ancient slags is affected by:

1. the ores and gangue,
2. the fluxes added,
3. the furnace construction material (including lining, bricks, temper added to clay for furnace lining),
4. the fuel
5. process conditions (furnace shape and size, heat distribution, air supply, time)
6. cooling conditions
7. weathering (according to time and climatic conditions, and deposition).

Ancient bloomery slag usually consists of the following phases (=areas of homogenous structure and composition):

Wustite, FeO, the most common iron oxide in ancient bloomery slags. It is shaped like the branches of a tree, which are called dendrites.

Fayalite, an iron silicate, $2FeO.SiO_2$, which melts at about 990^0C and usually takes the form of broken, elongated grey laths in silicate based slags. It is the most common component of ancient slags, often occurring as slag stringers in wrought iron.

Interstitial Glass, non-crystalline solid phase. In bloomery slag it is the phase with most of the impurities.

Before ending this section is necessary to briefly discuss the possibility of the production of cast iron in antiquity. Evidence (slag, white and grey cast irons, artefacts) from Messemvria-Zone suggests that the bloomery furnace also produced cast iron as early as the 5th century BC, but most probably accidentaly. The following section will address the issue in order to provide the background to better understand the complications of this discovery that are fully discussed in chapter V (analytical data) and chapter VI (interpretation).

Cast iron

Cast iron is the name used to describe an impure iron with 2-5% carbon in modern terms (Craddock 1995, 239) or with up to 1.9 % carbon and other impurity elements such as silicon and phosphorus in antiquity (Tylecote 1962, 312). Cast iron has two forms, white and grey. They both describe the appearance of the surface exposed after fracturing. In white cast iron most of the carbon exists as cementite (iron carbide with the formula Fe_3C) and it appears white under the microscope (as in Fig. 28a-c in chapter V); in grey cast iron most of the carbon exists in the form of graphite (as in Fig. 28d-h in chapter V), a soft plate or flake- like structure (Tylecote 1962, 312-314). Cast irons are more brittle than wrought iron or steel and most cannot be worked without farther processing (Craddock 1995, 239-240). Another name used to describe cast iron is liquid iron or liquid steel. In this work the traditional terms (grey and white cast iron) will be used because it is beyond the purpose of this research and the author's expertise to re-evaluate and provide clarification in a series of terms used by metallurgist.

The production of cast iron in Western Europe has been associated with the introduction of the blast furnace generally used from c.1500 A. D. (Tylecote 1962, 300) and transmitted to Europe from China (Craddock 1995, 251). The earliest use of cast iron is well documented in China as early as the mid 1st millennium BC (Rostoker and Bronson 1990, 17). Cast iron can be produced in several ways and it can be produced in a bloomery furnace by increasing the fuel to ore ratio. Some African smelting furnaces produced iron rich in carbon (Craddock 1995, 249). The principles of the smelting are the same, but a higher temperature and thus a stronger blast are required. What happens is that iron under 1523^oC is completely liquid and can be almost perfectly separated by the impurities of the ore. With the progressive addition of carbon this temperature can fall to a minimum of 1143^oC (Rostoker and Bronson 1990, 16). Under highly reducing conditions, the iron, in a perfectly liquid state, gathers in the bottom of the furnace. The impurities of the ore along with the fluxes and charcoal also form a liquid slag, which since it is lighter sit on top of the iron. The smelter opens a hall on the furnace wall and first taps out the iron into halls where it is left to solidify. The same happens with the slag. The product of this operation is an iron containing up to 4% carbon. Iron on that stage is in a cementite form and surrounds island of pearlite. Holes are indicative of the casting process. This is called white cast iron and it is not ready for use because it is hard to work with. What follows is further re-heating of the mass of white - cast iron. During re-heating, iron absorbs more carbon that after cooling takes the form of long graphite flakes. The produced mass is ready to be used by the smith. Further re-heating will force the graphite flakes to expel more carbon. The by-product, slag, is different from the bloomery slags from wrought iron production. It contains almost no iron oxide since all iron has been removed completely. Mainly silicate phases are present and some round iron prills indicating a good control of the temperature.

II. 2ii. The structure of iron and iron alloys.

Iron alone has rarely been used in the manufacture of weapons and tools since it is a soft metal. What has been used widely though is the iron-carbon alloy known as steel. Steel displays characteristic properties, considered as metallic. It is a good conductor of electrical current and has a characteristic 'metallic lustre' when it is freshly cut. It is strong and it can be readily formed into practical shapes. Finally, its permanent deformability (ductility) is an important asset in permitting small amounts of yield to sudden and severe loads (Shackelfold 1992, 3). Its metallic properties, with the exception of electrical conductivity, made it an attractive material for many uses in antiquity. Nevertheless, metallic properties are dependent on the microstructure of the material and they can change when the structure changes. Factors such as temperature, time, and pressure during the manufacture of the artefacts can change the structure of the material and thus its properties.

Carbon in iron comes from the use of charcoal during the smelting process and can also be added in later stages while hammering in hot. Carbon also can reduce the melting point of iron (from 1538°C to approximately 1200°C) and give extra strength. For a better understanding of both the metallographic results presented in this chapter and the manufacturing techniques employed in the production of the material under study, it is important to explain the role of carbon in iron. The Fe – C phase diagram is an essential tool. It depicts a number of phases at a range of (Fe-C) composition (relative percentage of carbon and iron) and temperatures. The iron-carbon diagram is not representative of what really happens because carbon forms cementite with iron (Fe_3C) and thus we have the Fe-Fe_3C diagram. In this phase diagram slow cooling conditions, usually in room temperature, are only considered. Microstructure and thus properties also change with different heating and cooling rates.

What makes the control of the process really difficult is the fact that iron changes structure at different temperature. Iron, element no 26 in the periodic table, crystallises in the body centred cubic (bcc) system and its melting point is 1538 ° C. As can be seen in diagram (Fig. 6) under 910°C iron crystallises in the bcc system (see glossary) and it is known as a-iron or ferrite. A typical bcc unit cell can be seen on the site of Fig. 6. There is one iron (Fe) atom in the centre of the cube and one-eighth atom at each of eight unit cell corners. Above 910 °C, a-iron transforms to austenite, known as γ-iron, which has face centred cubic system. As displayed in Fig. 6 in a typical FCC unit cell there is one half atom (i.e. one atom is shared between two unit cells) in the centre of each unit cell face and one-eighth atom at each unit cell corner (Shackelfold 1992, 87-89).

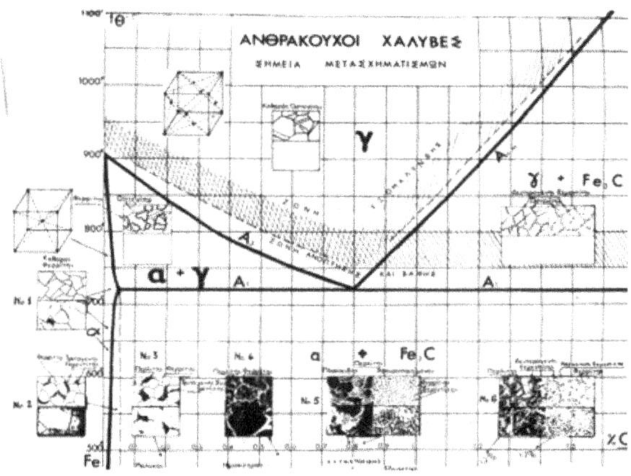

Fig.6. The iron carbon (Fe-C) phase diagram after Varoufakis (1979, 17).

This is a crucial transformation because the amount of carbon that iron incorporates (and thus the quality of steel produced) is depended on its structure. Fig.6 shows the most important phases of this process. Carbon atoms are smaller in diameter than the atoms of iron. In the face centred cubic system (γ-iron), the centre of the unit cell is free and thus the carbon atom can take this place without causing changes to the rest of the structure (Varoufakis 1979; Shackelfold 1992). This is not possible when iron is in its body centred cubic phase (α-iron). A good smith should know, although empirically, how to control this factors in order to produce or repair a steel artefact, since differences in the amount of carbon, temperature and cooling conditions can result in the production of steel with completely different properties. The process of increasing the carbon content of the surface layers of a metal (often wrought iron) by heating the metal bellow its melting point in charcoal is called carburisation. The diffusion of carbon into iron indicates the degree of successful application. The diffusion of carbon in iron is very slow. The smith should keep the iron to be carburised in fine charcoal and at a small distance from the tuyère to avoid rapid oxidation of the charcoal, but not far away so a temperature up to 900°C-1100°C can be obtained with enough consumption of the charcoal.

Under these circumstances, an experienced smith will take five hours to raise the carbon content up to 0.45% at a depth of 2mm belllow the surface (Tylecote and Gilmour 1986, 15). Tylecote and Gilmour suggest that it is easier to do the carburising first (usually in thin layers of wrought iron) and then a piece of steel can be welded to iron (Tylecote and Gilmour 1986, 15). Here, only the phases at specific compositions (0.2 %, 0.8 %, 1.2 % of carbon in iron) will be considered, because ancient steel usually has these compositions.

In the first example (0.2% carbon in Fig.6) the carbon content in a-iron cannot be more than 0.025% (soft steel) in 723°C. The crystal structure is ferrite with large grains (indicating soft metal and slow cooling) and some pearlite between the grain boundaries (formed due to the incorporation of carbon by a-iron). Steels with the above composition are described as soft or hypo eutectoid.

In the second case at 0.8 % of carbon in Fig., equilibrium (eutectic point) is reached at 723° C (eutectic temperature) when the two elements are fully melted in the right amounts. Steel with this composition is very hard.

At 1.2% carbon and beyond and in 1130° C we have γ-iron (crystallized in the fcc system). A typical microstructure of this phase on the solid stage is the lamellar pearlite with secondary austenite between the grain boundaries shown in this kind of steel is known as hyper-eutectoid, is usually very hard and brittle and requires further heat treatment to reduce brittleness.

Another iron alloy used in antiquity is the iron-phosphorus alloy. Until recently, the presence of phosphorus in iron was considered by archaeometallurgists as a large disadvantage since it can increase the brittleness during cold working. Recently, both experimental and analytical work has shown that phosphorus rich iron was deliberately in use in antiquity. In two cases (Africa and England) phosphoric iron has been used for the manufacturing of curved blades such as sickles (Schmidt and Childs 1995). This seems to be the case with Thrace as well. One interpretation could be that phosphoric iron was chosen because it was too difficult and it took a long time to carburise the long sickle blades. In the iron-phosphorus alloys, phosphorus content up to 0.3% induces phase change from ferrite to austenite and back to ferrite (Fe-P diagram). Alloys with more than 0.5% phosphorus can stay in their ferrite structure after cooling. This means that some blades can show ferrite structure but with a higher hardness (Tylecote and Gilmour 1986, 7). Phosphorus has also been detected in the chemical analyses of under study artefacts and it is again discussed in chapter V. The microstructures and their abbreviations discussed in this work, according to Tylecote and Gilmour (1986, 4), are:

F	Ferrite
P	Pearlite
B	Bainite
F + P	Ferrite and Pearlite in more or less equal amounts
P +F	Pearlite with some ferrite
C	Cementite
M	Martensite
TM	Tempered Martensite

II. 2iii. Heat treatment of iron and steel.

In antiquity edged blades were produced mainly in four ways: by surface carburising, by welding iron to steel, by work – hardening or cold-working wrought iron. Because steel was expensive the idea was to combine the minimum of steel with the maximum of iron. As described in the previous section, the bloomery process in most parts of Europe and Mediterranean produced iron. Steel was produced with a number of methods (a) directly from the smelting process (as cast iron); (b) by solid-state carburisation of ferrite iron; (c) by a number of methods producing liquid steel, the most famous being some African furnaces, the Catalan method in Europe and the Tatara process in Japan (Craddock 1995, 249).

The traditional diffusionist view that the method of producing liquid steel was transmitted to Europe from China is currently being challenged with the discovery of new archaeological finds. Today more archaeologists argue for the local development of iron production (multiple inventions) and for local production of liquid steel earlier than the Middle Ages (Gassman 2001; Mack *et al.* 2000). The Messemvria-Zone cast iron finds (Fig.28a-h in chapter V), probably accidental, coming form a well-sealed context dated back to the 5th century BC, add new light to this argument. Therefore it is important to describe how steel was produced and treated. The commonly used method was carburisation, meaning re-heating of iron in a bath of charcoal so the carbon could be absorbed. The alternative way was by de-carburising cast iron. These issues will be discussed with some of the Messemvria-Zone samples in chapter V.

The heat treatment of steel involves re-heating the steel in charcoal at temperatures up to 750°-850°C and then plunging it into a liquid (water, oil) to cool it rapidly (quenching). The final structure of the blade will depend on the amount of carbon content and the rate of cooling. The temperature is critical no hardening will be successful if the temperature is 700°C, as explained in the previous section. Above this, a steel with 0.85% carbon will be hardened at 750°C while steel with less carbon, for example 0.4%, will need to be heated up to 850°C and then quenched.

This treatment will result in a martensitic, needle like structure, which is practically unusable. Martensite though is brittle and thus the blade could break or crack during forging or use. Re-heating the blade further to a lower temperature and forging again can eliminate this effect. Analyses of knives from Roman Britain (Tylecote and Gilmour 1986, 17) showed that this process can take place in temperatures between 200°C -600°C and can last from some minutes to half an hour. The result is tempered martensite, a structure hard and less brittle than martensite. In most of the cases the complete blade and rarely only the edge was plunged into liquid.

The time-temperature-transformation diagram (TTT) is used to demonstrate how the microstructure of a material with given composition can change in relevance to temperature and time. This is specifically important with regard to the heat treatment of steel where controlled temperatures and cooling conditions are crucial. Grains of a specific phase change shape and size corresponding to different conditions. As can be noted in the Fe-C diagram (Fig. 6) slow cooling produces large grains. Hammering iron in cold produces large deformed (usually elongated to the direction of hammering) ferrite grains. Fast cooling of steel, on the other, hand can produce grains with more angular shapes or needle like as happens with martensitic structure, which has a brittle phase (see PLATE 20a in the appendix). This means that the blade produced can break more easily, and that is why further working is necessary to reduce brittleness. Martensite contains cementite and pearlite. Re-heating and hammering of martensite will result in a different structure, namely, the pearlite will absorb more carbon and will be spherodised and the cementite grains will brake due to hammering that usually follows (see PLATE 8 in the appendix).

Cooling conditions involve time (slow cooling versus fast cooling) and environment (cooling in the air usually at room temperature, or in water, or in oil). Further working of the material, i. e. re-heating the iron into charcoal to carburise it or re-heating steel to make it softer and then hammering it while hot to harden it again and finally quence it in water, leaves its fingerprints in the material. All the properties observed macroscopically (softness, hardness, brittleness) correspond to specific structures. Techniques such as carburisation, annealing, quenching, pilling, hammering etc. employed by the smiths in different sequences are responsible for the microstructures observed under the optical microscope today. By using metallographic studies the manufacturing techniques by which archaeological iron was produced can be described step by step contributing to a better understanding of why a series of technological choices were made.

Chapter II. 3. Aims and methodology.

Chapter II. 3i. Type of evidence and data used in this research.

The peculiarity of the area and of the subject of this research, along with the fact that Thrace does not play a large role in the existing literature, makes it necessary to clarify the kind of data used in this research.

The data used in the chapter I (Landscape) come from recent papers about the geomorphology of sites such as Avdera, Maroneia and Messemvria-Zone (Syrides 1996; Psilovikos and Syrides 1992; Xidakis *et al.* 1988). Some of the information is in preliminary form. With regard to the documentary evidence (chapter III), Greek and Latin sources have unavoidably have been used, given the absence of direct Thracian written evidence. Perseus library in the web (http://www.tufts.edu) was used to retrieve most of the data used on chapter III. Chapter IV (achaeological data) presents the data from both surveys and excavations in the area. Information is based mainly on preliminary reports (in *Archaeologiko Deltio -AD, Archaeologiki Ephimeris - AE, Praktika this en Athinaes Archaeologikis Etaireias -PAE, to Ergo tis Archaeologikis Etaireias - EAA,* from 1970 to 1996). Papers in publications of local conference proceedings (to *Archaeologiko Ergo sti Macedonia kai Thraki –AEMTh*, annually from 1987 -2001, Kavala and its Region 1988) or international meetings (2nd Symposium of Thracian Studies, Alexandroupoli 1992) have been used since no final publications exist. The volume 'Thrace' (published in 1994) is used frequently in this work, because it is the only existing publication covering the history of the area from the early Iron Age to modern times, according to recent archaeological finds. Nevertheless, all the information about the artefacts and the industrial waste under study comes directly from the excavation reports and the related publications.

Chapter II. 3ii. Data collection and sampling strategy.

Laboratory processing of the material under study is the final in a long series of steps including locating, identifying, and recording, photographing and sampling. In the case of Aegean Thrace, locating and identifying the iron-related industrial waste was the most time consuming for a number of reasons worth noting, since they influenced the methodology undertaken. Kalyva and Messemvria-Zone, partly because of their small size and their limited chronological span were easier to deal with. Avdera caused considerable difficulties during sampling. The fact that several people and institutions have been involved in the excavations in different parts of the city meant that the distribution of the archaeological material to different Museum collections, namely Avdera, Komotini and Kavala. Collection and recording of the material took even more time and effort. Since a new archaeological museum and new displays in Avdera were under construction, all the archaeological finds have been split up to even more categories than before (i.e. knives for exhibition or not, for conservation or not etc) which made the collection and review of all the material more complex. Messemvria-Zone is smaller than Avdera and it has been excavated systematically by only two people (A. Vavritsas who dug the enclosed part of the 4th century settlement; and T. Tsatsopoulou who works in the rest of the 6th, 5th, and 4th centuries BC city.

When I started sampling the Kalyva archaeometallurgical material, I did so following the different typological categories, i.e. tools, knives, weapons, nails, keys, etc., having a long list of different artefacts. The industrial waste did not have this variety. After analysing all these samples, I found that not all types of material are appropriate for extracting conclusions about the technological knowledge applied during their manufacture. It would seem so far that technology (in terms of knowledge of properties of the material, chosen techniques of manufacture and successful application of them) could better be considered by studying the technology of edged blades (whatever they are tools, knives or weapons). I decided to limit the types of artefacts to be sampled from the other sites to mainly edged blades. I further divided all the edged blades into two main subgroups, the straight and curved, because the study of analogous material from Kalyva showed a potential differentiation of material and manufacture between these two groups.

After the systematic recording, measuring, photographing and drawing of the objects, pieces (< 0.015m) from slag and blooms and (0.005m) from the blades, see fig. 7) were removed by a TS 8/ E Trim saw using the saw cat soluble cutting oil (typical dilution 3-10 % in water). All the sections were mounted in a FT fast setting resin that needs one to four hours to cool. The mounts were then wet grinded using three grinding papers (no 220, 500, 1200) and polished using the standard metallographic techniques of diamond polishing. A Rotosystem 250 machine (Stuers 1992: 55) has been used for both grinding and polishing of the samples. The polished sections were then etched in nitric acid in alcohol (NITAL with 2% nitric-acid and 98% methanol) and examined by both binocular and optical microscopes. The main body of the analytical work undertaken was contacted in the Metallographic Suite of the University of Strathclyde under the supervision of Prof. A. Hendry and assistance kindly provided by the lab-technicians Mr. J. Kelly and Mr. D. MacMahon; mounting and processing of polished sections also took place in the laboratories of the department of Archaeology in the University of Glasgow under the supervision of Dr. R. Jones and Dr. E. Photos-Jones and under the scrutiny and loving care of Mrs. L. Farrell.

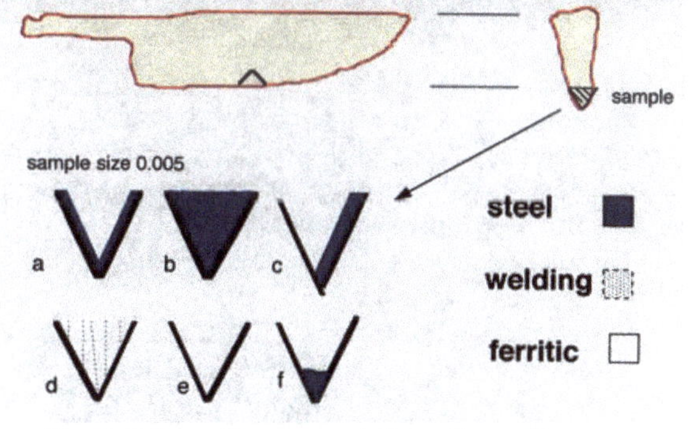

Fig.7. Schematic representation of sampling technique (by M. Kostoglou).

Hardness measurements, according to the Vickers system, were made in order to check the phases present and to evaluate the heat-treated blades (Tylecote and Gilmour 1986: 7). The way to measure hardness is to indent the specimen with a pyramidal diamond indenter under a controlled load. This leaves an impression the size of that is a measure of the hardness of the material, expressed with number, i.e. 150, the type of the system used, in this case Vickers Hardness (Hv), and the load used, i.e. 100p.

In order to define the chemical composition of the blooms, slag, as well as of some inclusions in the blades, SEM analyses and examination were carried out in a "Leo microscopy" Stereoscan S360 SEM instrument with integrated "Linch Analytical" AN10855 energy dispersive EDAX micro analyser. The analytical results are in their normalized form and in oxides. When an element is not present, this is indicated with the abbreviation nd (not determined).

Summary: the first part if Chapter II sought to provide an insight into the study of culture and its practices. Groups of people can share the same cultural identity that is carried into material culture in a symbolic manner. But culture is a complex concept that includes other sub-structures and sub-groups which could be ethnic, political, legal, sexual, technological, religious and so on. Each one has its own identity. Boundaries between those identities are flexible and changeable, as this thesis will demonstrate with respect to Thracian identity over the centuries.

Material culture (related to iron technology in this study) carries social significance as well as information about the cultural identity of the people by whom it was produced. In order to decode the meanings of material culture we have first to identify and then study the different components on a very small scale (both in time and in place). A major component in the present research is the process of iron making.

Nevertheless, the cultural aspects of iron technology could only be considered if first technology is understood in its own terms (Lechman and Steinberg 1979). The second and third parts of chapter II described the processes and the problems involved in the study of iron in order to create the background for a meaningful integration of the analytical data into archaeology. This integration has always proved difficult:

'the chemist is fully aware that ancient slags are not always well defined, homogenous materials, easily placed in particular categories, nor can he do without the terminology and the reference to physico-chemical principles when explaining his results to the archaeologist' (Bachmann, 1982, 1-2). This work aims to provide an archaeological understanding of the physical processes and analytical studies related to iron technology, and to demonstrate that recently developed analytical techniques are the best tool when they are used in an archaeological context and not on their own.

However, before moving to the discussion of the analytical data (chapter V) and archaeological (chapter IV), it is necessary to briefly evaluate how issues related to culture, identity and iron technology, with reference to Thrace and its inhabitants, were understood and presented in ancient literature and iconography. The following chapter (III) focuses on ancient Greek and Latin sources because these are on the core of the culture-historical approach (chapter II) used in the construction of conventional history of the region (chapter I).

Chapter III. Text and iconography.

This chapter attempts to re-evaluate the use of ancient sources in the conventional historical interpretation; ancient literature has been used in the construction of arguments about the Hellenisation or Romanisation of the Thracians in Aegean Thrace (already presented in chapters I, II and under discussion in chapters IV and VI). It is argued that ancient writings do not necessarily present definitive evidence on the existence of a distinctive Thracian nation with a unified national identity in antiquity (see chapters I and II). They rather imply that the existence of the Thracian ethnic identity was based on distinctiveness from the Greeks or the Romans. This ethnic identity was initially constructed by the Greeks in terms of the difference of these people and not on any particular knowledge of their characteristics.

The second part of this chapter focuses on texts related to the metal production in the area under study. This group of bibliographic references became an important part of this work after noticing that almost half of the texts referring to Thrace or Thracians were in connection to metalwork. Particular references to iron are given in an attempt to establish the existence (or not) of a metal working tradition in the area, and the indigenous involvement in it. Ancient references to iron metallurgy and steel making techniques in general are not included in this section because they are part of existing publications (Pleiner 2000 and 1969; Photos 1987; Conophagos and Papadimitriou 1981; Halleux 1974).

Thrace and Thracians in ancient literature.

In the authors of the 5th and 4th centuries BC the Greeks use the term "Thracians" as a collective name for a series of tribes with a strong sense of honourable birth. Their clothes, weapons, and occupations are in accordance with their mountainous environment. They live in isolated communities in the highlands between the Aegean Sea and the River Danube and they keep their socio-political structure in forms of the past (table 5 in the appendix). For purposes of clarity the first section of the present discussion is organised

Name	Title(s)	
Late 6th and 5th cent. BC.		
Herodotus	Histories	30
Euripides	Rhesus	20
Thucydides	Histories	69
Late 5th and 4th cent. BC.		
Xenophon	Anabasis, Hellenika, Kyropaedia, On Hunting, Ways and Means	32
Plato	Laws	2
Isokrates	Panegyricus	7
Aristotle	Constitution of the Athenians, Politics.	1
Demosthenes	On the false Embassy, Philipic, Against Aristocrates, On the Chersonese, Against Pantaenetus, Against Timotheus.	44
2nd cent. BC to 1st cent. AD.		
Pausanias	Description of Greece	22
Diodorus Siculus	Historical Library	55
Strabo	Geography	70
Plutarch	Alkibiades, Cimon, Nicias	18

Table 1: Authors and works used in the text listed in chronological order (summarised from Warrington 1961. R. no = total number of references after frequency test in Perseus library on 5.4.1999).

towards three major issues related to the subject of this work:

a) the knowledge of geography/topography of Thrace and Aegean Thrace;
b) the knowledge of the socio-political structure of the people living in the territory;
c) the identity of the people under study.

a) First, the name "Thrace" occurs often (see table 5 in the appendix) in the ancient literature but most of the time without specific meaning. According to Herodotus (*Histories V,* 17) and Thucydides (*Histories* II, 97. 1-2), the River Strymon was the southwestern boundary between Macedonia and Thrace (see Fig. 1 in chapter I). Different tribes occupied this area and Herodotus (*Histories* V.3) mentions a lot of tribal names. Herodotus' understanding of names and places has been proved problematic more than once. Along with the well-known example of Naukratis (Snodgrass in Shanks 1997, 122), we should add the ancient topography of the Thracian coastline. A series of excavated Greek colonies (*emporia*) on the Thracian coast were named after the list given by of Herodotus describing Xerxes' passage to Greece (*Histories*, V.3).
According to new archaeological data (amongst them inscriptions and coins) the site, previously thought to be Messemvria, has being identified as Zone changing the topography of the rest of the Thracian coastline (Tsatsopoulou 1987). This example, along with the dispute over the location of the royal cemetery of the capital of ancient Macedonia (Kotsakis 1998; Faklaris 1994), illustrates well the problems arising from taking ancient literature as ancient reality.

With regard to ancient Thrace we find that no boundaries between tribal territories are known. Although it is mentioned that some tribes used stones to mark their boundaries in the 4th century BC (Xenophon *Anabasis* 7.5.13), no archaeological evidence verifies this tradition. At that time (5th century BC), all who lived South of the River Istros (Danube) as far as the Aegean, the Propontis and Macedonia were called "Thracians". The Greek population who lived on the colonies along the coast were referred to as 'the Greeks of Thrace' (Herodotus, *Histories* 7.185.1).

In Roman times there are more generalisations and more confusion regarding territorial or regional boundaries. Strabo (table 1 and table 5 in the appendix) says that the Greeks inhabit the seaboard; that Thracians occupy the entire region between the river Strymon, the Pontus and the Haemus Mountain; and that some people attach the area between the Rivers Strymon and Nestos (Fig.1) to the Macedonians (Strabo, *Geography*12.4.4 and in 7.7.4). This confusion might be due to the creation of the first Macedonian Meris (the territory stretching between the rivers Strymon and Evros; see Fig.1) in AD 46 by Emperor Claudius (see chapter I and table 1 in the appendix). Nevertheless, no clear territorial evidence exists and boundaries between Thracian tribal lands, Thrace and neighbouring lands are often confusing.

The constant migration of people (Strabo, *Geography* 7.3.10) and changes in names should also be added to the above evidence. Most of these changes seem to be related to the administration policy of the Romans. Plutarch talks about settlers sent to Thrace (*Perikles* 11.5) and we learn that 'Aelius Gatus governor of Macedonia, transplanted fifty thousand people from the far side of the river Ister into Thrace, some of them Getae, but all called now Moesi' (Strabo, *Geography* 7.3.10)

Summarising all the above, we could say that between the 5th century BC and the 1st century AD political changes affected territorial tribal boundaries as well as regional boundaries. Constant movement of population caused considerable confusion on the use of topographic names and *ethnonymes* that is evident in Greek and Latin literature.

b). Regarding the knowledge of the socio-political system of the region, 5th century BC authors agree that the socio-political structure of the Thracians was based on tribal organisation and kingship when the rest of Greece was experiencing more democratic regimes (Herodotus *Histories* 7.111.1, Thucydides *Histories* 2.2.9.2-3, 5.6.2). Relationships with other people took place often *via* diplomacy and gift-exchange (Xenophon, *Anabasis* 7.3.16, Thucydides *Histories* 2.29.4-5, Strabo *Geography* 7.3.8).

The attempts of the Odryssian tribe to unify all the tribes under one king were not always successful (tables 1 and 5 in the appendix). Xenophon reports on free tribes under no king (*Hellenica* 5.2.17). Tribal organisation and kingship survived even under Roman domination suggesting an element of continuity in the local culture. Strabo reports that two tribes existed in his times and that Thrace, although devastated, could afford fifteen thousand cavalry and two hundred thousand infantry-men (*Geography*, fragment 47). This is striking if compared with Herodotus' description where Thrace is mentioned as the second most populated area of the world after India (*Histories* V.3).

Devastation in rural areas under Roman control is a common literature topic in Srtabo. As suggested (Alcock 1996, 27) people turning to pastoral activities in Roman times were regarded as taking a retrograde step in their social-economic lives. According to survey data, this does not seem to be the case, at least for the province of Achaia. No survey data exist for Aegean Thrace, but according to the literature (from Herodotus and Thucydides to Strabo and Diodorus Siculus) the Thracians were always described as a people engaged in pastoral and other mountainous activities living in villages, often fortified, known as *kome* (Thucydides *Histories* I.5.1). This rural character, however, should not be uncommon in most of the areas of the provinces of Roman Greece and Thracia. Pausanias' description of a city could hardly be found in Greece since most cities, even Athens lacked clearly defined urban characteristics (Alcock 1996, 29; van Andell and Runnels 1987). The above evidence suggests that the rural character of the economy in Aegean Thrace and its different socio-political structure were over-emphasized by the Greek and Latin authors in an attempt to create' the other'.

c). With reference to the third issue of the present discussion, namely the identity of the people of Aegean Thrace, references are made to three distinctive groups: the Thracian tribes, the Greek colonists and the Romans, mainly military units. We should add the Macedonians, the people of Macedonia under Philip II. Collective characteristics are attributed to the Thracians in terms of clothes, music, language, religion eating and drinking habits (see appendix table 5 in the appendix). All these characteristics are opposed to the Greco-Roman ones and often described as barbaric.

Fig. 8. Attic red-figured vase depicting the death of Orpheus by Thracian women (reference no: Boston 90.156 in http://www.perseus.tufts.edu on 7.4.1999).

There are two particular prototypes of a Thracian female or a Thracian male, and they can be seen both in literature and iconography. The image of a Thracian woman entered Attic iconography in the late 5th century (table 3 in the appendix) and became common motive in the 4th century BC. Thracian women are either slaves or maenads. From Herodotus again we learn about Thracian women sold to Athens as slaves. The maenad image, on the other hand, could be related to the impact of Euripides' Bacchae produced in 406 BC (Warrington 1961, 226), which although composed in Macedonia, refers to Thracian customs and landscapes.

The male prototype of a Thracian is a warrior known as *peltast*. These warriors were often used as mercenaries by the armies of the Greeks and Romans. Thracian warriors were famous for their fighting ability, a concept that has characterised them from Classical to Roman times (see table 5 in appendix). They had organised forces of foot soldiers, archers, scouts, targeteers, and swordsmen, which were often used as mercenaries by the Greeks or the Romans. The Thracian *peltasts* influenced Greek warfare and images of these warriors should be more common in the Greek world after their introduction to the Athenian army by Iphicrates in the 4th century BC (Connolly 1981, 49; Warry 1980 50-51; Best 1969, 7; Warrington 1961, 391).

Iconographic evidence on pottery (Fig. 9, and table 3 in the appendix) suggests that light armour was the characteristic of the Thracian *peltast,* at least according to the Greek interpretation. The *peltasts* were equipped with a crescent-shaped shield, the *peltae.* As it has been suggested it had Thracian origin but the Greeks used it as well (Snodgrass 1967, 78). It was usually made of sheep or goatskin stretched over a wicker frame. *Peltasts*' armour consisted of two short javelins or one long thrusting-spear, and a dagger. The long thrusting spear has been identified as the Thracian weapon *'romphaia'* (table) some examples of which, made from iron, have been found in Bulgaria (Kolev 1985; Secunda 1983) and are dated to the Hellenistic period (c.4th -3rd centuries BC). Long spears are also mentioned by Xenophon as a characteristic Thracian weapon (*Kyropaedia* 6.2.10). They remained the typical weapon of the fighting Thracian tribes up to the Roman period since they are depicted on Trajans' column in Rome. In Greece there are only four examples of this weapon and they all come from the Evros region. A long iron spear is currently on display in the archaeological museum in Komotini, and it was found along with another two iron spears and a bronze one in the area of Polyanthos. They are all dated to the Hellenistic period.

The light armour mentioned above, along with the long coat - *zeira* - usually decorated with geometric motives (table 2, no 9, Harvard 1959. 219 and no 3, 1969. 14), the low boots with a flapped, turned down border and a fox-skin helmet, completed the military attire of the Thracian *peltast* (Best 1969; Connolly 1981). This kind of clothing was unusual

Fig.9a. Attic red-figured cylix depicting Thracian peltast on the *tondo* (Robinson 1931, Corpus Vasorum Antiquorum 2, plate III).

for the Greeks. Nevertheless, it was appropriate for the way of living and fighting of the tribes from the mountainous north where heavy snow covered their places for most of the year (Xenophon *Anabasis* 7.4.3-4). *Peltast* became a part of the Greek army after general Iphicrates' reform in 4th century BC, when Athenians might have come in closer contact with more Thracians (Warry 1980, 50-51; Warrinton 1961, 391). It might not then be coincidental that the depiction of the Thracian *peltast* became more common in that period in the red-figured vases of Attica (see appendix table 3). Nevertheless, the Greeks perceived Thracians as a 'non- Hellenic' nation. Because they are perceived first as 'non-Hellenic' or 'barbaric' people and secondly as Thracians, they are often givecharacteristics shared by other barbaric people such as strange looking clothing and bloody customs. In most of the cases they are described as hostile not only to the Greek colonists but also amongst each other (see table 5 in the appendix).

Summarising the previous evidence with respect to the classification and discussion on ethnic and national identity in chapter II. 1, one could argue that the Thracians are not qualified as one nation with the same ethic identity.

- The numerous tribes did not have the same or similar language, religion and customs.
- They did not occupy specific territories.
- They lived in constant movement caused either by dislocations and migrations of some tribes or later by the administrative policy of the Romans.
- They were fighting against each other and against the Greeks and the Romans. The Odryssians (the most powerful tribe) never succeeded to unite all the tribes.
- And most of all, we do not have evidence

about what these people bellieved themselves to be; nor about how they called themselves, their *ethnonyme*.

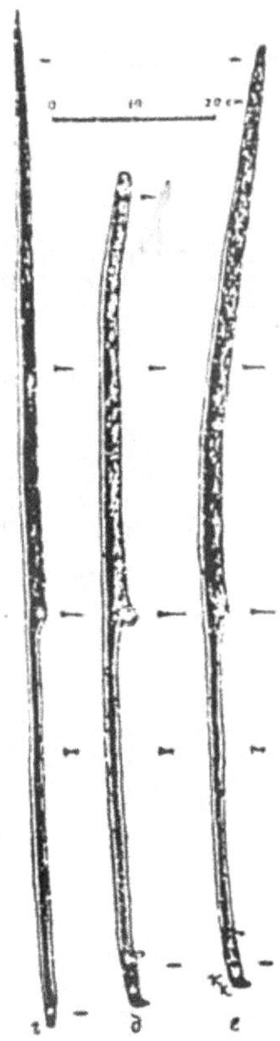

Fig. 9b. Hellenistic iron *romphaeae* (Kolev 1985).

These characteristics although they do not support the argument about the Thracians having a national identity, as defined in chapter II.1, they could support the existence of a tribal society including smaller groups with distinctive ethnic identities. These ethnic identities were often competitive with each other but the Greeks grouped them together under the same name 'the Thracians'.

The indigenous involvement in metal production.

Moving to the next aspect of this discussion, namely the indigenous involvement in iron production, it is interesting to note that the Greeks and the Romans often differentiated the Thracians from the different cultural groups that shared the same area on one issue: their acknowledged working and managerial skills in the running of mines due to the exploitation of the gold, silver and iron mines located in their own territory. The broader region has had a particular significance in the myth about the origin of metalworking. Fire genies and mythical metal workers such as *Cabeiri* in Lemnos and Samothrace are often placed in volcanic regions (Aldington and Ames 1959, 129).

There are particular references to the rich gold and silver mines in Thrace with special reference to Mount Pangaion or to specific tribes and territories that specialised in mining (table 4 in the appendix).

Xenophon (*Ways and Means* 4. 22) reports a case in Laurium silver mines where a Greek called Nikias had a thousand men and let them out to Socias the Thracian who had to pay an obol a day per man. Moreover, a mining area in Laurium was called Maroneia because the miners who lived there came from Maroneia, a city on the Thracian coast (Demosthenes *Against Pantaenetus* 37. 4) where silver mines were also located (Triantafyllos 1973; Papastmataki 1996).

Control over the exploitation of the gold and silver mines of Thrace was the subject of political conflict. The indigenous population of the area, namely the Thracians, often went to war with others such as the Athenians, and the Thasians (Thucydides *Histories* 1.100.2) or the Macedonians, over control of the mines.

While the relationship between Thrace/Thracians and gold or silver mining is emphasised by the ancient authors, this does not seem to be the case during the Roman Period. The Latin authors refer more often to the issue of the Thracian's involvement with metals working and extraction but this time with particular reference to iron. Strabo (VII. fragment 45) reports that two Thracian tribes, the Sintoi and the Odomantes (they lived in the area east of the upper Nestos River valley) were good metal smiths who practised their trade well before the Romans arrived in the area.

Titus Livius (XLV, 29, 10-11) mentions that the Romans kept the exploitation of gold and silver mines of the First Macedonian Meris (the territory stretching between the rivers Strymon and Evros; see Fig.1) under their control. According to the same text, at this period the exploitation of the copper and iron mines was left to the local population, presumably the Thracians.

Furthermore, during Roman times iron ores were exploited in Nicopolis, a site "beside Nestos" in the upper River Nestos valley. Also armouries known as *fabricae* were set up in the Thracian City of Handrianoupolis, in the east upper Evros valley.

Finally, the Roman province of Thracia (an extended territory roughly coinciding with

Herodotus' Thrace) was one of 44 armour production centres recorded in *Notitia Dignitatum,* a document compiled in the beginning of the 5th century AD (Randsborg 1991, 94-101; Robinson 1975, 7). Also Thrace is reported to supply the Empire not only with iron weapons but with iron ores and iron as well (Velcov 1980, 187).

The references examined thus far suggest that Thrace indeed did have a metal working tradition. The experience of the Thracians in the running of mines, especially of gold and silver, is well testified for the Classical and Hellenistic Periods. The Romans turned their attention to the iron ores, mines and workshops of the area. This change in material choice could also signify a cultural change in terms of social-economic organisation or even in value systems. The socio-cultural implications of technological changes cannot be fully understood before the conclusion of the analytical study.

It could be argued that the conventional interpretation of the history of the area (as presented in chapter I) was constructed with emphasis on the interpretation of a small selection from ancient literature. References were drawn mainly from Herodotus and they appear to be painting a uniform picture. The previous discussion did demonstrate that the ancient authors do indeed speak about Thrace and the Thracians on various occasions but they draw a rather incomplete picture. In an attempt to emphasise uniformity between the Thracian tribes and to juxtapose them to the Greeks or the Romans as the barbaric 'other', several issues were ignored or misinterpreted.

Further interpretative limitations are imposed by the 'double hermeneutic', taken from Giddens and the sociology and applied to Classical archaeology by Shanks (1997, 121). Archaeologists should try to understand both a world of meaning and interpretation (society) and the role of the individual who attempts to do so. It is beyond the scope of the present chapter to go deeper into the problems of the 'double hermeneutics' but this factor could explain the problems that ancient historians fase in studying or interpreting their material.

I will illustrate the biased selection and use of ancient literature (in modern and ancient times) with a final demonstration.

A frequency test for the word "Thrace" was carried out in the Perseus library (http://www.perseus.tufts.edu, 5.4.1999). The results are depicted in table 1. Since Herodotus is used extensively in the conventional history (chapter I) of the area (especially on describing the barbaric customs of the Thracians), it could be expected that he is the one who refers most frequently to the region and its people. According to the data in table 1, Herodotus comes sixth after Strabo (with 70 references), Thucydides (69), Diodorus (55), Demosthenes (44) and Xenophon (32). It could be argued that the frequent use of a word in a textual context is not always meaningful. Thus quantity may not indicate better knowledge of a subject. This seems to be the case when information from Herodotus is compared with information from Demosthenes, for example. Thrace in Demosthenes text is only a geographical region of strategic importance and political implications for the interests of Phillip II of Macedon. Thucydides, on the other hand, was of Thracian origin himself and the Athenians used him as general in the war against Amphipolis (their former colony in the River Strymon) due to his vast knowledge of the area and his good relationships with the indigenous people (Warrington 1961).

Xenophon also had a personal knowledge of the area and the people since he crossed Thrace (Warrington 1961) with ten thousand soldiers in his way back to Greece. He focuses on the landscape because he is interested in military practices and strategies. He describes warriors and he refers to their culture only if related to war (table 5 in the appendix). He knew how the Thracians fought and lived because of the mercenaries that took part in the same expedition under his command. Thucydides, Demosthenes and Xenophon, although they appear to be more 'scientific' and thorough in their knowledge of the area, they are used often in the reconstruction of past historical events such as battles and less in the reconstruction of socio-cultural aspects of life. Finally, the authors of the Roman period, Strabo, Plutarch, Cicero and Pausanias are trying to be more specific in the information they give. Nevertheless, they do refer to their predecessors and despite their effort to be critical they often repeat the same information. This is the case with both Strabo and Pausanias who are concerned with the accuracy of the geographical features and the names of people and places. Here one could comment on the confusion of primary and secondary sources (Shanks 1997, 121), i.e. Strabo of the 1st century AD is used as a source for evidence and reality for the 4th and 5th centuries BC.

The picture portrayed in the two sections (Thrace and Thracians/indigenous metal working) of this chapter illustrates that the information obtained by the ancient authors is quite subjective and biased. It depends on the focus of the particular research, and the personal experiences and knowledge of the researcher (both in modern and ancient times).

Text and iconography on the one hand, and archaeology on the other, describe two different realities. Conceptions and preconceptions of geographic knowledge, local traditions and identities over long periods of time are subject of change. Who and what is "Thracian" in Aegean Thrace from Classical to Roman times is discussed in chapter IV in the light of the archaeological data and forms part of the final interpretation in chapter VI.

Chapter IV. Archaeological data.

The first section of this chapter presents archaeological evidence related to the iron production in North Greece. The data used are coming either from already published works or from recent surveys in the region. As it will be demonstrated, a rich multi-period industrial landscape, including mining and smelting activities, is clearly displayed in North Greece, especially in the area between the rivers Strymon and Evros (Agean Thrace). The data used in the first section will provide the regional, technological and cultural background for the discussion of the archaeological evidence from excavated sites presented in the second part of the chapter. After comparing and combining published and unpublished finds from both surface and excavated sites, this chapter concludes with a suggested model of economic and social patterns of iron production in Aegean Thrace. The conclusions of this chapter will be compared with the interpretation of the textual evidence (chapter III) and of the analytical data (chapter V) in order to portray the social archaeology of iron technology in Aegean Thrace as proposed in this thesis (chapter VI).

IV.1. Iron working in North Greece.

The Iron Age in Greece falls between the 12th and the 8th centuries BC, a period called the 'dark ages' (Snodgrass 1980). Iron technology in Greece and elsewhere, has been associated with major socio-cultural changes. Population decrease, discontinuity of settlement occupation, disappearance of writing and monumental architecture, and change in burial customs from multi-burials to single burials (Snodgrass 1987) are few of the changes occurring during and after the coming of iron. Possible explanations for the introduction of iron to Greece vary from the mythological legend of *Dactyles*, who discovered iron smelting, and the introduction of the new material either from the East or from the Balkans (Pleiner 1969), to the bronze shortage hypothesis due to the collapse of trade with Near East (Snodgrass 1987) and to the independent development of iron production in Macedonia (Sherratt 1994). Current research emphasises the lack of adequate data for the social study of the Iron Age in Greece (Morris 2000). Nevertheless, most scholars accept that iron technology was diffused to the Greek mainland from the Levant via Cyprus and other islands, amongst them Crete and Rhodes, circa 1200 BC (for a summary of the different suggestions see: Pleiner 2000 and 1969; Pigott 1996; Cradock 1995, 259-261; Snodgrass 1987; Photos 1987).

The problem of the introduction of iron in Greece and the discussion about its origin, although of major importance, is out of the chronological framework and the scope of this work (see chapter I). I have briefly referred to this problem for two reasons: first, to draw attention to the fact that most of our knowledge of early iron lies on amlargeuous documentary evidence, limited archaeological finds and even fewer analysed iron finds. Second, to clarify that the present work does not answer the questions of when, how and why iron was introduced to Greece. It rather explores the uses of iron technology seen in the archaeological record during a period of time when iron is fully incorporated into society (stage 3 according to Snodgrass 1980).

During the last twenty years numerous remains of past mining and metallurgical activities, such as mining works and metalworking sites, have been reported and data of chemical analyses referring to a number of slag occurrences have been published (Wagner and Weisgerber 1988; Papastamataki 1975, 1985 and 1986). Although the list is very short compared to other European countries, a number of relevant archaeometallurgical studies have already been undertaken in Greece.

Silver and gold mining dominated the research interests following the work in the silver mines in Laurium (Conofagos 1980). Silver or gold mines have been, thus, located on Siphnos and Kythnos (Gale and Stos-Gale 1984), on Thassos (Wagner *et al.* 1981), in Nikisiani on the Pangaion Mountains (Papastamataki 1986), and in Palia Kavala (Vavelidis *et al.* 1996). The mining areas in the Pangaion Mountains and in Palaia Kavala fall within the territory of Aegean Thrace. According to archaeological, epigraphic and numismatic evidence (Samsaris 1987) the gold and silver mines of this region were under Imperial control. Especially in the 2nd century BC, the silver and gold mines in Attica and Thrace provided major supplies to the Roman Empire under the control of procurators appointed by the Emperor. Mineral rights, taxes, legal conditions and technological obligations are defined by regulations and legislation enforced by provincial governors (Healy

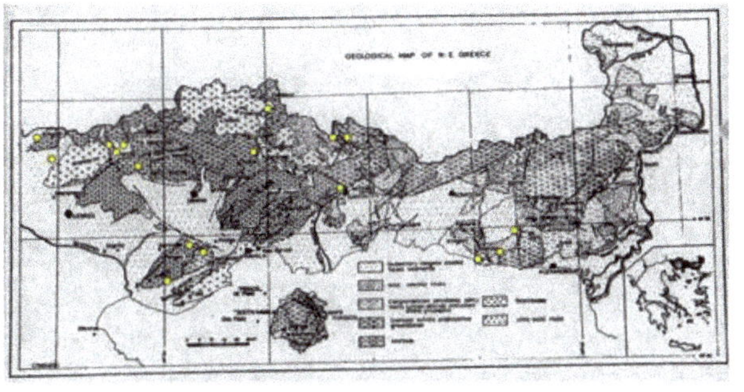

Fig. 10. Map adapted from Photos 1987. Dots indicate iron working sites (slag heaps).

1978). Specific strategies were also employed towards the industralisation of gold and silver mining in order to increase production. Under the Roman administration control, mechanical devices, such as waterwheels, or practices, such as *hushing* (breaking up mineral deposits with a huge and sudden volume of water) and *cross-cutting* (manipulating the natural terrain to drain mines), are employed to increase efficiency in underground mining (Curchin 1991, Healy 1978). Evidence for these activities exist in Spain (Curchin 1991). Also, evidence for cross cutting has been found in Mademochoria (an area in Chalkidiki, Greece) and in Rio Tinto, Spain (Davies 1935). Smaller mines or mines of secondary importance, such as these for copper and iron, were left under the control of conductors or individuals whose undertakings were checked and taxed by the local governors (Healy 1978). The use of the silver and gold mines for the extraction of iron ores should not be excluded. This was the case in Attica where the steel used in the Erechtheion was produced from iron coming from the silver mines in Laurium (Conophagos and Papadimitriou 1981). There is, however, an obvious differentiation between the administration and operation of silver/gold mines and of copper/iron ones during the Roman period. Aegean Thrace is often mentioned as an important mining region in the major bibliographic references (Courchin 1991, Heally 1978, Davis 1935) along with other well-known mining areas, such as Rio Tinto in Spain. Nevertheless, no systematic work has been undertaken so far to study the ancient exploitation of the mining areas listed in this section, despite the existence of significant archaeological indications.

Photos (1987) has given a very detailed review of the iron metallurgical activities in Greece with particular reference to Macedonia, from prehistory until the Ottoman period. Chiotis (Chiotis *et al.*1996) recently published analytical results on iron working from Aggistron, Serres, dated to the Ottoman period (Fig. 10). Papadimitriou also examined material such as iron shells of wooden poles, from the ancient pier found in river Strymon close to the ancient city of Amphipolis (Papadimitriou *et al.* 1996). None of the aforementioned studies, however, treats metallurgical activities in their socio-cultural context. They are all focused on the study of limited groups of finds from isolated sites.

Published archaeological and /or analytical data from excavations, related to iron or other metal finds from the region of Aegean Thrace are very limited. The only direct evidence for iron mining so far is the mining tunnel found accidentally in an area called Ktismata, southeast of Maroneia (Fig. 2 and Fig. 10) dated in the 3rd – 2nd centuries BC according to the pottery found inside (*AD* 1973, Melphos 1995).

Slag-heaps have been located in Kimmeria, north of Xanthi (Stavropodis 1983) and Thermes (Papastamataki 1985) and they are discussed later in this section. The geographical location of these sites is shown in Fig. 3 and 10.

Iron finds have been discovered in almost all the excavated sites in Aegean Thrace. The earliest group of iron finds is dated to the 9th century BC. It was found in a burial close to an Iron Age dolmen in the Rodhope Mountains northwest of the river Evros in the modern village of Roussa (Fig. 2). The group included mainly jewelry, one iron ring and a chain, supporting the idea of the symbolic use of iron in early times (Pleiner 1969, 2000). It also contained a single-edge iron knife that was not available for sectioning. A group of knives are also found in the excavation of the archaic cemetery in Lissos on the coastline south of Komotini in Fig. 2 (currently on display in the Museum of Komotini). Finally, an iron sword was found accidentally in Vafeica (AD 1999, in press) 1km north of the modern town of Xanthi (Fig. 2). The current state of the iron and the importance of the finds did not allow sampling. It was impossible to make any deductions on their manufacture and composition and it remained unknown if they had a steeled edge. These early iron finds from Aegean Thrace are included here because the list of early iron finds in Greece is very limited. The fact that utilitarian iron artefacts, such as knives of the type described by Sherrat in 12th century BC Cyprus (Sherrat 1994), are included amongst prestige/display items, indicates their considerable significance for the study of the early development of iron technology in the region. Compared to iron artefacts, industrial waste and related finds (i.e. slag, blooms, furnace lining, tuyères) from excavated sites are rarely part of the record. There are few exceptions where excavated material was studied by archaeometallurgists (from north Greece Papadimitriou *et al.* 1996; Photos 1987; from south Greece Conophagos and Papadimitriou 1981; Varoufakis 1979). In most of the cases presented here, there is a generally acknowledged difficulty in providing a safe date for the activities studied. An earlier date (pre-Roman) is almost always assumed. Radiocarbon dating is possible only if charcoal has been preserved in the slag, but no such data for North Greece exist so far. It is, therefore, necessary to establish an archaeological and analytical framework for the study of the morphology and composition of the iron-related products and byproducts per period in order to:

a) define the metal which was obtained during the smelting activities in each site;
b) characterise and classify the smelting activities associated with the Classical, Hellenistic, Roman, Byzantine and Ottoman periods; and
c) increase our knowledge on the smelting techniques employed in each period.

Changes in technology can then be appreciated in a safer chronological framework and the results can be better correlated with the rest of the archaeological information.

The slag-heaps in north-east Greece have first been located and systematically recorded in the frame of a research program for the investigation of uranium exploration, undertaken by "Demokritos" Research Center (1975). Finds include slags, fragments of furnace walls, lining, tuyères, sherds and pieces of coal (loose or entrapped in the slags). In Fig. 10 all surveyed sites, areas with slag depositions and other relevant metallurgical remains, are indicated. It can be noticed that the areas with traces of iron and possibly other metals are located in the same regions with the areas in Fig. 3, indicating a rich industrial landscape with traces of activity from Classical/Hellenistic to Ottoman periods. The descriptions that follow refer to iron working sites mentioned in Stavropodis (1983) and Photos (1987) and re-visited by Bassiakos Pournis and Kostoglou (Bassiakos *et al.* 1999). All the sites mentioned here are included in Fig. 10

Faia Petra 1

At an area about 7 km NE of the little town of Sidirocastro (=iron castle) two sites with slag-heaps situated west of the village of Faia Petra (=brown stone) have been recorded. The first site lays 300m west of the village. Remains of a slag-heap are apparent on the surface while less amount of slag is scattered around in a distance less than 50 m. The geologist S. Manolarakis (IGME, Serres) who discovered the site reported that the slags have been transported here from their original position (Faia Petra 2). The slag is radioactive and ropy on the surface, indicating flowing outside the furnace. Fresh fragments reveal dense, dull black interior and limited pores.

Faia Petra 2

The second site is about 100m west of the last houses of the village. It is composed of two adjacent slag-heaps, originally placed on a slope overlooking, to the east, a nearby spring. The heaps were covered by a soil stratum (30cm to 50cm thick) and appeared due to the construction of a road, which cut them. At least a few tens of tonnes of slag remain buried under the soil layer. High values of radioactivity were measured on an artificial section by contact measurements.

Some also contain small pieces of entrapped charcoal, enough for radiocarbon dating. In the lower slag-layer, near the ground we found fragments of furnace walls, which consisted of backed light-grey clay with highly vitrified interior surfaces. Cone-shaped clasps for joining tuyères, diameter 4cm to 7cm, of the same colour (light grey) were also observed in some fragments of furnace walls. Close to those clasps we noted three cylindrical imprints of smaller diameter (2,5cm), probably corresponding to pipes once used for blowing air into the furnace.

Ano Vrondou

In the village of Ano Vrondou, about 30km NE of the city of Serres there is a slag-heap situated in a yard of a homestead, in the south-eastern fringe of the village. The lower layers of the heap seem undisturbed. The slag fragments are radioactive and ropy on the surface. Among the slags we found fragments of tuyères very similar to those observed at the site Faia Petra 2. Furnace linings as well as fragments of furnace walls, 10cm to 15cm thick, were also apparent. Three fragments of temperature-resistive rocks, such as granite gneiss and schist, each of them with one side partly vitrified and glazed were also found. They undoubtedly constitute parts of metallurgical furnace walls and seem to have been transported from the nearby hills, where these formations occur (IGME 1982).

Panorama

Stavropodis (1983) has briefly mentioned a radioactive slag-heap situated in that site, without giving any more details. A small slag heap, as well as pieces of slag are scattered around the heap in a distance of about 500m is situated near the northern yard of the Aghios Dimitrios monastery, 30km north-east of Serres and 2km south-west of the village of Panorama (Fig.10). The main slag-heap is radioactive. A thick soil layer covers the heap, so estimation of the slag reserve without excavation seems difficult. The slag fragments have collapsed ropy appearance and their outer surface is dull, dark brown to dark grey coloured. The interior is dark brown to black, dense with very few pores. Most of the slag pieces are rather small. Charcoal is scarce and very few fragments, less than 0.5cm, are entrapped in the slags. No furnace walls or any other relevant remains were observed at that site.

Katafyto
The site lies 7km north of the Ano Vrondou village and 800m SW of the Katafyto village. It has been briefly reported by Stavropodis (1983) and studied by Photos (1987). The slags cover the slope of a little valley, in the bottom of which flows a stream. Their radioactivity is high. The reserve is estimated around 5,000tons. The fragments of the slags are weighting more than 10kg. Their external surface is ropy but not glassy. In their interior the slags are non-dense and very porous. Many fragments of furnace walls, 15-20cm thick, which consisted of reddish backed clay, were found among the slags. The colour of some tuyère fragments found in the same heap was also reddish.

Vathytopos
Remains of a sizeable slag pile, at least 2.5m high, exist 3km east of the village of Vathytopos, at the southern side of a bridge, already mentioned by Stavropodis (1983). The pile is close to a stream, partly constituting its riverbanks. An intra-layer of fluvial material near the top, as well as limited slag resetting of the upper layers are indicators of past river-flood(s). The radioactivity of the pile is high. Slag removal has badly disturbed the original stratification. No sherds or furnace bottoms or walls were observed. Freshly fragmented slag-pieces reveal very porous and dull-black interior. A massive iron working is assumed here, indicated by the large amount of slag.

Domatia
The slag pile at the village of Domatia was known since the late sixties as has been reported by Stavropodis (1983). The village is situated 24km southwest of the city of Kavala, on the main road to Thessalonica. Only a small part of the pile is still in place. The radioactivity of the slags is very intense. Many of the slag pieces are voluminous, some of them weighting over 10kg. Their external surface is ropy, irregular and non-glassy. Their interior is very porous, due to the existence of gas vesicles, which have diameters over 2cm. Pieces of charcoal, 1.5cm or less, are entrapped in the slags. Remains of furnace walls and bottoms, as well as conical clasps for joining tuyères were also observed. A few non-diagnostic sherds were found among the slags.

Vrondou valley
At the 24th kilometre of the road from Serres to the villages of Vrondou (Ano Vrondou, Kato Vrondou) lie the remains of a little radioactive slag-heap reported by Stavropodis (1983). The road crosses a cultivated valley at the site where the radioactive slags were detected. Only an amount of slag mixed with the cultivated earth can be observed today. Any estimation of the slag reserve is very difficult. The radioactivity of the slags is medium.

Most of these sites correspond to a series of mining areas and slag-heaps with evidence for possible exploitation in early times. In a recent work (Bassiakos *et al.* 1999), the detected radioactivity of the slags has been associated with the smelting of radioactive magnetite iron sands. The original hypothesis that the radioactive elements in the slag come from the use of radioactive coal as fuel has not been proven.

Aiming at re-measuring radioactivity, another five sites with metallurgical residues were revisited as part of the same project. None of the bellow mentioned sites revealed any radioactive anomaly.

Aggistron
A group of smelting sites, which have left large slag piles and heaps, all located around the modern village of Aggistron (close to the Bulgarian frontiers, 33km northwest of the city of Serres), has been studied by Chiotis *et al.* (1996). One slag-heap, 2.5km west of the village has been reported as radioactive (Stavropodis 1983). Chiotis *et al.* (1996), based on an indirect radiocarbon result and on their own analyses, ascribe the whole massive activity (probably for noble metals production) to a post-Roman period, presumably Byzantine or later, at a time when water-driven belllows were applied in metalworking.

Kimmeria
2km north of the city of Xanthi, on the eastern riverbank of a stream, lies another already known large slag-heap. Part of it has already been removed and the original stratification of the rest is disturbed. A reserve of about 1,500 tons of slags is estimated here. The measured radioactivity is low. The slag fragments are voluminous, irregular ropy and non-glassy in their upper surface. Their interior is rich in pores, some of them extended also up to the surfaces. From this site, Stavropodis (1983) had published a photograph of a simple shaft furnace, almost integral (unfortunately not giving dimensions or any other details), which does not exist any more. About ten thick fragments of furnace walls and bottoms, as well as conical clasps for joining tuyères were observed; they consisted of reddish baked clay, similar to those seen near

Katafyto. Among the slags an intact tuyère was found, fitting well into a terracotta clasp. A few sherds, found in the heap, bellong to the Ottoman period.

Thermes
Two slag-heaps, located 500 meters apart north of the village of Thermes close to the Greek-Bulgarian border, were reported by Papastamataki (1985). The total amount of slag estimated to be 1500m^3. No evidence for furnaces has been reported although Papastamataki identified the area for the charcoal preparation. The slag from Thermes comes from smelting of galena (Papastamataki 1985) suggesting the production of silver as the main activity in the site.

Komaros
The site is located approximately 5km North of the ancient town of Messemvria-Zone. The geomorphology of the area suggests mining tunnels. Local people talk about slag-heaps close by and in the site of Petrota. Komaros and Petrota will form part of a larger survey project in the future.

In an attempt to put the metallurgical finds in the archaeological context of the region they are located, the above mentioned surface finds are grouped geographically as follows:

Vrontou Mountains
Aggistro, Phaia Petra, Ano and Kato Vrontou, Kataphyto, and Vathytopos (Fig.10). All the sites with the exception of Aggistro revealed radioactive slag-heaps. The archaeological evidence related to the area is limited. There are finds from the area of Sidirokastro where a Byzantine castle has been discovered (AD 36, 1981, B2 Chronica, 346) but the slag-heaps are not close by.

During the fieldwork, surface pottery was recorded from the sites in Aggistro and Faia Petra but no pottery or any other archaeological material was apparent in Vrontou, and Vathytopos. The pottery fragments recorded in Aggistro cannot be dated safely since they are mainly coarse wares with no dateable characteristics. The site, however, is located close to a hamam building of the late Byzantine early Ottoman Period. Pottery fragments from the Ottoman period were scattered in Kataphyton. In Faia Petra pottery fragments where often visible between the slag and the furnace lining. Amongst them there is a fragment from the lip of a medium size open pot (*lecanis*) of grey colour. The fragment can be dated in the late Hellenistic period, around the end of the 3rd century BC (Drougou 1991). If magnetite sand was the primary ore source in the area (Bassiakos *et al.* 1999, Photos 1987, Papadakis 1975) then these open pots could be used for the washing of the sands before smelting.

Menoikion Mountains
Panorama, Mikropoli and Kalithea, all with radioactive slag (Fig.10). The broad area of the Menoikion Mountain has revealed plenty of archaeological evidence. In the area of Alistrati there are many graves from the Hellenistic to the Roman period and it has been suggested that a small rural settlement *(kome)* was located in the area (AD 36, 1981, B2 Chronica, 346). Pit graves from the 2nd and 1st centuries B C have also been found in the area of Prosotsani (AD (39) 1984, B2 Chronica, 273). An ancient city of the 5th century BC called Gazoros has been excavated in the area of Nea Zichni (Poulios 1995).

Pangaion Mountains
Nikisiani and Palaiochori (Fig.10). The rich gold and silver mines of Thrace where well known and there are particular references to the Mount Pangaion often in connection with the interest of Phillip II of Macedon in the area (Xenophon *Hellenica* 5. 2. 27; Thucydides 4. 105. 1; Plutarch *Cimon* 4. 1; Demosthenes *On the Hersonese* 45).

The foundation of colonies such as Philippi (by Philip II of Macedon), Amphipolis (by Athenians) and Fagris (by Pierians) and a series of Thasian colonies along the coast demonstrate the importance of the area in antiquity. The sites continued their life both during the Roman and the Byzantine period.

Archaeometallurgical research has been concentrated on the northwest slopes of the mountains. An ancient smelting furnace has been discovered in the area of Ninisiani in a site called Libadia in close proximity to two fortified sites of the Classical and Roman times (AD 37, 1982, B2 Chronica, 326) but it was impossible to locate it.

Rodhope Mountains
In Kimmeria, Thermes, and Komaros (Fig.10) no radioactive slag was found (with the exception of Kimmeria). The pottery found in Kimmeria (mainly coarse and glazed wares) bellongs to the Ottoman period. No archaeological material has been noted in Thermes. However, in the broader area there are ancient castles situated on strategic positions in order to control routes of communication between the Aegean and the Thracian hinterland. In some cases, as in Kalyva-Kastro, a settlement is also included. The relationship between these fortified settlements and the mining areas have still to be investigated. The site in Komaros lies immediately north of the walls of the ancient Samothracian colony of Messemvria-Zone.

All the above areas have been inhabited continuously from the Classical (even earlier) times up to the present. Furthermore, these areas are identified with the territory of ancient Thrace (see also chapters I- III). The lack of systematic research in the area made the identification of any possible connection between mining areas and specific archaeological sites impossible. A research project between IGME and the Department of Antiquities of East Macedonia in the Pangaion Mountain in the 1980s is still unpublished. Poulios suggested that the late Byzantine pottery collected from different mining areas, amongst them Nikisiani and Palaiochori, points to the exploitation of the mines during the Ottoman Period and this corresponds with Photos' dating (Poulios 1988, 344; 1987).

Dating of the aforementioned sites was based on macroscopic characteristics of the slag, surface pottery and on comments by the 16th century traveller P. Belllon (Belllon 1553, chapter 57). In my opinion there is not enough evidence to accurately date the sites. Furthermore, by ascribing most of the mining-smelting sites between the rivers Strymon and Evros to the Ottoman period (Poulios 1988, Photos 1987) most of the researchers excluded *de facto* the possibility of earlier exploitation. What all the previously mentioned sites - both radioactive and not - have in common, is that all historically bellong to the same geographical and political unit, the first Macedonian Meris (see chapter I) where iron/copper mines were left under the control of the indigenous population (see chapter III). In this region, as mentioned earlier, archaeological, epigraphic, and documentary evidence (summarised in Samsaris 1987) strongly indicates that although gold and silver mines were under Roman imperial control, the exploitation of the iron and copper mines was under indigenous people. An arranged tax was paid to the Roman authority of the area (located probably in Philippi). This possibility along with the rich, multi-period industrial landscape testifies to the importance of metal production in the region.

Nevertheless, it is possible that the mines of the Ottoman period continued the exploitation of previously existed mines. Excavation in one of the smelting sites could provide evidence for the history of exploitation in the region since it seems that it covers more than 2000 years. The analyses of excavated finds in this thesis (see table 2), includes artefacts and industrial waste from the 5th century BC to the 3rd century AD and aims in creating a safe archaeological, analytical and chronological framework for the regional study of iron in North Greece.

IV.2i. Messemvria-Zone (end of 6th to 4th centuries BC): typology and distribution of industrial waste and iron finds.

Messemvria-Zone is a "commercial colony" (*emporion*) and was founded on the Thracian coast in the 7th.century BC by the inhabitants of the island of Samothrace. It lies on the coast, at the South end of the Zonaia Mountains. The archaeological finds cover mainly the period from the 6th century BC to 4th century BC, but there are also indications, mainly in the form of coins, that the occupation of the site was continued up to the 5th century AD (Tsatsopoulou 1997). In order to better understand the distribution of the iron material, the three main phases of the city and their features will be briefly described.

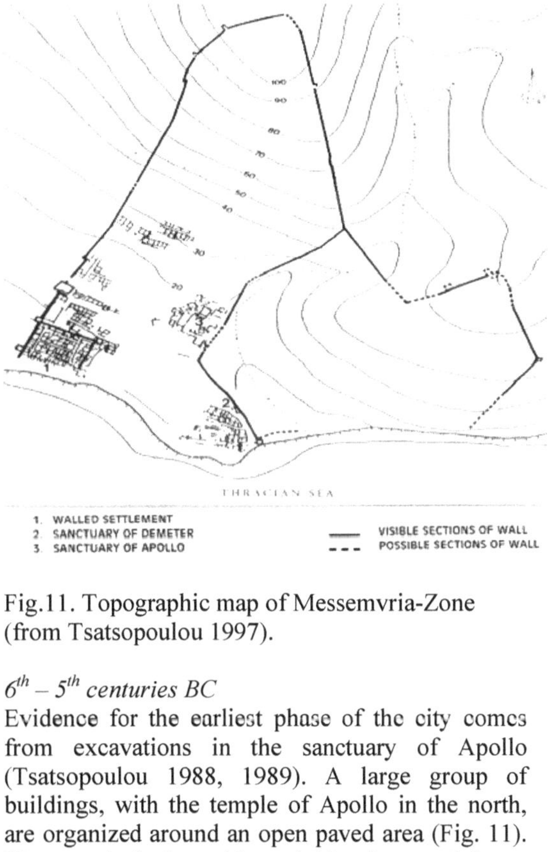

Fig.11. Topographic map of Messemvria-Zone (from Tsatsopoulou 1997).

6th – 5th centuries BC

Evidence for the earliest phase of the city comes from excavations in the sanctuary of Apollo (Tsatsopoulou 1988, 1989). A large group of buildings, with the temple of Apollo in the north, are organized around an open paved area (Fig. 11). The temple is 9m wide and 15m long and has a north-south orientation (Fig. 13). The complex occupies a total area of 35 x 45 square meters. The buildings are arranged in a Γ-shaped complex around the open area. The architectural complex of the sanctuary of Apollo is the only building dated to the 6th century in the whole-excavated area. There is no clear evidence about the size, extension and the general layout of the city in the 6th century.

Fig. 12 (a). Pottery with incised inscriptions in the local idiom from the sanctuary of Apollo (from Tsatsopoulou 1989).

Most of the finds were discovered both inside and outside the temple of Apollo (Fig. 18). There are many marble and clay architectural members, fragments of marble inscriptions, parts of archaic statues (*kouroi*), and many fragments of pottery. They are all dated in the 6th and 5th centuries BC. The pottery is comprised of Attic black-glazed and black-figured vases, of mainly table-drinking wares (*skyphoi, pinakia, cylix*) and local hand-made pottery known as Thracian. Half of the 6000 pottery sherds (usually Attic black-glazed or black-figured wares) discovered in the temple and its vicinity bear incised inscriptions with the name of the god Apollo written in a local idiom resembling the Aeolian dialect. One third of these inscriptions are written in the archaic way from right to left (Tsatsopoulou 1989). This evidence leads to the safe assumption that the sanctuary was functioning in a local/regional scale. Since Messemvria-Zone is one of the commercial colonies of the island of Samothrace it could be suggested that the sanctuary of Apollo was under the protection of the great temple of Apollo and Kabeiroi in Samothrace.

Weapons (PLATES 5a-b, 6a-b,and 7a in the appendix).

Spearheads (MES 22, 24, 25, 26, and 27, the first five from left in Fig. 12a) and one arrowhead (MES 29, second from left in Fig. 11c) have been discovered during the excavation of the Apollo Sanctuary (Fig. 13). According to the pottery found in the same context (Fig. 12a) they are dated at the end of the 6th century BC. In the same context bellong a group of iron spits (MES 31, 32, 34, 35), and the iron currency bar (MES 21) pictured in Fig. 14a-b; also the knives (MES 17, 18), and the small key (MES 20) probably bellonging to a wooden chest. The deposition of these artefacts, north of the wall T4 on the external northwest corner of the temple (Fig.13), suggests that they were probably placed there after they were removed from display in the temple.

TYPE OF FIND	MESSEMVRIA-ZONE		AVDI	
	Recorded	Sampled	Recorded	
KNIVES	15	5	34	
SICKLES	2	2	2	
SCISSORS			6	
SPEARHEADS	11	5	6	
ARROWHEADS	12	4	6	
AXES			6	
SCISSELS			2	
WEDGES			7	
SPITS	10	4	9	
BARS	1	1		
SAWS			1	
PLOUGHS	1			
HORSE-BITS				
STRIGILS	2		21	
KEYS	2		7	
PINS			9	
RINGS	1		8	
SPATULAE			8	
BELLS			3	
NAILS	6		145	
OTHER (hooks, sheets, chain rings)	3		24	
TOTAL	**66**	**21**	**158** (+145 nails)	
SLAG	**36**	**16 (all SEM)**	**72**	

Table 2. List of artefacts and industrial waste (recorded and sampled) from the three sites

(b)

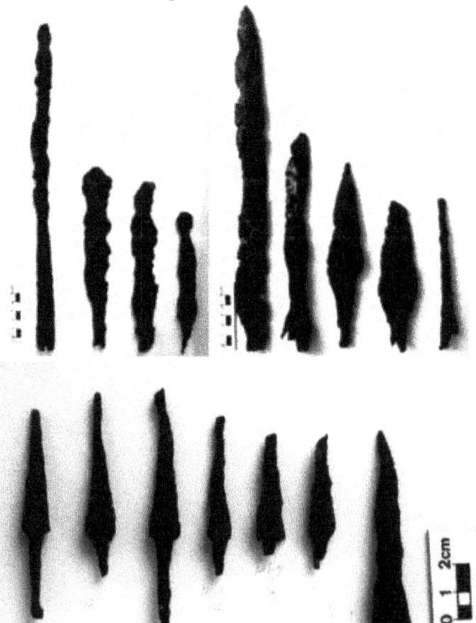

(c)

Fig. 12. Messemvria-Zone. (b) Spearheads from the Sanctuary of Apollo (end of the 6th century BC); and (c) arrowheads (4th century BC)

Fig. 13. Messemvria-Zone. Aerial photograph of the Apollo Sanctuary. The arrow indicates the approximate location of the iron artefacts (from Tsatsopoulou 1997).

The small chest key and the majority of iron finds, as well as the very good quality of the pottery found within, indicate that this was probably the place where precious artefacts where kept after being removed from display in the temple.

Two spearheads (MES 23, 24) were found in the southeast part of the city, close to the sanctuary of Demeter and they are dated in 5th century BC. One spearhead (MES 55) was found in the northwest part of the city during the 1995 excavation season and it is dated to the 5th century. Arrowheads (MES 47, 62) were found scattered close to the west fortification wall. Spearheads (MES 7, and 10, PLATE 7ain the appendix), one dagger (MES 16) and arrowheads (MES 1, 2, and 3, PLATE 5b in the appendix) were found in the enclosed settlement (Fig. 16), and they are dated to 4th- 3rd centuries BC. Spearheads and arrowheads cover a typology of the late 6th and 4th centuries BC, and their manufacturing techniques are discussed in chapter V.

Spits and iron bar (PLATE 7b in the appendix).

Although iron and steel bars are often found in Central and North Europe, they are not a common find in Greek and Mediterranean excavations. There is no evidence for currency iron or steel bars in the Mediterranean during the first millennium BC (Tylecote 1987, 255). The elongated bar of iron (Fig.14b) found in the sanctuary of Apollo, is thus a unique find because its morphology, metallography and archaeological context suggest that it is the earliest and only iron currency bar found in the Mediterranean.

Various objects of different shapes and sizes were used in different parts of the world to give to the bloomery iron (produced in a smelting furnace after the reduction of ore and with the use of fluxes) a marketable form. Thus bars, flattened balls or plates were used in Europe, Asia and Africa (Rostoker and Bronson 1990, 96). The best-known analysed examples are the sword type bars of pre-Roman southwestern England (Tylecote 1987, 206-208; for a general review on Great Britain see Crew 1994). Most of these finds are in use in the middle ages and they are made of steel. Quality control takes the form of a deep incision to expose the metal in the interior of the bar.

The absence of evidence rather than evidence itself had forced scholars such as Pleiner (1969) to suggest that iron was circulated in the ancient Greek world in the form of iron spits. I had the opportunity to revise this statement by studying the spits and bar found in the same archaeological context.

The Messemvria iron bar is 0.26m long, 0.45m wide, 0.017m tall and weights 615.6gr (Fig. 14b). A cross-section was removed from the left edge of the bar and was examined under the metallographic microscope and the Scanning Electron Microscope (see chapter V).

The shape (elongated and flattened in order to be handled easier), the weight (almost that of a single bloom) and the context (votive, in the same group with the spits) suggest that this is not a simple iron bar but most likely a currency bar.

14 (a)

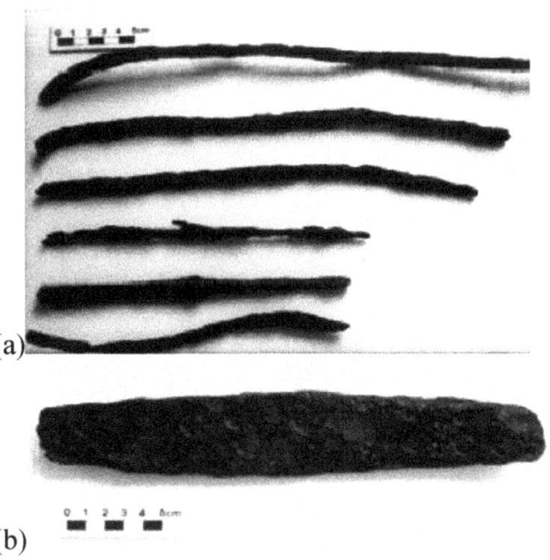

14 (b)

Fig.14. Messemvria-Zone. (a) spits and (b) iron bar found in the Sanctuary of Apollo (end of 6th century BC).

A total of 6 spits (Fig.14a and PLATE 7b in the appendix) were also studied. All the samples examined displayed steel microstructures (see chapter V). The spits analysed were made by welding and then twisting low carbon steel to high carbon steel. In order to check if this was the case with all spits I sampled another four from the ancient settlement of Avdera, a Greek colony on the Thracian coast (Fig. 2 and 21). The Avdera spits were found in a domestic context during the excavation of the Classical settlement. They are made not from steel as the Messemvria ones but from ferrite iron with a lot of slag inclusions (chapter V).

This evidence, although contradictory, is very important. The Messemvria-Zone spits seem to be of special value and use because they are found in a ritual context and they were made of steel. The Avdera spits found in a domestic context were made of ferrite iron because their use –roasting- did not require any of the properties of the steel.

These new finds suggest a clear differentiation in terms of the quality and use of spits. Further research is required to establish whether the shape, weight and carbon content of iron and steel currency bars was standardized.

On the basis of the above evidence it may be concluded that the iron bar, probably forged from single blooms, was the marketable form for the circulation of ferrite iron in ancient Greece at the end of the archaic period (6th century BC). The spits, more easily carburised to steel due to their long and thin shape, were used as steel currency bars.

5th – mid 4th centuries BC

The fortification walls, most of the excavated parts of the city, and the cemetery bellong to this phase. The excavations revealed the fortification double wall (1.37km long) that starts from the sea and climbs up to the top of the hill, 110 m high, where the acropolis is located. The fortification wall divides the town in two parts. The west part of the town occupies 73.3 acres and the east 49.3 acres (Fig. 11). The towers are located 30-35 meters apart. Most of the existing archaeological data comes from the west part of the city since trial trenches on the east did not provide any archaeological traces. The buildings are arranged according to the Hippodamian system very well known from ancient cities in Greece and Italy as well as from the excavations in the Macedonian city of Olynthus (Ward-Perkins 1991, Fig.11 and 11a; Robinson 1930). Parallel to the west city-wall, roads 2.50-3.50m wide and north-south streets 2.50m wide form-building quarters (Fig. 11 and 16). The steep slope on this section of the city (ΔT_4) led to the construction of oblong buildings with narrow corridors between them. The levels of the floors vary significantly.

Houses are typical of the 5th century, often having a second floor, with rooms arranged around a courtyard and facing due south to the sea. Most of the finds indicate domestic spaces but there are also at least two cases where a co-existence of domestic and industrial activity could be argued. This is a house 162m^2 in area close to the west city wall where large storage facilities and finds, such as clay moulds for figurines, indicate a small workshop. The other is located north of the first North-South street and it contains two rooms full of amphorae placed upside down; their abundance and typological variety suggest that the owner had practised extensive trade activities within the Aegean network (Tsatsopoulou 1995).

The part of the city in the southeast corner (AT_2) is not built according to the Hippodamian plan (Fig.15). Roads follow the natural relief along the coastline. Part of the settlement and the south city wall are missing. Houses are similar to the ones described earlier, but they are smaller. During the excavation of this part (Tsatsopoulou 1988) extensive destruction layers were discovered under the layers of the 5ht century BC indicating major re-constructions. Finds include spindle-whorls, fishhooks, storage and tablewares, Attic black and red-figured vases. Local hand-made wares, such as these pictured in Fig. 15b and 18c, were found in all layers. Also, water wheels and storage arrangements, and some bath facilities were found *in situ*. A small temple dedicated to the goddess Demeter was discovered here in 1973 (Vavritsas 1973). It comprises three rooms and it is oriented towards the east city wall. Finds include clay, marble and metal figurines, mainly female. Fragments of Attic vases and of hand-made pottery, fragments of statues, one gold medal, one silver vase, a red-figured pelike full of gold and silver sheets decorated with either eyes, body parts, animals, or people bearing gifts to the goddess (Vavritsas 1973) were amongst the finds.

Although a new town emerges in the 5th, there are some elements suggesting continuity with the 6th century. These are handmade pottery and burial customs (discussed in the following section). The city planning of the 5th century BC differentiates two areas: one with the sanctuary of Apollo dominating the area. The other quarter is located with reference to the temple of Demeter on the east wall. The 'Apollo' area is built according to a well-defined Hippodamian system. Houses and workshops face the streets and the sea (Fig.11 and 13). According to the excavation diaries (1986-1996) high quality imported pottery, mainly drinking and transport wares predominate in the material record.

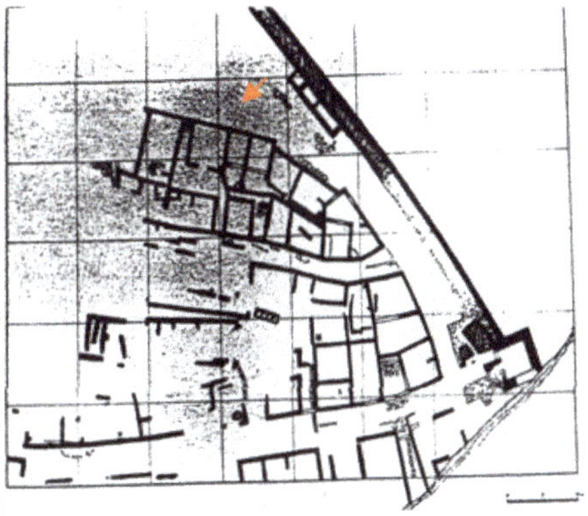

15 (a)

15 (b)

Fig.15. Messemvria-Zone. (a) map showing the planning of the southeast part of the city with the temple of Demeter (3 rooms) oriented towards the east wall (from Tsatsopoulou 1988); (b). The handmade jug approximately 0.10cm tall and is typical of the pottery found in the southeast section of the city (Vavritsas 1973).

The 'Demeter's' quarter (Fig. 11 and 15) is smaller. It is not build according to plan. Streets follow the natural formation of the area. Houses are smaller and look packed. Large storage vessels, presumably for cereal storage *(pithoi)* were discovered in some houses and handmade local pottery is noticeably more than the imported. Iron slag is predominant in the material record (excavation diaries 1987).

Differences in architecture and pottery distribution could reflect social differentiation within the town. For example it could be suggested that entrepreneurs, traders, rich craftsmen inhabited the area of 'Apollo', along with an elite, probably priests who were responsible for the temple. The 'Demeter' quarter could bellong to people dealing with agriculture as the storage facilities and the depictions on the silver offerings indicate.

Most of the industrial waste comes from the area north and north-east of the enclosed settlement, the destruction layers of the 5th century city in the 'Demeter quarter, and the area out of the west fortification wall (Fig.16).

Most of the material is slag, either shaped or small amorphous fragments. Some scattered fragments have been found in the foundation layers of the streets as well (excavation diaries 1992 - 1995). It also seems that iron slag is found more often in layers before and during the 5th century BC. This is the case with the slag found between the years 1987-1996. This also seems to be the case with the earlier excavation in the enclosed settlement. Vavritsas, in his reports (*PAE* 1966-1983) and in the rare excavation diaries often refers to slag fragments found in the layers under the foundation layer of the 4th century enclosed settlement. However, he did not collect any slag from 1966 to 1983. Many fragments of slag are today apparent in the dumping area of these early excavations. Slag was also found north and west of the sanctuary of Demeter, and in the area of houses 1-3 of the enclosed settlement (Vavritsas 1984 in Tsatsopoulou 1987, 72). The end of the 5th century BC could be suggested as a *terminus ante quem* for the date of the slag found in the destruction layers of the city. The southeast part of the city, namely the area outside of the west wall of the enclosed settlement, is suggested as the location of the main industrial activity in Messemvria-Zone (Fig. 16). To check this hypothesis, the author conducted a small trial excavation on the foundation of the west wall of the enclosed settlement, as part of the restoration work in 1996 (Kostoglou 1996). The wall is based on a 0.20m thick layer of red soil, which contains a lot of stones and fragments of pottery. Under this, there is another layer almost 0.20m thick of gray soil with a lot of gravel. The soil looks burned. In 1987 Vavritsas excavated a furnace-like structure on the corner of the northwest tower of the enclosed settlement. The structure is still *in situ* but there is no recording of the excavation.

Fig. 16. Messemvria-Zone. Aerial photograph showing the layers of the 4th –3rd century BC enclosed settlement and the layers of the 5th century BC underneath (from Tsatsopoulou 1997). Arrows indicate the area with the higher density of slag.

The so-called furnace is located 2.50m outside the west wall of the enclosed settlement. It is 0.90m high and has an external diameter of 2.40m (internal 1.50m). The structure is based on the same thick layers of red and gray soils that were described earlier and seems to be in use contemporary with the enclosed settlement, namely 4th century BC. The walls are 0.35 wide; they are built of bricks and are covered with clay lining inside. The floor of the structure is again made from clay. There are no indications to imply involvement in pyrotechnological activities. The brick walls, the floor, and the clay lining are free of every deformation or discoloration suggesting contact with fire.

Hence, this structure, seems to have been used as a solid container, a kind of storage facility, unless it was not used at all. The drawings and photographs from this excavation are courtesy of the 19th Ephoreia of prehistoric and Classical antiquities in Komotini. Nevertheless, the structure is built on a very well isolated layer, which might be indicative of some industrial activity involving firing, such as smelting, that pre-dates the structure. This has been proved to be the case after a small trial section was carried out on one of the sides of the structure where some stones indicated the existence of another, possibly earlier, wall.

The clay structure was founded on top of a structure that might be a small smelting furnace. What survived is part of two walls (0.68m and 0.50m thick and 0.70m high) forming a circle with an internal diameter 0.60m. The walls were constructed with schist stones (a material known for its property to stand high temperatures). The space between the two walls was covered with the material from the construction of the clay structure, but signs of firing were obvious. The slag number MES101 was found in contact with the one of these walls. As it will be shown in chapter V, this is a furnace bottom slag and provides the evidence for the identification of the structure as a smelting furnace. It is the only excavated furnace so far.

Domestic and/or agricultural tools such as knives (MES 65, 68), sickles (MES 44, 45, 67), and falces (MES 53, 67) were found in the northwest-excavated part of the city. Knife MES 9 comes again from domestic context in the enclosed settlement. Although knives are used for a number of purposes, tools such the falces (MES 53 and MES 97) are known as 'falx arboraria' and they are used for pruning trees and vines (White 1967, 86, Fig.59).

Knives, sickles, falces (PLATES 7c and 8 in the appendix).

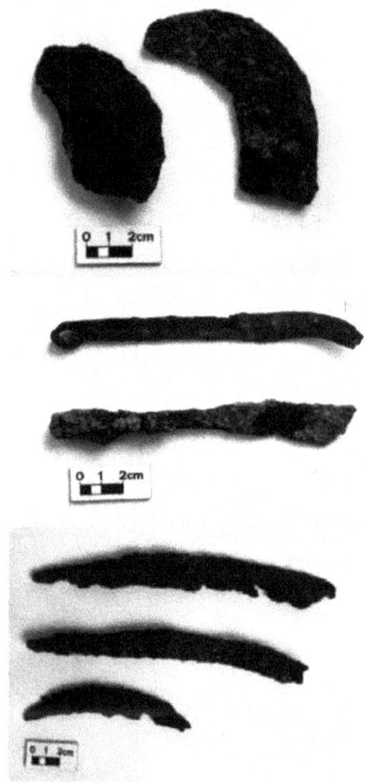

Fig. 17 (a). Messemvria-Zone. Sickles, billhooks and knives from the northwest part of the 5th century BC settlement.

4th – 3rd centuries BC

An unknown event caused the destruction of the 5th century BC city clearly pictured under the enclosed settlement in Fig. 16. After that, the city is concentrated on the enclosed settlement. This settlement is founded on top of layers bellonging to the city of the 5th century BC and it has a slightly different orientation (Fig.16). In the small area of the enclosure houses, small workshops (one for wine making) and administration buildings are located. Walls and towers are repaired with the same stones used in the buildings of the 5th century BC. The city in the west and east is destroyed but he sanctuary of Apollo is still in use. Inscriptions of citizens such as ΑΔΑΜΑΣ ΒΟΣΤΑΔΟΣ (Adamas Bostados) appear on the external surface of the northwest tower. Finds include red-figured pottery, local handmade pottery, coins, figurines, arrows, and iron tools (Vavritsas PAE 1966-1983). Messemvria-Zone is the only trade centre in the Thracian coastline having its own coinage. A large number of coins, more than 4000, were discovered during the excavation of the city. They are representative of cities in close proximity, such as Maroneia, Aenos, and Avdera. There are also coins from the Aegean Islands and the Ionian Islands, as well as from Thessaly, Macedonia, and Asia Minor. Coins of Maroneia represent 30% of the total. They bellong mainly to the 4th century BC and there are three types all depicting the head of Apollo on the first side (Galani-Krikou 1992).

Zone is also recorded as paying high tax to the Athenian League. Zone pays two talants, as much as its metropolis Samothrace and only one talant less than Maroneia, which is a far largeger colony. On the basis of this evidence it has been suggested that in the 4th century BC Zone was independent from the metropolis Samothrace (Tsatsopoulou 1997).

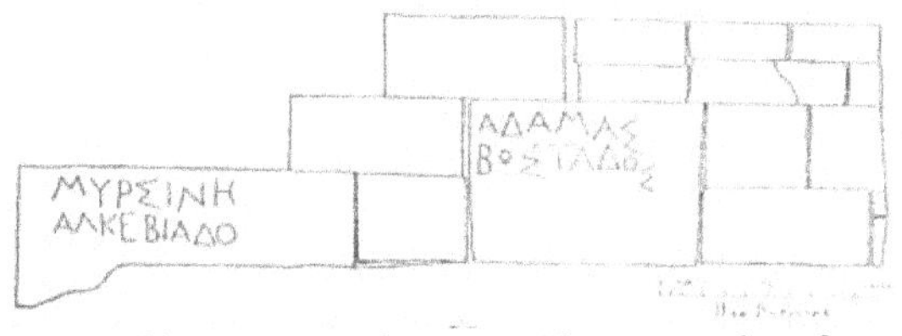

Fig. 17 (b). Messemvria-Zone. The external surface of the south wall of the NW tower of the enclosed settlement with inscribed names (Vavritsas 1977).

The finds of the 4th and 3rd centuries BC suggest a major demographic decline in the town although the reasons are hard to trace. Evidence for military action can be seen on the multiple repairs of walls and on the iron weapons found inside the settlement (discussed later in the section). Strong discontinuity in the settlement patterns is evident. Whatever caused the destruction of the town has also put an end in the iron production as it is discussed later in this section. Change to a monetary-based economy is also evident in the 4th century BC when the city has its own coinage. Elements that survived are the local handmade pottery (Fig. 15b and 18c), and Hellenised names of Thracian origin such as Adamas Bostados (Fig.17) (Tsatsopoulou 1987).

Cemetery

The cemetery of the city covers mainly the 5th and the 4th centuries BC although it was in use since the Iron Age as three excavated tumuli dated in the 9th –8th centuries BC (Vavritsas 1969-1973). It is located in close proximity, less than 500m, out of the west fortification wall. The burials include a variety of materials and practices. Burials are oriented from east to west and they are located in two rows 10m apart from each other suggesting that the cemetery, as well as the city itself, were built according to a plan. Burial customs include both inhumations and cremations. Sarcophagi, chest or pit burials, handmade pithoi and Greek amphorae are all in use with no clear distinction in the quality or quantity of the grave goods. Three stone circles were found including different types of burials. Stone circles form the base of tumuli and they seem to be an indigenous tradition. For example, the tumuli in Messemvria-Zone and other ones in Aegean Thrace, thought to bellong to Roman veterans of Thracian origin, have such a basement (Triantafyllos and Terzopoulou 1987). On the contrary the tumuli in Avdera from the 6th to the 3rd century BC do not.

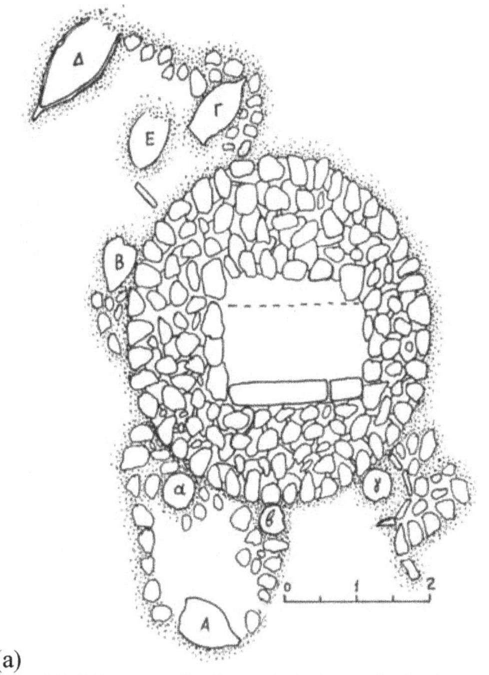

(a)

Fig. 18. Messemvria-Zone. (a) stone circle the basement of a tumulus with multiple burials (Vavritsas 1970);

(b)

(c)

Fig. 18 (b) typical group of red-figured vases and handmade pottery (c) from the cemetery dated in the mid 4th century BC (Vavritsas 1983).

A typical tumulus for Messemvria-Zone (Fig. 15) contained different burials in its periphery. Both amphorae and handmade local pithoi were used as containers of the cremated bodies. The finds are red-figured vases, gold jewellery and handmade vases. Iron finds from burial context consist only of one type of find, strigils.

Three monumental burials were also discovered. One of them had one cremation buried inside and a second outside. The cremation outside covered an area 2-3m in diameter and included 9 red-figured vases (*pelikae*) of the 4th century BC, as well as bronze vases, iron strigils, and small votive clay vases. It was probably a mass burial according to the size and the finds (Tsatsopoulou 1991). There is no evidence to connect the mass burial with the destruction of the 5th century town. There are more inscriptions from the cemetery with both Greek and Hellenised Thracian names (Tsatsopoulou 1987).

The cemetery of Messemvria-Zone is in continuous use from the Iron Age to the 4th century BC. This suggests strong cultural continuity. In the author's opinion, the material record of Messemvria-Zone points to a small community with very distinctive character and a distinctive cultural identity when compared with other sites of the same period (i.e. Avdera). The material cannot simply be characterised as local or Greek. Burials as this pictured in Fig. 18 present a mixture of customs known from both the Greek (amphorae used as containers for cremations) and local traditions (cremations in tumuli). A mixture of both imported and locally made handmade vases is included in the grave goods (Fig. 15).

The only iron finds from the cemetery are two strigils and two iron rings (Tsatsopoulou 1987). Strigils are a common find in Classical cemeteries. They are often found in the cemetery of Avdera (see next section) but they are not common in Messemvria-Zone.

Although elaborate rings from precious metals and stones have been found, the iron rings mentioned above are of special interest because they bellong to a specific type: a simple ring with oval shaped upper part made only from iron. The earliest rings of this type are dated as early as the late Mycenaean period and they are found in Peloponnese (Pleiner 1969, 8). There is no technical information obtained from the study of these rings, but they seem to be of symbolic significance. A type of iron ring known as 'samothracia ferrea' sold on the island of Sampthrace is mentioned in Roman authors of the 1st century BC (Pleiner 1969). Initiates of the cult of the Great Gods of Samothrace were wearing such rings (Blakely-Westover 1999) but it is hard to make any assumption about their meaning and the invocation of the metallurgical act. Although it cannot be ruled out, there is no direct archaeological or analytical evidence to suggest that the iron rings found in Messemvria-Zone and Avdera were all in use by participants of the cult.

IV.2ii. The archaeology of iron in Messemvria-Zone: conclusions.

Focus on the process of consumption of imported material may provide a particularly useful path for tracing indigenous agency. The anthropology of consumption (Appadurai 1986, Bourdieu 1984) shows that demand is never an automatic response to the availability of goods, especially not in colonial situations. Indigenous people show selective preferences. In an archaeological context Dietler (1998) demonstrated this in the study of Greek imports to south France. Dietler sowed that the archaeological study of consumption involves: a) consumption context and patterns of association with imported goods; b) the relative quantitative representation of imported goods; and c) their spatial distribution (Dietler 1998, 300-301). The analyses of ethnographic parallels for the social role of the material demonstrated that this should take place on specific temporal and spatial dimensions (Schiffer 1996, Thomas 1991).

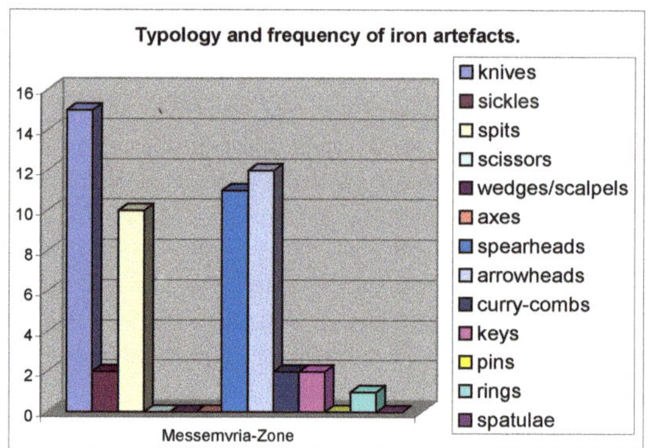

Table 3. Typological frequency of iron artefacts from Messemvria-Zone (based on table 7 in the appendix).

This model suggests that in a colonial situation imports should not be regarded as prestige goods as it often happens in the case of the core – periphery model (see chapter II.1). In this context the amount of imported pottery in Messemvria should not be regarded as indicative of the identity of the people living there. The typology (drinking cups-*cylikae* and *skyphoi*) the ritual and burial context and its spatial distribution (Apollo sanctuary and cemetery, fewer in the settlement) point to a different explanation. If these elements are combined with the fine Attic glassed wares bearing local idiomatic inscriptions and architectural characteristics (double walls, different quarters, stone base tumuli) an interesting picture emerges: the local society is appropriating Greek material according to its own cultural traditions. Two of these traditions can be safely named: (a) ritual drinking and (b) cremation burials in stone base tumuli.

As it can be noticed in table 3, the majority of iron finds that survived in Messemvria-Zone is knives, spits, spearheads and arrowheads. The rest of the categories are represented with less than 5 examples.With regard to their deposition no differences are noticed between iron finds in domestic and ritual context indicating that most of the finds found on display in the sanctuary of Apollo were utilitarian objects. This is well indicated by the deformed (presumably in battle) spearhead MES 23 (chapter V). Although weapons such as spearheads and arrowheads are found in both ritual and domestic context, no weapons were found in burial context (table 4), although such a ritual is known to occur since the Iron Age (Snodgrass 1980, 1967). This depositional pattern should reflect the reality of the 5th and 4th centuries BC in the site because the Messemvria-Zone cemetery and indeed most of the city were found intact. This reality is hard to explain but it could mean that iron weapons were precious objects often inherited to the next generation, thus complicating their depositional patterns. There is no evidence to suggest that the weapons were kept in a city-controlled storage, as is the case in Greek city-states (Snodgrass 1987, 1980). The absence of spatulas, pins and strigils, groups known from Greek graves for example in Attica (Morris 2000), indicate another area of differentiation from Greek burial customs.

Finally, the unique iron and steel currency bars found in ritual context signify that iron had not only economic value but also a high symbolic meaning. The meaning of spits found in Greece has intrigued researchers for long periods because they could not explain why such objects are deposited in the burials and in sanctuaries (for a recent discussion see Haarer 2001). As it will be demonstrated in chapter V, the fact that some of the spits are actually made of steel and could be used as currency (Kostoglou 2003) opens a new way for a new interpretation of such objects.

With regard to iron production it should be reminded that almost all debris of smelting and smithing activities in the town are coming from the southeast part of the town and they are dated in the 5th century BC or before. The mode of production seems to be small-scale, household or community workshop where industrial and domestic activities co-exist. There is no evidence to indicate large industrial activities. The iron and steel currency bars also suggest that iron production was important in the economy of the site before the 5th century BC. No evidence exists for the following centuries. In the 4th century numismatic and epigraphic evidence on taxation by the Athenian League leads to the conclusion that a monetary type of economy based on coins and talants of gold or silver was practised.

Political and economic changes might have forced a change in mining /smelting activities. Since iron is not on demand by the Athenian League the Messemvria-Zone mining turns to precious metals. The study of this change is beyond the scope of this research. It should be noted though that it contradicts the general interpretation of the mining/smelting activities in the area (see chapter III).

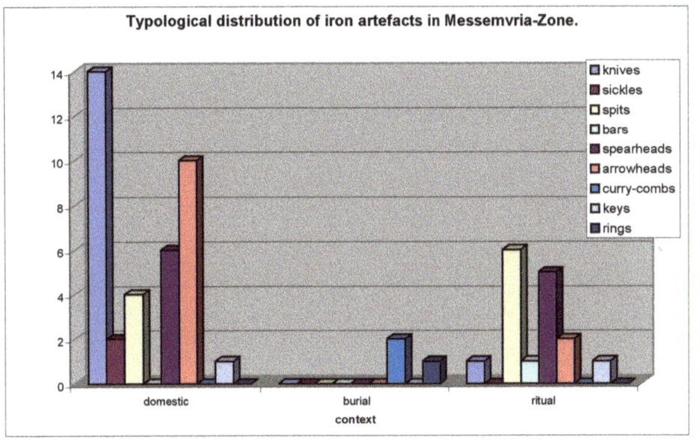

Table 4. Distribution of iron artefacts from Messemvria-Zone (based on table 7 in the appendix).

Finally, it should be emphasised that indigenous elements are strongly apparent on the archaeological record and consist of locally made pottery, burial customs and the use of a local idiomatic language. This community has a sense of bellonging and a sense of cultural continuity manifested in everyday practices and rituals taking place in the temples and the cemetery of the town (Kostoglou 2009).

IV.3. Avdera (5th century BC to 2nd century AD): typology and distribution of industrial waste and iron finds.

The city was first founded first by Klazomenai (c.650 BC) and second by Teos (c.550 BC), in the East Nestos valley, on the cape Baloustra. Peculiar geomorphologic changes of the area influenced the development of the colony (Syrides 1997). Industrial waste and iron finds are described chronologically with reference to their archaeological context.

7th – 6th centuries BC

According to new archaeological data (Koukouli-Chryssanthaki 1988) the first colonists founded a fortified city (Fig.19) using the natural protection of the foothills. The excavation discovered part of the archaic wall in the north of the city, an archaic temple of Demeter, and burials (Koukouli-Chryssanthki 1984). The temple of Demeter was in use from the end of the 6th century BC to the end of the 4th century BC when the Hellenistic cemetery of the city took its place. Pottery from Corinth, the East Aegean islands and the Ionian coast dominates the archaeological record. Pottery of Attic workshops is rare in the early phases of the city indicating that the first colonists kept stronger relationships with the region they came from.

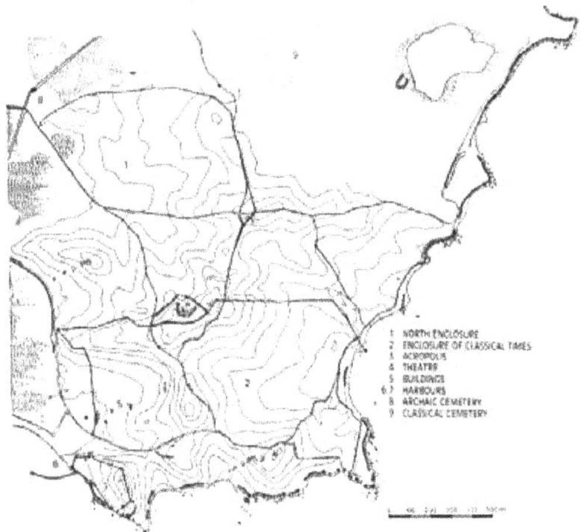

Fig. 19. Map of Ancient Avdera (from Kallintzi 1997).

No iron slag or iron finds were found in any of the excavations of the 7th and 6th centuries BC indicating that iron was not produced locally and that iron artefacts were a rare and precious commodity in this period.

5th-4th centuries BC

The first small colony became a city-state during the Classical times having all the typical characteristics: city walls (5.340m), acropolis, theatre, temples for the city's Gods and Heroes, two harbours, political autonomy, and independent coinage. Imported pottery from Athens, Corinth, and the Ionian coast, silver coins of Avdera found as far away as Egypt and Syria, and the finds from the city' s cemetery indicate that the city flourished during the Archaic-Classical times (Koukouli-Chryssanthaki 1988; Isaac 1986, 73-123, Boardman 1980).

The city walls cover an area of 10 acres and they include areas without buildings in the east, presumably for cultivation or for herding animals (Koukouli-Chryssanthaki 1988, 41). If this is compared with the double walls of Messemvria-Zone then it could be concluded that the land for used agro-pastoral activities was fortified. This evidence suggests that the *chora* of these two colonies was limited. This is a very important realisation for the discussion of the impact of the Greek colonies on hinterland, and the discussion of the Hellenisation process at the end of this chapter.

In Avdera as well as in Messemvria-Zone, the same co-existence of domestic and industrial space is noted and houses of the 5th - 4th centuries BC are built according to the Hippodamian system of planning (Koukouli-Chrysanthaki 1983).

No slag was found in any the areas excavated suggesting again that iron smelting probably did not take place in the city and most of the iron was imported in finished or semi-finished forms. Iron artefacts of the 5th and 4th centuries BC include axes, wedges and spits.

Knives (PLATE 11a, 11b, 12a and 12c in the appendix).

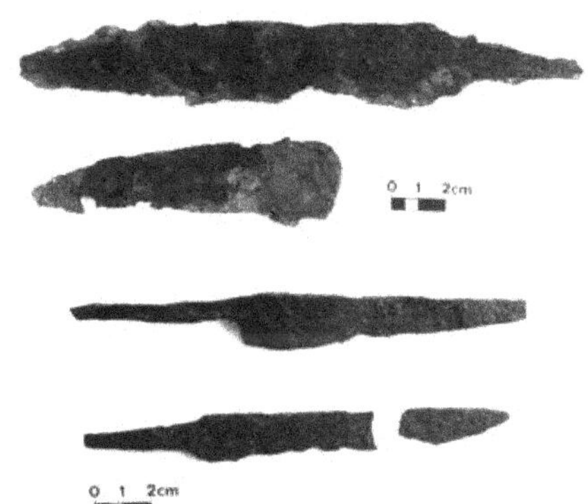

Fig. 19 (a). Avdera. Knives dated in the 5th –4th centuries BC.

These tools (samples AVD 11, 15, 51 and Sickle AVD23 in) are used for a number of domestic and agricultural processes. Their typologies can be traced back to the beginning of iron technology in Greece (Pleiner 1969).

Wedges (PLATE 13 in the appendix).

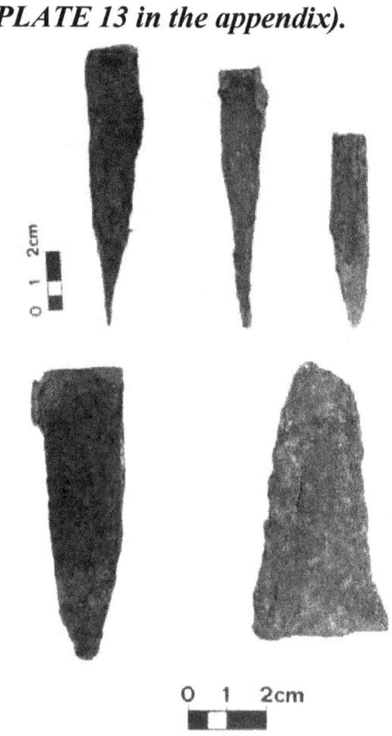

Fig. 19 (b). Avdera. Wedges found in the west gate of the Classical wall.

Their discovery in connection with the Classical masonry wall in the west gate (table 9 in the appendix) indicates that they were probably used to work the stone *in situ.*

4th – 3rd centuries BC

The city walls were destroyed and never re-built at the middle of the 4th century BC and the city's plan is still the Hippodamian (Koukouli-Chrysanthaki 1984, 5-6). Amphorae from the island of Thassos are predominant in the record indicating strong trade relationships with the North Aegean (Kallintzi 1994, Koukouli-Chryssanthaki 1988). Houses (Skarlatidou 1989) continue on the type known from the previous period and domestic and industrial activities are sharing the same building. Archaeological evidence indicates that pottery (Koukouli-Chrysanthaki 1983), clay figurines (Lazaridis 1960), and bronzes (Koukouli-Chrysanthaki 1992) were produced at local workshops during the 4th century BC. None of the excavated workshops produced evidence for iron making. Iron artefacts of the 4th and 3rd centuries BC are found mainly in the cemetery.

Archaic and Classical cemetery

The extensive cemetery of Avdera covers the period from the 7th century BC to the Roman times. The burials from the 7th to the 6th centuries BC correspond to the early phases of the colony and pottery of Ionian workshops is often included in the grave goods (Kranioti 1984, 1987).

There is no clear differentiation between burial customs and type of burial: clay vessels of everyday life such as amphorae and pithoi, clay sarcophagi, chest or pit burials, and often free burials include inhumations or cremations of adults, children, or new born babies. In the case of cremations the places where the pyre took first place were also discovered. Close to the burial the remains of deliberately broken eating and drinking vessels indicate that after-burial rituals took place (Kranioti 1987, Skarlatidou 1984). Tombs are covered with tumuli and some of them have been identified to bellong to the same families (Koukouli-Chrysanthaki 1988). Similar cemeteries are known from other places in Greece such as Attica (Morris 1996). The study of the cemetery is not yet complete but it has been estimated that a striking 64% of the burials bellong to children and new born-babies (Skarlatidou 1987). The cause of death has been identified as malaria (Agelarakis 1997) due to extensive marshlands (Syrides 1997). There are no iron artefacts amongst the grave goods of the archaic and Classical cemetery.

2nd century BC-3rd century AD.

The settlement of the early and late Roman times is smaller than the earlier phases and occupies only the area of the Classical acropolis (Fig. 19). The decline of the city in 3rd century AD was caused by malaria (Koukouli-Chryssanthaki 1997, Syrides 1997). Roman houses (Samartzidou 1984) have been excavated in the area of the west gate (Fig.21).

Most of the industrial waste and iron finds come from trench Γ (Fig.20) which is situated outside the Classical west gate and it is an area with houses and workshops.

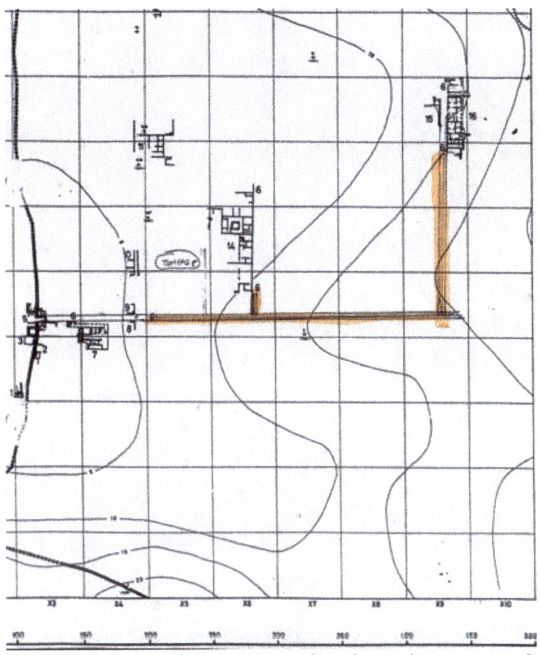

Fig.20 (a). Avdera. Map showing the area of Trench Γ where most of iron artefacts and industrial waste found in Avdera (courtesy of the 19th Ephoreia of Prehistoric and Classical Antiquities).

Axes (samples AVD30, and AVD39, PLATE 11b in the appendix).

This implement (*pelekis* in ancient Greek, *securis* in Latin) is used either simple or fitted with a hatchet (White 1967, 60-61). In its double form it has a vertical cutting edge on the one arm and a horizontal one on the other. Axes are heavy-duty tools. Heavy long-handled types were used for working standing timber, lighter types for chopping roots, making stakes etc; in agriculture they were used to root out surface roots or to prepare ground previously under timber for cereal cultivation or to root out old vineyards (White 1967, Fig. 35, 36).

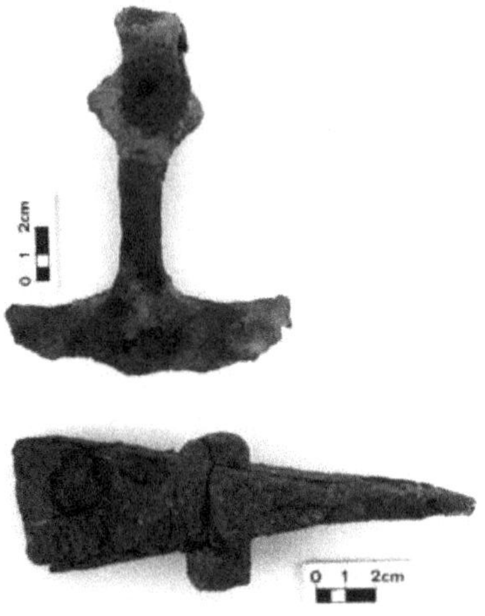

Fig. 20 (b). Avdera. Axes found in the excavations of the 5th and 4th centuries BC.

Two pointed axe (sample AVD48, PLATE 13 in the appendix).

Fig. 20 (c). Avdera. Two pointed pitchfork found in Trench Γ.

The heavy two pointed axes, as literary and monumental evidence prove, were commonly used for breaking the soil in stony ground, especially where olives and vines were to be planted (White 1967, figs.26-27). Although such finds were also collected in Po valley in north Italy (White 1967, 50) this is the first time that such an implement is found in Roman Greece.

Wedge (AVD61)
Similar to the one mentioned earlier from Classical times.

Sickles and billhook (PLATE 11b).

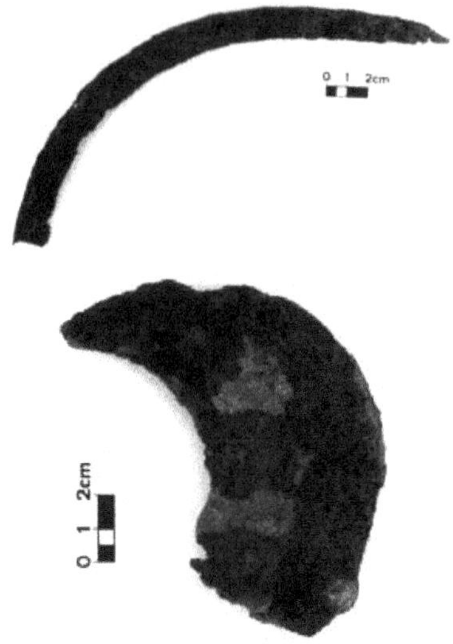

Fig. 20 (d). Avdera. Sickle AVD 23 (left) and billhook AVD40 (right).

This type of agricultural implement consists of a smooth curved blade, equipped with a smooth serrated cutting edge, attached to a short handle, which lies in the same place as the blade. The curvature of the blade varies from a shallow arc to a semi ellipse. It is generally accepted that this implement is a Roman invention known under the general term *falx* and having many variations (White 1967, 80, Fig. 52, 53).

Spits (samples AVD43, 44, and 50, PLATE 11a in the appendix).

Fig. 20 (e). Avdera. Iron spits from domestic context.

Spits are found in domestic context during Classical times. The fact that none of them is made of steel (see spits in Messemvria-Zone) suggests that spits in Roman period had only a domestic use and they were not used as currency.

Saw (sample AVD26, PLATE 12b in the appendix.)

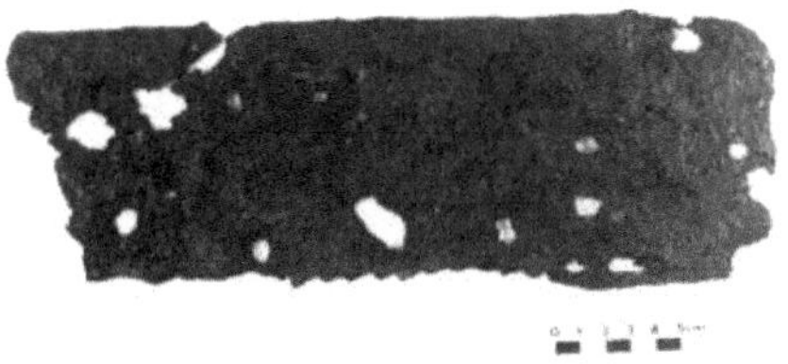

Fig. 20 (f). Avdera. Carpenter's saw dated in the Roman period.

The earliest example of a saw is found in Fortetsa, Crete and it is dated in proto geometric period (c.10th century BC, Pleiner 1969, Fig.5). Saws can be either one or two handled and they were used for pruning trees or working timber (White 1967, 116, Fig. 92).

Hellenistic and Roman cemetery
The cemetery of the late phases of the city occupies the area of the archaic city, covers an area of approximately 1100m^2, and displays the same variety of burial types and after-burial rituals known since the 7th century BC (Samiou 1988). Grave goods include small clay vessels, oil lamps, bronze mirrors and jewelry. The iron artefacts from the cemetery are: two iron spearheads, one dagger, one arrowhead and a most of the strigils (see table 9 in the appendix and Fig. 23 and 24).

Scissors/shears (sample AVD12)
Often used in operations such as sheep shearing where one hand is required to work the tools (White 1996, 119-120, figs. 95-96).

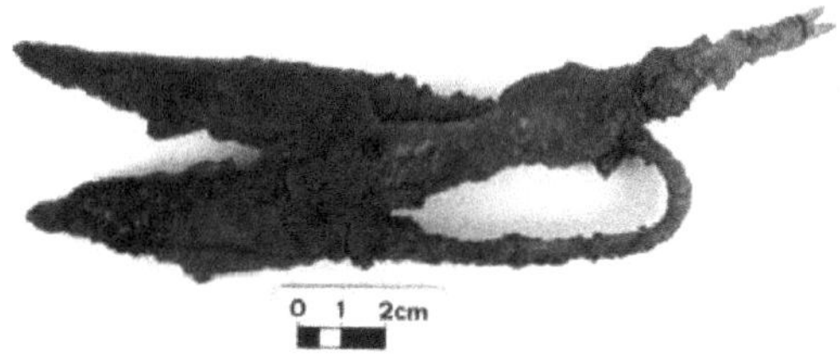

Fig. 20 (g). Avdera. Scears and knife from the Hellenistic cemetery.

Strigils (samples 1, 5)

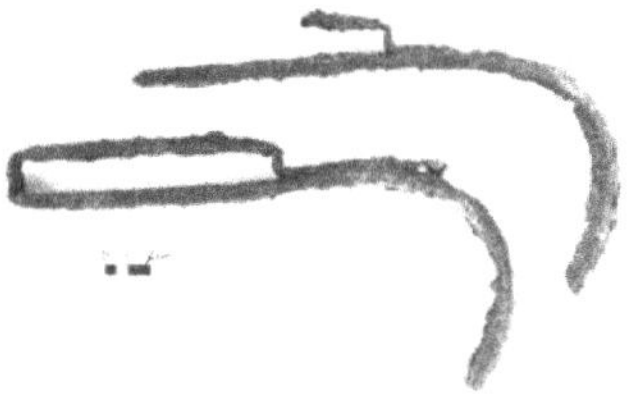

Fig. 20 (h). Avdera. Strigils, Hellenistic cemetery.

As it can be noticed in table 9 more of the strigils found in cremation burials and they are dated in the 4th century BC.

Nails
A number of 145 nails were discovered during the excavation of trench Γ, outside the west gate of the Classical city wall where Roman houses and workshops were discovered (excavation diaries).

IV. 3ii. The archaeology of iron in Avdera: conclusions.

As discussed earlier in this section, iron artefacts and industrial waste are not apparent in the archaeological record of the 7th and 6th centuries BC. On the basis of absence of evidence it should be concluded that iron was not produced and commonly used during the first phase of the colony due to either lack of access to resources and/or lack of the technological knowledge. There is not enough archaeological evidence for the study of iron in Classical Avdera. However, a few finished iron objects such as wedges, sickles and knives are found in domestic context (table 6). These are all small portable artefacts easily transported and there is no evidence that they were made locally.

Small portable finds used for cosmetic operations such as strigils, pins, and spatulas (tables 5 and 6) were found in both domestic and burial context of the 4th century BC. Iron finds are not included in the 'rich' burials of the cemetery where gold jewellery, figurines, and imported pottery are found (Kallintzi 1997). Strigils and spatulas, however, are part of cremation burials (table 6 and table 9 in the appendix) that are considered expensive (Morris 2000, Kallintzi 1997). Iron rings were found in stone sarcophagi (AVD 243, 245 in table 9 in the appendix) and one ring was found in a cremation burial (AVD 20 in table 9 in the appendix). Weapons such as spearheads are found in domestic context and not in burials (table 6) supporting the depositional pattern noticed in Messemvria-Zone.

Industrial waste is again missing from the record, indicating that iron was not smelted locally during the 4th century BC. Slag is found only in Roman context suggesting that iron smelting/smithing took place locally only during the Roman period.

Heavy-duty iron tools and implements occur for first time in the record of Avdera. Axes point to operations in order to clear stony ground cultivation and to work timber. Knives, axes, a saw, a billhook, wedges, nails, and slag (tables 5 and 6) come mainly from the excavation of a domestic area with workshops outside of the west Classical gate dating in the Roman period. This might mean that land under timber was needed for cultivation. It is assumed that major deforestation of two extended forests (Kotza-Orman=great forest and Domuz-Orman=boar forest) took place in Roman times and only traces of the forests are found in the area today (Kallintzi 1997). The iron tools found in Avdera provide evidence for agro-pastoral activities such as clearing stony or rooted ground for cultivation, pruning trees, and shearing animals. Finally, spits occur only in domestic context, they are all dated in Roman times and none of them is made of steel (chapter V) indicating that they were used for roasting.

To conclude, the archaeological evidence indicates that Avdera was a colony with Greek ethnic characteristics and mainly Greek ethnic population. Although it should not be ruled out, no indigenous elements exist so far in the archaeological record of Avdera. There is no local hand-made pottery of the types found in Messemvria-Zone and Kalyva-Kastro. The locally produced pottery of the 4th century is wheel-made and imitates common Hellenistic types. The imported pottery is predominantly coming from workshops in Asia Minor and Corinth, and after the 5th century BC mainly from Attica. The city is destroyed in the 4th century BC as Messemvria-Zone but there is no evidence for repairs.

All iron slag and the majority of artefacts come from the Roman period where agro-pastoral activities are predominant. Almost all the debris and most of the iron finds come from Trench Γ where small workshops are located. A co-existence of domestic and industrial activities is also evident.

There are no mineral resources within the area of the ancient city-state of Avdera (extending from the River Nestos in the West to the lake Vistonis in the East, and from the Aegean Sea in the South to the foothills of Rodhope mountains starting in Toxotes). Mining areas such as Thermes and Kimmeria with evidence for possible exploitation in antiquity, are located almost 30km north of Avdera (see Fig. 3 and 10) and fall within the territory possibly controlled by indigenous people. The source of iron used in the city will be discussed in chapter V. Metal supply is a major problem in Avdera, especially in early times and it seems that it was solved due to the Roman administration policy.

Tables 5 and 6.Typology, frequency and distribution of iron finds in Andera (table 9 in the appendix).

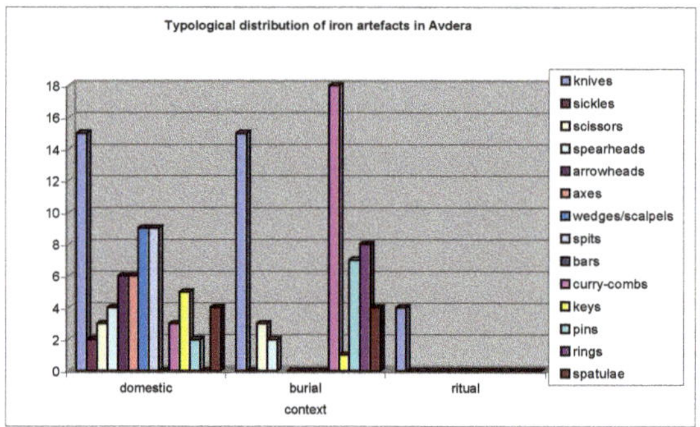

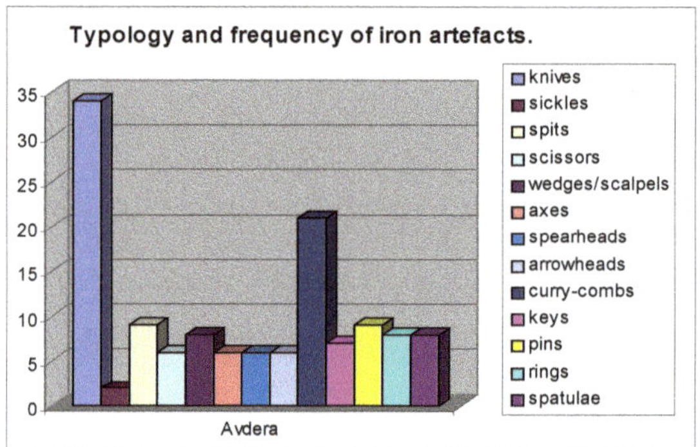

IV.4i. Kalyva – Kastro (2nd and 3rd centuries AD):

Typology and distribution of industrial waste and iron finds.

Philip II built the fort of Kalyva (c. 340 BC) in the Highlands of Xanthi (map 2) during his second campaign in Thrace (346-342 BC). Philip II founded the fort in order to ensure the strategic route next to the River Nestos by the attacks of the Thracian tribes. Local hand-made pottery dated in the 9th century BC indicates that indigenous Thracian tribes occupied the site before the coming of the Macedonians. Local hand-made pottery of the type often found in Messemvria-Zone (Fig. 15b, 18c) is found at Kalyva-Kastro as late as the 2nd century BC (Triantafyllos *AD* 1991). Macedonians, Thracians and Romans have occupied the fort between the 4th century BC and the 3rd century AD. Kalyva-Kastro is strategically situated in the upper valley of the River Nestos in order to control natural routes of communication between the Aegean and the Balkan hinterland and to command access to natural resources. During the last years excavations conducted by D. Triantafyllos (Director of the 19th Ephoreia of Prehistoric and Classical Antiquities) revealed evidence for iron making in the form of slag, blooms, agricultural tools and other finds coming mainly from the

Roman Period (2nd and 3rd centuries AD). The evidence from the excavation at Kalyva included the plan of the irregular trapezoidal perimeter wall with two rectangular and four circular towers (Fig. 21) and also recorded four building phases:

Phase I. 4th- 2nd centuries BC. This phase is dated after the destruction of the Odryssian State by Philip II. Four of the six towers were built of irregular and trapezoidal stones, the twin gates, the wall of the gates and the large stone cistern bellongs to the earliest phase of occupation.

Phase II. 2nd century BC - 2nd century AD. The Romans demolished Macedonia and decided that all the cities, villages and the forts east of the River Nestos should now bellong to the first Macedonian Meris. The end of this phase may have been caused by attacks by Thracian Tribes.

Phase III. 2nd - 3rd century AD. Fast repairing and partial rebuilding of the fort.

Phase IV. 3rd century AD. A settlement (houses, storerooms and workshops) was created within the boundaries of the fort. Most of the finds under study bellong to the last two phases.

During the last decade, excavations unearthed more than 300 iron artefacts, covering a large variety of items dating mainly to the 2nd and 3rd centuries AD. They include knives, axes, sickles and other agricultural tools, arrowheads, door keys, jointers, nails, rings. Sixty-nine objects have been chosen for examination, and 32 have been analysed during this first stage of investigation (Kostoglou 1997).

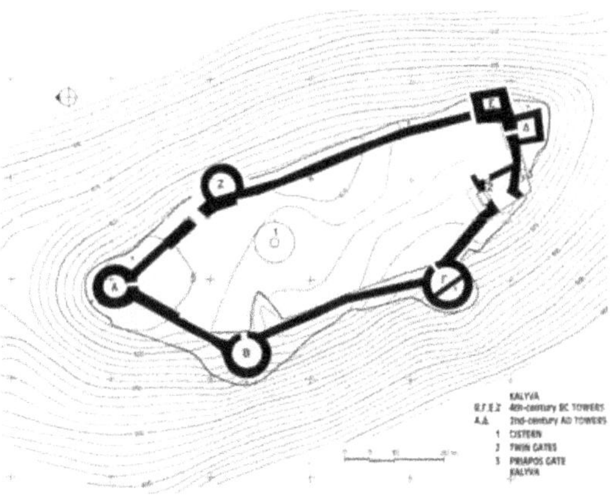

Fig. 21(a). Map showing the area where most of iron artefacts and industrial waste found in Kalyva-Kastro (from Triantafyllos 1994).

Iron/steel artefacts from Kalyva-Kastro can be classified according to the following groups:

Weapons : Three samples from the sockets of spearheads and one of arrowhead.

Knives (PLATES 17-20 in the appendix).

Fig. 21 (b). Kalyva-Kastro, knives.

Four different types of knives can be discerned

Type 1: Straight, single edged blade (*KL8, 46, 39, 41, 61, 16, 20, 62, 48).*

Type 2._Curved, single edged blades. (*KL 4,12, 23, 65, 14).*

Type 3. Long, triangular, single edged blades *(KL 66,67).*

Type 4. Small triangular, single edged blades probably part of scissors such as found on Avdera *(KL59).*

Axes (sample KL 55)

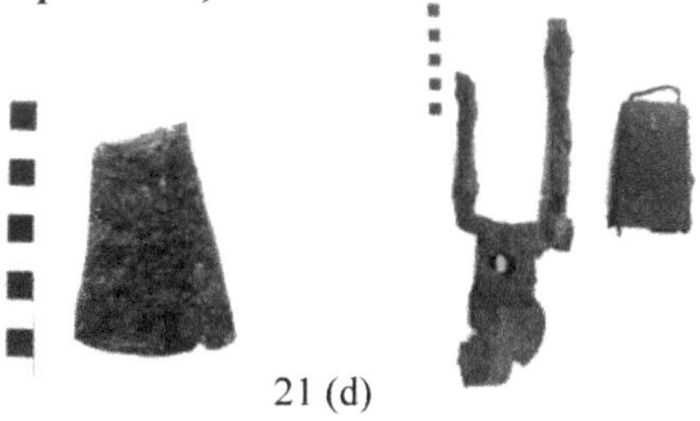

21 (c) 21 (d)

Fig. 21. Kalyva-Kastro. Iron tools: (c) single axe (KL55); and two pointed axe (d).

This blade is wedge-shaped in section, its thickness decreasing from the narrow cutting edge to the broader haft end (which in our example is missing). In its Roman form it might be either single or double-edged, having a vertical cutting edge on the one arm and a horizontal on the other. Its Roman name was *" securis"* (White 1967, 60, Fig. 34). The double-headed axe pictured next to KL55 is known as *ascia* or *rastrum* and it was used to prepare stony ground for cultivation (White 1967, 67, figs. 43-47).

Sickles, billhooks (Kl 3, 33, 64.)

This type of agricultural implement consists of a smooth curved blade, equipped with a smooth serrated cutting edge, attached to a short handle that lies in the same place as the blade. The curvature of the blade varies from a shallow arc to a semi ellipse (KL 3, 33). It is generally accepted that this implement is a Roman invention known under the general term *falx* and having many variations (White 1967:80, Fig. 52, 53).

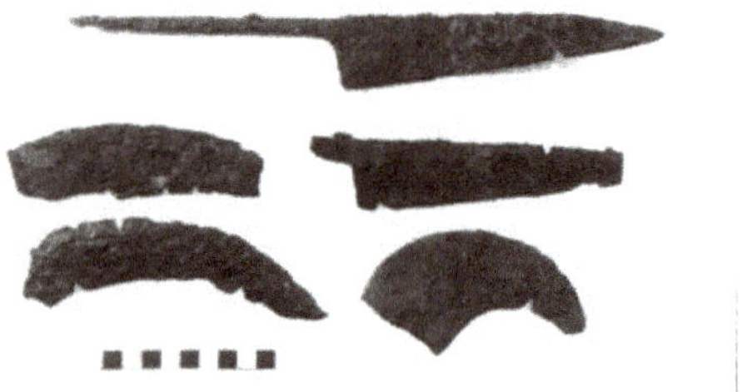

Fig. 21 (e). Kalyva-Kastro. Sickles (two on the left) and billhool (bottom right).

The blade Kl 64 bellongs to another subdivision of the *Falces*. Its Roman name was *fax arboraria* (billhook). The term billhook is generally used to describe knives used for lopping and pruning trees. The term "billhook" indicates that the blades are "hooked" (i.e. curved) in profile and in addition they have the characteristic projection at the top of the curved blade. The implement consists basically of a long flat blade with a single cutting edge curved outwards at the end to form a right angle and ending to a point (White 1967, 86).

Other
One iron hoe, (for typology and function see Reis 1976), a belll for sheep, and an iron horse-bit all indicate pastoral activities within the settlement.

IV. 4ii. The archaeology of iron in Kalyva-Kastro: conclusions.

It has been proposed (Triantafyllos 1994) that during the first centuries, namely from the 4th to 2nd century BC, only soldiers (of Macedonian or of Thracian origin?) lived in the fort in order to control the passage from the River Nestos, and to give information about the enemies' movements. From the 1st - 2nd centuries AD veterans of legions and allied troops with their families (probably of Thracian origin) settled in the fort and thus a village was created with economy based on agriculture and pasturalism. This type of rural settlement - usually in the mountainous zone - known as *kome* (mentioned by Thucididis I. 5. I) was the main model of economical and social life in the Roman Province of Thracia (Triantafyllos 1994). Small-scale workshops for smelting and smithing iron are located within the fort that has a predominantly rural character. The nature of the finds ensures the rural character of the settlement: clay figurines loom weights, lamps, querns, and coins are typical finds. Typical Roman material was found often and it is found with locally made indigenous ones such as the clay figurines and hand-made pottery depicted in photograph 21 (b). It is notable that the vast majority of the archaeological finds consist of iron artefacts. They bellong mainly to the 2nd and 3rd centuries AD. The absence of iron finds from the earliest phases, could be explained as the result of recycling or the continuous repairing of the fort.

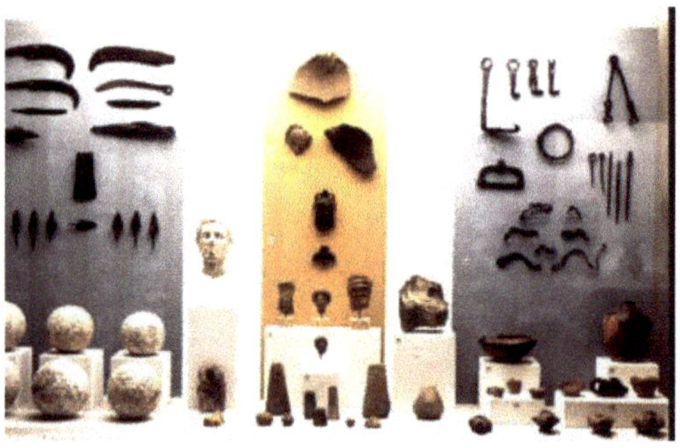

Fig.21 (f). Typical group of finds from Kalyva-Kastro. The arrows indicate one marble head made of Roman style, one locally made clay figurine and hand-made pottery (on display in the Museum of Komotini).

These finds are knives, axes and other agricultural tools (table 7), arrowheads, door keys, champs, nails, and chain rings. All the iron artefacts and industrial waste come from the settlement since no burials were excavated in Kalyva-Kastro (Fig. 8). The iron artefacts are mainly knives and sickles for domestic purposes, and heavy-duty tools such as axes for woodworking and land preparation for cultivation. They are similar with the Roman tools in Avdera. No strigils, spatulas and pins are found supporting again that these finds are more included in burials.

Indigenous elements in Kalyva-Kastro could be seen in local hand-made pottery found in the 9th century BC and in 2nd AD (Fig. 21b), in local figurines (Fig.21b), and in names of Thracian origin (Triantafyllos 1987). No Roman weapons or distinctive Roman finds (with the exception of the marble head in Fig.21b) are found, indicating that the Romans were mainly mobile military units of soldiers.

IV. 5. Discussion and interpretation: economic and social patterns of iron production in Aegean Thrace.

The economic and social patterns of iron production as they were practised on local and regional level and the visibility of identity in the settlement patterns, burials and archaeological finds, are discussed and conclusions are drawn.

This chapter has discussed finds from the mining/smelting landscape on the one hand and from the excavated sites on the other. The surface data have been discussed in the first section of this chapter. It is suggested that these data should be kept in the background of the present research, despite difficulty in dating and contextualising the activities they represent. It will be demonstrated in chapter V that the evidence from the sites presented in the first part of this chapter (included in Fig. 10) points to a diverse model of exploitation. Diverse, local developments in iron technology strongly question the assumed standardisation of iron production under colonial or imperial situations as the 'Hellenisation'and 'Romanisation' theories presume (see chapter II.1). The study of the excavated material showed clearly that the development of iron technology in each site is unique and can only be understood if the analysis is context-specific (Kostoglou 2009, 2008). Iron in each site follows different patterns of production, consumption and deposition.

In Messemvria-Zone iron is produced locally before and during the 5th century BC levels. It has a high socio-economic and symbolic meaning as it is indicated by the iron votives in the sanctuary of Apollo. Iron production stops after the destruction of the city and there is no evidence for iron production in the 4th century BC when numismatic and epigraphic evidence suggest the gold and silver was produced in Messemvria-Zone.

In Avdera iron has not been found in any of the 7th and 6th centuries BC. A few portable iron artefacts are found in domestic and burial context in Classical times. Iron finds in Hellenistic times are not included in rich tombs. No iron votives are found in any of the cities' temples from 6th to 3rd century BC. Iron workshops in the city appear only during the Roman times. Iron artefacts in Roman Avdera testify to a number of agro-pastoral activities.

In Kalyva-Kastro the majority of the archaeological record consists of iron industrial waste and finds. They cover a small variety, mainly knives, sickles, and axes. The distribution of strigils is different. As it can be seen in Fig. 9, strigils are often part of the record to in Hellenistic Avdera. On the contrary, only two strigils are found in Messemvria-Zone during the 4th century BC, suggesting that these objects were not commonly used in this site. The absence of strigils from Kalyva-Kastro is probably due to the fact that they seem to be found in burial context or because their use in Roman times was not extensive. The distribution of strigils might reflect different patterns of social practices towards the body within the sites.

Artefacts in all three sites testify for the practice of agro-pastoral activities from Classical to the Roman times. For example, billhooks for pruning, and axes (Fig. 9) for wood working and preparing stony ground for cultivation are found. Knives are predominant in the record (Fig.9) and they are found in both domestic and burial context (Figures 4, 6, and 8). Weapons such as spears and arrows are found in streets and houses (Fig. 4, 6, and 8). This conclusion agrees with the rural socio-economic model proposed as being dominant in the Greek countryside from Iron Age (Snodgrass 1987) to Classical (van Andel and Runnels 1987) and continued during the Roman times (Alcock 1996). However uniform this conclusion might seem, there are other finds such as the currency bars in Messemvria-Zone, the strigils of Hellenistic Avdera, and the implements of Roman Avdera and Kalyva-Kastro that provide interesting patterns in the study of iron consumption on local scale. The analysis of the iron finds and industrial waste in chapter V provides more evidence for the socio-cultural role of iron production and consumption. What comes across the archaeological record generally is a sense of continuity with local traditions that has been overlooked by the conventional historical interpretations (chapter I). Here I would like to summarise the main points of the discussion of the archaeological data of each site comprising the second part of this chapter:

1. Despite major destruction in the 4th century BC, probably due to the Thracian campaign of Phillip II in Thrace (Hammond 1972), all three sites are in continuous occupation since at least the 6th century BC.
2. Cemeteries are continuously used and re-used. Burial customs and rituals do not change from 6th century BC to the Roman times. Strong family and community identities emerge from the study of burials in Messemvria and Avdera.
3. Local pottery traditions continue in Messemvria-Zone and Kalyva-Kastro producing hand-made wares, mainly drinking cups and small containers. The local tradition in Avdera produces wheel made tablewares.

Finally the mode of iron production in all three sites seems to be the house-based or community-based workshop.

Table 7. Typology and frequency of iron finds in Kalyva-Kastro (based on table 11 in the appendix).

Table 8. Distribution of iron finds in Kalyva-Kastro (based on table 11 in the appendix).

Table 9. Typological distribution and frequency of iron finds in all three sites (based on tables 7, 9, 11 in the appendix).

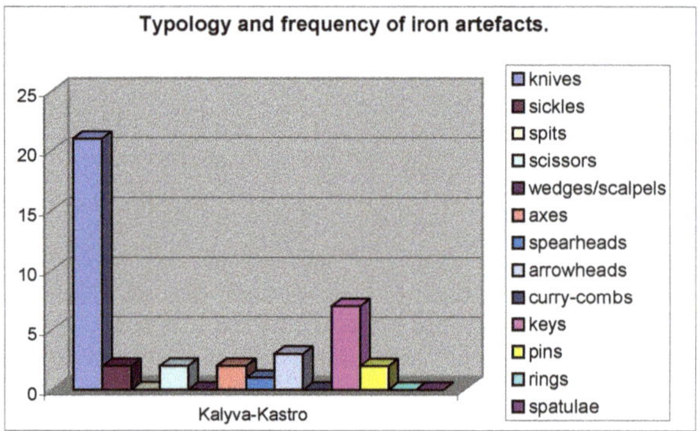

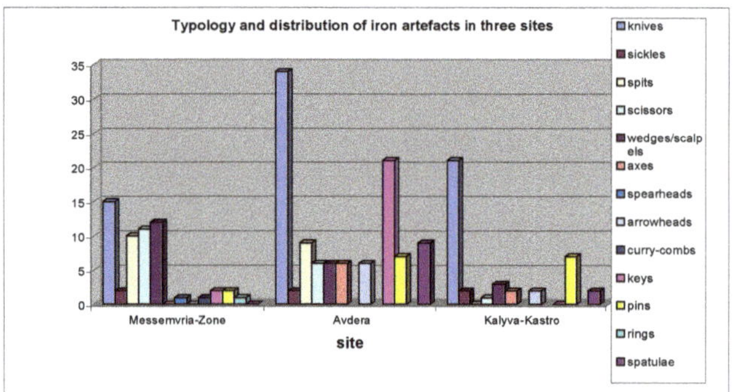

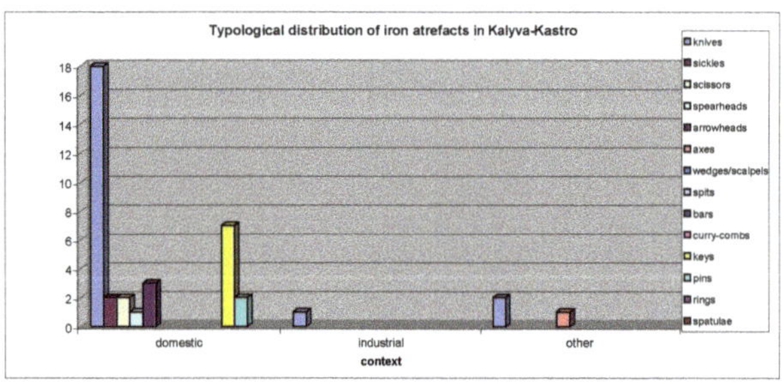

This picture strongly contradicts the conventional history of the region as presented in chapter I. Therefore a number of questions arise: what was the impact of the Greek and Roman presence in Aegean Thrace? How is the study of short-distance relationships between societies with different patterns of social or economic organisation possible? In the context of this research these questions involve the study of Greek or Roman culture versus the indigenous culture in order to see if such asymmetric interactions have the potential to bring about major transformations of the social relationships.

As noted in chapter II,1 (theory and methodology) the problem with such studies is how archaeology and the interpretation of the material culture can contribute to the understanding of such complex and multivariant relations. The relationship between the Greeks and the indigenous populations, which came in contact with each other during the colonisation of other lands, has long been understood and interpreted under the centre - periphery model. Colonies organised in cities or other urban centres have thus been regarded as foci of civilization (Andersen *et al.* 1997; Aufrecht 1997). On the other hand, the countryside frequently inhabited by the indigenous population was regarded as primitive and uncivilized.

The concepts of centre and periphery also include other opposing ideas such as city - country, urban-rural, civilised-barbarian; innovative -primitive. The contrast is not only a spatial one but also cultural, implying a contrast between an innovative, dominating region and others which are backward and thus subjected.

Supply and demand (in crops, fuel, minerals, and slaves) is another opposing idea since the centre was thought to exploit the periphery economically. Urban centres consequently cause the acculturation of the underdeveloped areas and scenarios such as 'Hellenisation' or 'Romanisation' are created to explain social and cultural changes (see chapters I and II).

According to this model and due to stylistic similarities and documentary evidence, it has been argued that Thracians came in contact with the Greek culture through the Mycenaean. This was the first phase in their integration progress. The second took place during the two periods of colonization of the coast (8th and 6th cent BC) and the final was after Philip II of Macedon (4th cent BC). The history of the area has been created without any reference to the indigenous population.

The core-periphery model has been criticised both by post-colonial archaeologists (Rich and Wallace-

Haddrill 1992; Champion 1989; Rowlands 1987) and by anthropologists of consumption (Dietler 1998, Appadurai 1986, Bourdieu 1984). It is generally accepted that this model is dominated by cultural and economic (market) determinism and does not consider the input of indigenous populations in the developments. Therefore an alternative model is required.

In the relationship between the Greeks/Romans and the indigenous population two possible alternatives to domination can be seen:

a. Political, territorial and/or military control of the indigenous by the Greeks or Romans; and
b. Peaceful coexistence with occasional episodes of hostility and warfare.

Colonies are founded in order to cover certain colonial needs (economic, political, military) and usually they present urban features such as public buildings, fortifications, and town planning. The urban appearance of the colony, however, does not mean that the entire region has been involved in the process. The urbanisation processes can cause gradual nucleation or, as regional processes, they can cause change in the settlement pattern that encompasses the emerged town and other minor settlements in a long period of time. One way to study the colonisation progress on a regional level is to ask if the colony has been integrated into the region or kept its distance (van Dommelen 1997). It is, thus, relevant to investigate if the foundation of the colony respects the existing indigenous settlements. Colonisation does not mean by definition urbanisation as in the case of Sardinia; each of the two can have its own characteristics and rate of change depended on the colonial interests and the indigenous reactions (van Dommelen 1997). The contribution of the indigenous values and systems was recently appreciated due to the parallel consideration and study of the material culture (Sardinia, van Dommelen 1993; Sicily, Antonacio 1997; Etruria, Arafat and Morgan 1994; and Gaul, Royman 1992).

The first difficulty in applying this model to Aegean Thrace is to locate the indigenous settlements and the indigenous material characteristics. According to the culture historical approach presented in chapters II, III, and I the indigenous characteristics disappeared due to assimilation or acculturation processes. To find indigenous elements in the existing archaeological record of Aegean Thrace one should look beyond the present preoccupations and ideas. The existing literature argues that indigenous elements have been recorded so far only in the Iron Age. The Early Iron Age lasted in the area from c.1050 BC until the 8th century BC, the time of the first Greek colonization. The Thracian tribes, on the other hand, continued the Iron-Age type organisation until the late Classical period as it is described by the ancient authors and is evident by excavations in rural settlements.

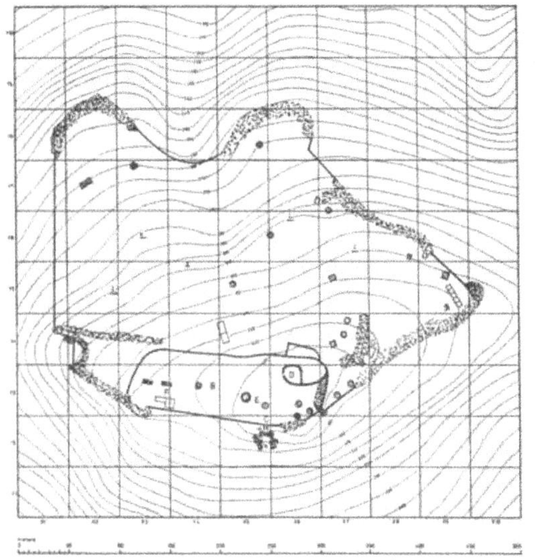

Fig. 22a. Hill fort in Asar Tepe (from Tsmpidis-Pentazos 1971)

In Aegean Thrace, Iron Age settlements have been found mainly in the upland and highland zones. The information given about the sites in the existing literature (Triantafyllos 1994, Triantafyllos 1972, Tsimpidis-Pentazos 1971) appears incomplete, as none of the sites have been systematically excavated. Nevertheless, the recorded sites are about 100 in number, and the majority of them are dated to the Early Iron Age according to the pottery found *in situ* (Triantafyllos 1994).

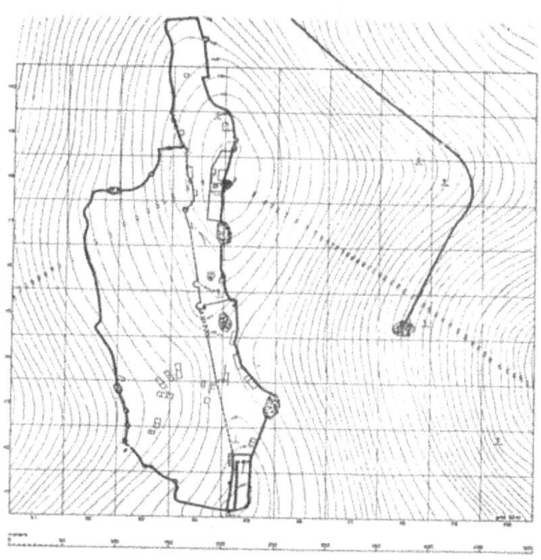

Fig.22b. Hill fort in Ayios Georgios, Maroneia (from Tsimpidis-Pentazps 1971)

These sites are situated in the Highland zone (Rodhope Mountains and Ismaros) and are fortified settlements with defensive walls. The best-known sites of this type are: Ayios Georgios-Maroneia, Ergani, and Koila (Tsimpidis-Pentazos 1971). In the small-scale excavation, handmade pottery dated to the Early Iron Age has been discovered.

Fortified settlements have been also recorded in Tsouka, Nymfaia, Roussa, and Kotronia. All of them are situated in the Highland zone, in strategic positions, near rivers and water sources, and on tracts of land suitable for cultivation. The main occupation of the settlers was herding sheep and goats, as well as hunting (Triantafyllos 1994, 42-44). This model is also supported by the archaeological data of the excavated settlements and the ethnographic record. As in villages today, cultivation would also continue on the mountain sites and around the settlements. The foothills are appropriate for either cultivation (mainly grains) or pasture. The first results of an ethnoarchaeological project (Eustratiou 1987) taking place in the upland settlements in Rodhope Mountains (Fig. 2 in chapter I), testify that both agriculture and pastoralism were being practised in the area.

The fortified settlements are characterised as Iron Age. However, none of them has been excavated properly. It is interesting though, that 15 out of 25 have continuous occupation from the Iron Age to the Roman times as suggested by surface pottery (Triantafyllos 1973). On the other hand, almost all the important sources of minerals and fuel (Fig.3 in chapter I) as well as routes to the interior (Fig.4 in chapter I) are located beyond of the territory of the ancient Greek colonies. This leads to the assumption that a network of close interaction between the Greeks and the indigenous population, involving trade and exchange of goods, should exist. The same network was used by the Romans to move military units and supplies to the Balkan hinterland. The relationships between the Aegean and the Balkan networks change over the time.

It seems that in the 5th and 4th centuries BC metals, timber, slaves, organic materials were exchanged for luxury and prestige items from Greece such as these found in the rich warrior tombs in Bulgaria from the same period. Archibald (1983; 1998) proposed the following model to explain the Greek imports in the Balkan hinterland.

5th century

Greek pottery and metalwork began to appear in the interior during the second quarter of the 5th century BC. Attic style pottery indicative of trade with Athens (via the colonies) is deposited in rich male or female burials (tumuli). Pottery found along the rivers suggests their use in early times. Thracian burials-cremations- in tumuli covered by a stone cairn or ring appear with 'Hellenised' burials-chamber tombs some with ashlar masonry.

This practice continued up to Roman times. In Aegean Thrace indigenous veterans are buried in their homeland in tumuli with stone bases. Tumuli in Thrace are located along the main routes of communication (map in chapter I) and most of them in the Diagonal and Via Egnatia (Samsaris 1984). Most of the tumuli cover the Hellenistic and Roman times. Considering that this burial practice is known in the region from the Iron Age we can safely conclude that it represents, along with the handmade pottery, the indigenous cultural tradition in the area.

4th century

Greek imports are found further north to the Haemos Mountains and to the hinterland of Odessos and to the South of the foothills of the Rodhope possibly via the river Nestos. Silver *phialae* in a warrior ashlar built tomb, north of the Danube estuary. The five *phialae* were the only non-indigenous element. Other tombs north of the middle Evros had native harness equipment and the same group of bronze vessels, and imported pottery.

After Philip II (346-341 BC).

Closer similarity between Bulgaria and the Aegean (finds from Nikissiani, Sedes, Derveni, Stavroupolis) is noticed in the archaeological record (especially in metalwork). Indigenous versions of jugs are known from Nikissiani, Arzos and elsewhere. Stylistic and technical similarities are noticed with Macedonia and Chalkidiki (bracelet with the snakes common motif). No sites are known prior to the 4th century BC other than hill-forts. After the 4th century the first cities, such as Kabyle and Phillipopolis (Plovdiv), appear in the hinterland both founded by Phillip II of Macedon. This evidence also suggests that hill-forts are part of the local tradition.

According to the present discussion, it seems that during Classical-Hellenistic times the Greek colonies and the indigenous settlements formed a network of small rural subsistence economies either complimentary or competitive. Control over territories, natural resources and/or people were often a cause of war as it is documented by epigraphic and archaeological evidence (chapters III and IV). The Greek city-states such as Avdera and Maroneia had more or less well-defined boundaries and were self-sufficient. Indigenous people lived in the towns such as Messemvria-Zone and in villages such as Kalyva-Kastro.

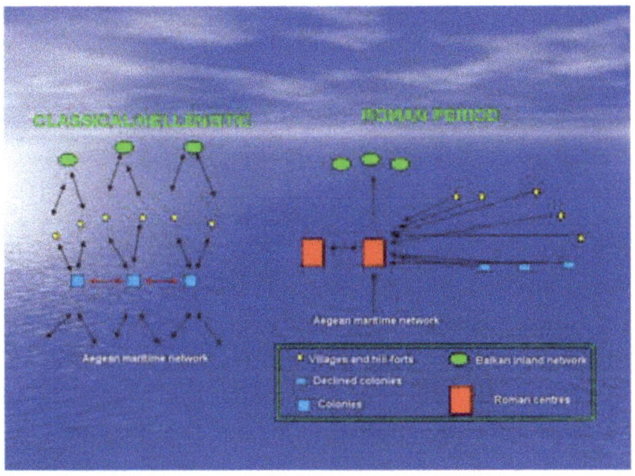

Fig.23a. Schematic representation of the socio-economic model based on trade and exchange between the Balkan and the Aegean trade networks during the Classical/Hellenistic to Roman times (based on chapter IV).

Under Roman control, production was intensified although there is no evidence of industrialisation that could be expected following the Latin texts (chapter III). None of the characteristics and strategies followed by the Romans towards the industrialisation of the production can be observed in the record of Roman Aegean Thrace. Known strategies include a) the use of vast quantities of water as energy source to manipulate the mineral ore in order to increase productivity and efficiency; b) the use of mechanical devices and methods to tackle flooding and to mine bellow ground level; and c) policies for provincial mine administration i.e. procurators appointed by the emperor to run the mines in Spain, regulations for the running of mines, shifting of skilled personnel and slaves to mining areas (Curchin 1991, Healy 1978). New energy sources, new mechanics and centralised control via a procurator appointed by Rome resulted in the re-organisation of the province in order to magnify metal production. None of these characteristics is evident in Aegean Thrace. No evidence for a large imperial mine exists so far and most of the production in the region seems to come from small scale local mines, dispersed in the mountainous zone between the rivers Strymon and Evros (Fig.10). The road and market inland-networks have been shifted from north-south due west towards a city called Philippi (Fig.2), which was the centre of Roman administration. The colonies declined despite their autonomous status and became rural. This change in their character suggests that the name colony might not be appropriate to characterise these sites during the Roman period. Upland and highland settlements located close to mining and/or trading routes continue their lives. Both colonists and indigenous population is now Thracian in the eyes of the Roman administrative policy. To conclude, it seems that in the case of Aegean Thrace colonisation did not affect the pre-existing indigenous settlements. The Greek colonies remained the only urban centres in the region during the Classical and Hellenistic times until they decline.

Urbanisation did not occur in Aegean Thrace as indeed in the Thracian hinterland (Archibald 1983) prior to Roman times. There is no archaeological evidence for the absorption of large numbers of indigenous population in the colonies. Indigenous people continued living in their fortifying settlements in the upland and highland zones practicing economies based on agriculture, pasture, mining and interacting with the colonists via trade and exchange.

Roman administrative policy changed the above pattern by the creation of new urban centres in the region, which caused the decline of the old. For example the Roman city at Philippi is located close to inland passages to the Balkan interior and thus to Europe. It is also located close to vast mining areas. The new cities collected, via taxation (see chapter III), most of the mineral wealth produced in the mines of the newly created first Macedonian Meris. Fig. 23 summarises the main socio-economic changes in a schematic model.

The effects of these socio-political changes into the life and identity of the people of Aegean Thrace can be traced in the part of the archaeological record that testifies continuity with earlier cultural traditions. The identity of the communities living in Messemvria-Zone and Avdera during the early phases of the colonies was constructed around ideas of bellonging in a family and group of common origin, as well as sharing the same lands and ancestors. This ethnic identity (chapter II.1) was contained with religious rituals and festivals taking place in sanctuaries and cemeteries. Ethnic characteristics are not part of the archaeological record of the Roman period where identity seems to have a stronger socio-political component. This suggestion is further refined after the discussion of all data in chapter VI.

Chapter V. Analytical Data.

The first part of chapter V discusses analytical data related to open mining/smelting sites presented in the first part of chapter IV. The second part is dedicated to the discussion of the analytical work undertaken by the author on the study of the excavated iron-related material. The objective of the analytical study is to determine the type of smelting and manufacturing techniques in order to establish the level of technological efficiency in iron metallurgy practised in each site. Technological choices and change in the form of innovation or invention are also investigated. The final sections of this chapter compare the iron technology in all three sites in an attempt to establish if changes in iron technology correspond to indigenous developments or to colonial/imperial strategies imposed by the Greeks and Romans.

V. i. Iron in North Greece

The sites in Fig. 10 indicate areas where slag heaps and/or mining tunnels with possible exploitation in antiquity have been discovered. These units constitute simultaneously the richest in Greece metallogenic provinces in terms of abundance and variety of mineral deposits. The appearance of elements such as gold (Au), silver (Ag), lead (Pb), copper (Cu), zinc (Zn), arsenic (As), iron (Fe), manganese (Mn), nickel (Ni) as well as the relevant mineralisations (see Fig.3) have been repeatedly verified in the past, in various geo-mining projects or geochemical studies (Veranis 1989, Voudouris 1993).

The occurring polymetallic parageneses, mainly of porphyry and/or hydrothermal type (Maratos and Andronopoulos 1966), are tightly associated with radioactive anomalies. Radioactive anomalies, sites with radiation level more than 2.5 times the background level, were located in Central Macedonia and Thrace by systematic scintillometer car-borne (or on-foot) survey (Meyer 1971, Olsen 1974). More than 200 radioactive anomalies were detected and analysed. Some of the most intense radiometric anomalies are associated with the deposits of radioactive lignite, occurring in basins known as 'Serres basin' (Macedonia), 'Dipotana' (Thrace) and 'Aimonion-Kotyli basin' (Thrace) (Stavropodis and Pournis 1971). The above regions bellong geologically to the so-called Serbomacedonian, Circum-Rodhope and Rodhope geotectonic units (IGME 1983).

The radioactivity of slags in some of the numerous slag heaps/piles in Macedonia and Thrace was noted during the above mentioned scintillometer survey. Stavropodis (1983) first recognized the metallurgical importance of those technological residues and gave some semi-quantitative analyses of radioactive slags and limited data on their radioelement content. So far, approximately 15 sites with radioactive slags were known. They mostly correspond to iron-working sites of the past. The working hypothesis is that the radioactive elements in the slag come not form the use of the radioactive lignite (slag included charcoal fragments) but for the smelt of radioactive magnetite sands inn the Vrontou Mountains (Bassiakos *et al.* 1999, Photos 1987).

The tradition of smelting magnetite sands and the production of high quality steel is documented for the coast of the Black Sea (Tylecote 1981,Tsaimou and Papadimitriou 1993, Photos 1984). Magnetite sands are also occurring on the coastline of Aegean Thrace on many locations amongst them in Kavala and Alexandroupolis regions in Fig. 2 and 10 (Papadakis 1975). The titanium rich magnetite sands on the island of Thassos and the magnetite sands of the Kavala region have been smelted as early as the Hellenistic period (Photos 1987, 1984). The so far examined published material (Bassiakos *et al.* 1999, Photos 1987 and 1984, Tylecote 1981) testifies to the existence of the following mining /smelting traditions in Aegean Thrace:

1. Hematite in many locations;
2. Manganese/ferrous ores in Symvolon mountain;
3. Radioactive magnetite slag in mainland, Kavala region and Vrontou mountains;
4. Titanium rich magnetite slag and in slag inclusion of artefacts in the island of Thassos.

It should be emphasised that the smelting of iron sands seems to be a regional phenomenon occurring within the territory of ancient Thrace (Fig.1 and 10). There is no evidence for smelting sands elsewhere in Greece. Parallels exist in other traditions such as the Chinese (Wagner 1985). Although the ore deposits of Aegean Thrace east of the river Nestos in the Rodhope Mountains bellong to the same geological formation, radioactivity is not one of the major characteristics of the slag heaps. The lack of radioactive anomalies detected in the mining/smelting areas of Kimmeria, Thermes, and Komaros (Fig. 3) and on the slag from Messemvria-Zone, Avdera and Kalyva-Kastro (measurements taken by Bassiakos in 1999) was surprising. Since not all magnetite sands are radioactive, the possibility of using the magnetite sand deposits of the Aegean Thracian coast for smelting remains a question to be answered. The following section aims at determining the mining/smelting activities in the excavated sites where, as it will be noticed, there is no evidence for the smelting of iron sands in the material under study.

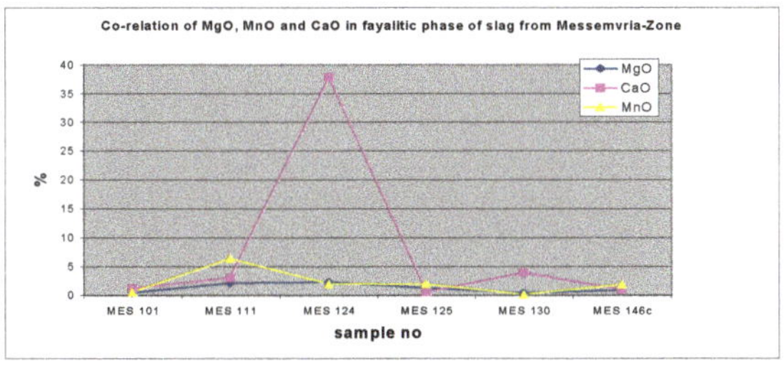

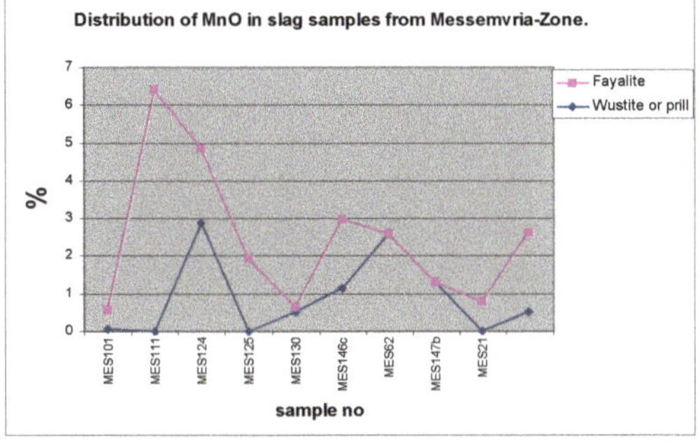

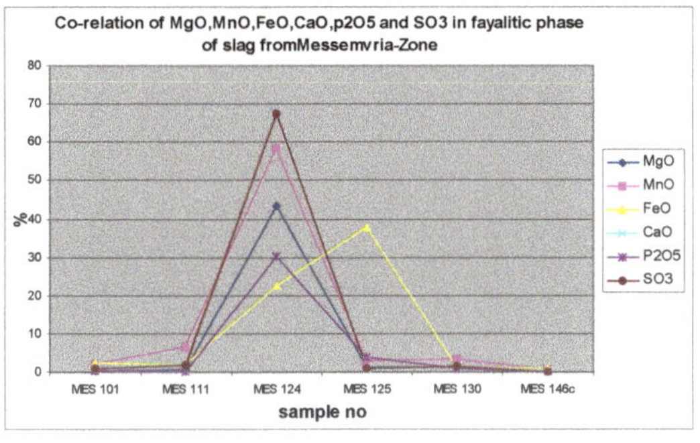

Table 10. Graph showing the distribution of MgO, MnO and CaO in the fayalitic phase of slag from Messemvria-Zone (based on tables 12a and 12b in the appendix).

Table 11. Distribution of manganese in the slag of Messemvria-Zone (based on tables 12a and 12b in the appendix).

Table 12. Distribution of major elements in Messemvria-Zone slag (based on tables 12a and 12b in the appendix).

V. 2i. Messemvria-Zone (5th and 4th centuries BC):

Metallography and SEM-EDAX analyses of industrial waste.

All the industrial waste that was found so far in the excavations of Messemvria-Zone consists of 50 finds of mainly slag. All of them where sampled for the purpose of this study. The total weight of the collected finds is 19.900 gr. We should add to this the fragments of slag still scattered on the site. 18 out of 50 samples have been analysed optically and 9 samples, representative of their categories, were analysed by a SEM-EDAX micro-analyser. The samples cover mainly the 5th century BC (see chapter IV).

Smelting slag

Group 1 (PLATE 1 in the appendix)

The first group of slag includes slag with moderate to low silica content, high lime content and rich in magnesia. This group includes samples MES 106, 124, 130, 132, and 137 (tables 12a and 12b in the appendix). In this group, the three main phases of a bloomery slag as described in chapter II.2 (wustite, fayalite, and glass) are clearly formed.

Silica is between 20-52% in area analyses (11.58% only in MES 124) and helps the formation of nice, long, needle-like fayalites ($2FeO.SiO_2$). Silica in the slag comes mainly from the gangue of the ore. The lime (CaO) content on the other hand, in most of the Messemvria-Zone samples is less than 10% and in the area analyses it varies between 4% and 13% with the exception of MES 130 (26.67%) (table 11 and tables 12a and 12 b in the appendix).

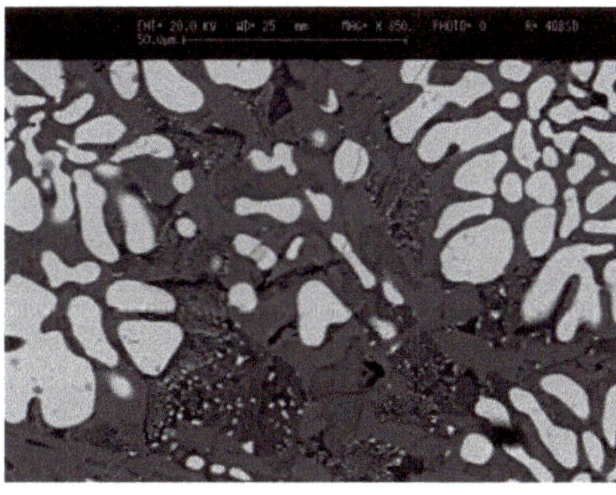

Fig. 24. Messemvria-Zone smelting slag, group 1: SEM photograph of sample MES130 showing globules of wustite (white), and plates of fayalite (light grey) in a glassy matrix (dark grey).

There are two possible reasons to explain the unstable variations of lime presence in the slag (Fig. 10). First, this could be due to insufficient control of the smelting process, thus, different smeltings could produce different products (bloom) and by-products (slag). Or, the smelter(s) were occasionally using lime as a fluxing material, which could contribute to a higher lime content in the slag. If we take into account the distribution of magnesia in MES 130 as well (table 10), then the chemical composition of the slag points out to the use of a magnesium-rich lime or dolomite as flux. If this is the case, a good degree of technological knowledge is assumed because the smelter tried to overcome the pure quality of the ores by using different material. In this category, sample MES 130 is typical of the slag produced from this smelting operation.

Manganese oxide (MnO) is present in considerable amounts in almost all the samples under study (table 10). Usually the amount of MnO present in the fayalite is between 1% and 3% with the exception of samples MES 111 where it is 6.40%, MES 62 with 3.22% and MES 53 with 4.58%. MnO is also present in the wustite (MES 124 with 2.88%, MES 146d with 2.57%, and MES 62 with 1.29%) and in the prills (MES 146c with 1.15% and MES 146d with 1.29%).

Group 2 (PLATE 2a in the appendix).

The second group includes samples where the silicate phases (fayalite and glass) are dominant and the presence of iron oxide (wustite and/or magnetite) or metallic iron (prills) are minimal. Samples MES 101, 111, 125, 146c, bellong to this group.

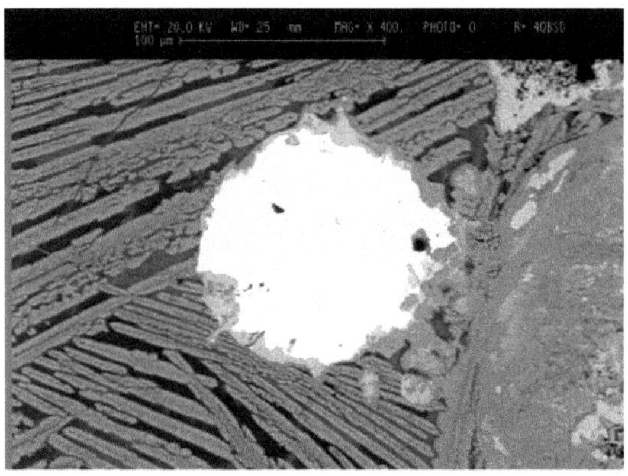

Fig. 25. Messemvria-Zone smelting slag, group2: Photomicrograph of sample MES111 showing needles of fayalite (light grey) and one prill of metallic iron (white) in a glassy matrix (dark grey).

The constant presence of manganese oxide in the wustite and fayalite phases of the slag (table 11) point the use of manganese as a safe tracer element for the industrial waste and artefacts produced locally at Messemvria-Zone is safe. This is very important for tracing the possible provenance of the artefacts analysed in the following section and it will contribute significantly to the discussion of iron technology of the area.

A problem arises because manganese can enter the metallic phase of iron during smelting (table 11) only if temperature exceeds 1500° C (Rostoker and Bronson 1990), which is not the case in the bloomery furnace (see chapter II. 2).

Nevertheless, there is another possible source of iron ores, which could explain the presence of manganese, and this is the use of manganese/iron ore for smelting. Such deposits are located in Komaros (see chapter II.2), approximately 5 kilometers north of the site. Bassiakos and the author, during fieldwork (summer, 1999) have located the site. There is evidence for possible exploitation of such deposits in antiquity, in the form of scattered slag and possibly mining tunnels (Bassiakos *et al.* 1999).

The manganese/iron deposits lie on the surface. One medium fragment (MES 129) was found in the same context with sample 111. MES 129 is mainly manganese oxide (37.08%) and silica (35.90%), and contains minor amounts of alumina (4.35%) and lime (4.33%). The iron oxide in the sample is low, only 2.16%, but this could be expected if the analysed sample comes from a discarded piece of ore that was not appropriate for smelting. An ore with similar composition is now termed as a manganese ore, but it would be treated as an iron ore by the ancient bloomery operators (Rostoker and Bronson 1990, 45). It would also be considered as a good quality primary source of iron because it could give useful quantities of iron and the manganese oxide would serve as flux for the slag. Therefore, no added fluxes were required.

It should therefore be assumed that manganese in the iron comes from the smelting of manganese/ferrous ores. The efficiency of the smelter to control the operation successfully is acknowledged.

Group 3 (PLATE 2b in the appendix)

Finally, in this category only one slag, the sample MES 101 (Fig.26) is included. It looks very similar to MES111 (Fig.25) which is an iron smelting slag. This slag was found *in situ* in the furnace debris (chapter IV). The most intriguing elements in this sample are the oxides of iron, copper and sulphur, which are present in the form of little prills, scattered on the silicate phase. These prills (little shiny dots under the microscope) constitute a typical matte that is a liquid or solid phase of sulphides usually FeS or Cu_2O. Matte formation in the slag is associated with copper production (see Craddock 1995, 149) but the matte present in MES101 is not enough to justify this. It is, thus, possible that MES101 is an iron smelting slag where iron and copper sulphides have been formed accidentally because a copper rich iron ore was used in smelting. Iron and copper are often associated especially in gossan deposits such as these found in Spain (Rostoker and Bronson 1990, 20) and in Cyprus (Knapp and Given 1996).

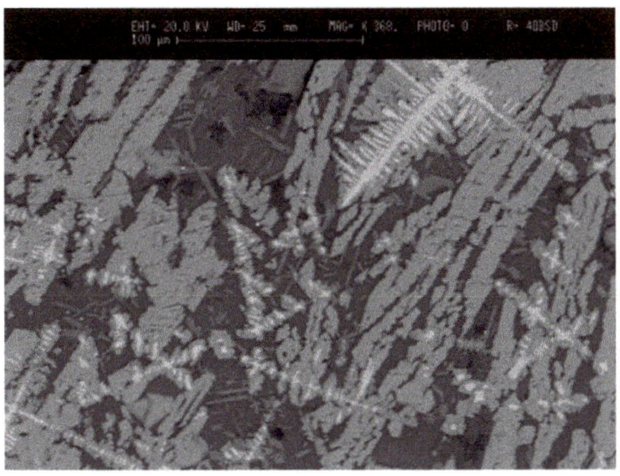

Fig.26. Messemvria-Zone smelting slag group 3: SEM photograph of MES101 showing iron oxide (white magnetite crystals), matte (little white dots), needles of fayalite (light grey) and glassy matrix (dark grey).

The chemical composition of MES101 (tables 12a and 12b in the appendix) with the high amounts of copper and sulphur points out to the smelt of mix sulphide ores such as chalcopyrite. This kind of ores are located approximately 15km north of Messemvria-Zone n the area of Kirke (see chapters I and II.2; also Fig. 3 and 10). The smelt of chalcopyrites could be possible only after roasting because this is a necessary step for converting the sulphides to oxides prior to smelting (Craddock 1995, Rostoker and Bronson 1990). No roasting furnaces or installations have been identified in the vicinity. Archaeolometallurgists are aware of the interaction between copper and iron smelting technologies and, in theory; they do not exclude the existence of cupriferous artefacts. On the one hand, the beneficiary impact of iron in copper smelting has been already acknowledged (Merkel and Barrett 2001, Craddock and Meeks 1987). On the other hand, it is well known that copper in iron smelting 'will alloy with bloom or cast iron during smelting and it will remain with it even when converted to steel. It has strengthening effect on the finished metal, and minor additions of copper to iron can be decorative' (Rostoker and Bronson 1990, 20). The use of copper rich iron in the manufacture of sickle MES44 (plate in the appendix) testifies that iron pyrites were smelt (MES101) and deliberately used for the manufacture of tools as early as the 5th century BC. It could be concluded so far that most of the samples from the first two groups testify for the use of manganese/iron ores in bloomery smelting (located probably at Komaros Fig. 3 and 10). The first group of slag represents an operation where considerable amount of iron is lost in the slag. The second testifies for a perfect operation with the minimum iron lost in the slag. Manganese oxide is constantly present in both the silicate and metallic phases of the slag. Although manganese/ferrous ores are considered to be self-fluxing, lime was also occasionally used. The current research also provided evidence for the smelt of iron pyrites (probably located at Kirke, see Fig. 3 and 10). No evidence for the smelting of magnetite sands located on the coast (Papadakis 1975) has been found so far.

Smithing slag (PLATE 3 in appendix).

Smithing slag is often represented by a lot of small, amorphous fragments, the so called hammer (Tylecote 1962). Their metallography indicates the presence of mainly iron oxide, which is present in a consolidated form with round shaped wustite (MES124, 143,147a). Smithing slag can also have a semi-spherical form as in the case of sample MES147b. This is known as smithing bottom hearth (Tylecote 1987).

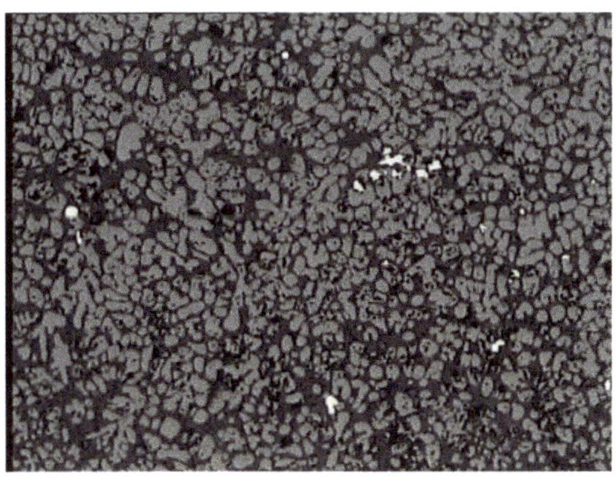

Fig.27 (a). Photomicrograph of smithing slag MES143 in magnification x264 showing mainly iron oxide (white).

Furnace lining (PLATE 4 in the appendix)

Two fragments of vitrified ceramic probably coming from the clay lining of a smelting furnace were also discovered (samples MES146c and 147d).

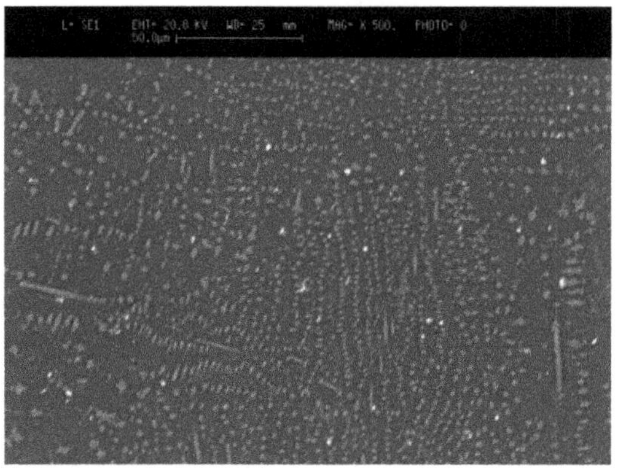

Fig. 27(b). SEM photograph of furnace lining MES146d showing iron oxide (white) dispersed in a silicate phase.

Material related to cast iron production.

The production of cast or liquid iron has been associated with the introduction of the blast furnace to Europe in the 17th century (Tylecote 1962). The technique was known in China since the 4th century BC and it was associated with the diffusion of Skythian tribes to the East (Wagner 1999). Descriptions of casting iron can be found amongst ancient authors such as Aristotle (for a detailed study see Halleux 1974) and some scholars (Photos 1987) see such evidence in ancient iconography as well. Authors in the past debated the possibility of cast iron production in ancient Greece, but the possibility is generally rejected given the absence of archaeological evidence (Papadimitriou *et al.* 1996, 271). This is the first time that archaeological material from a safe context indicates that cast iron, produced possibly accidentally, was the subject of experimentation by the smiths during the 5th century BC (Kostoglou and Navasaitis 2006). Although the discussion for the innovation of cast iron is, in general, still open, recent finds of Roman date from Britain (Mack *et al.* 2000) and others from 4th century BC period Germany (Gassman 2001) support the suggestion that the technique was multiply invented and not diffused to Europe from China. The finds from Messemvria-Zone are unique for the Mediterranean region and the earliest examples of this technique in Europe. The author understands the limitations of the current analytical study. Since the detailed technical study of these finds is beyond the scope and time of the present work, here I will briefly discuss the finds in their archaeological context and I will also present the first data of their metallographic analysis.

It is understood that cast iron is not necessarily produced in a blast furnace (Wagner 1999). Cast iron was one of the products of African iron smelting and it was also re-produced in modern experimental smelting in bloomery shaft furnaces (Killick 1991). High carbon in bloomery iron comes from droplets of cast iron falling into sponge iron during smelting in a bloomery furnace, so the same furnace can produce wrought iron, steel, and cast iron (Rostoker and Bronson 1990, 97-98).

There is also a direct method of producing steel as a bloom (Wertime 1962). Although the principles of the smelting are the same, a higher temperature and thus a stronger blast are required. What happens in the furnace is that iron bellow 1523° C is completely liquid and can be almost perfectly separated from the gangue of the ore (for general principles see Rostoker and Bronson 1990, Tylecote 1987, Craddock 1995). The gangue along with the fluxes and charcoal form an also liquid slag which sits on top of the iron, since it is lighter. The metalworker then opens a hole on the furnace wall and taps out the iron into pits where it is left to solidify. Then he does the same with the slag. The product of this operation is iron containing up to 4% carbon. Iron at that stage is in a ferrite form and surrounds islands of pearlite. Holes are indicative of the casting process. This is called white cast iron and it is not ready for use because it is not malleable. The sample MES127 (Fig. 28a, b, c) represents this stage of production, and it was probably discarded as unusuable after an initial attempt at heating and forging.

With the presence of carbon and under highly reducing conditions the iron is in a perfectly liquid phase. What follows is further re-heating of the mass of white cast iron. SEM analyses were not proved particularly useful because they gave almost 100% iron. Different analyses are required to establish possible trace elements and the exact amount of the carbon content. The technical implications of such a process should be part of metallurgical investigation. The analyses carried out at this stage of the research had as a goal to identify the metallographic structure and thus the type of the iron finds. The by-product, slag, is different from the bloomery slags. It contains almost no iron oxide since all iron has been removed completely, as indeed is the case with the slag of group 2.

Analytical work could verify if the slag and the cast iron finds bellong to the same kind of operation, i.e. accidental, or possibly experimental versus intentional production. I would argue on the second due to the fact that the 3 samples represent different stages. During re-heating, iron absorbs more carbon, which after cooling takes the form of long graphite flakes as in sample MES126 (Fig.28d-e).

To make iron workable further re-heating is required to help the graphite flakes burn out some of the excess amount of carbon (Fig. 28f). This process is known as *fining* in modern metallurgy (Craddock 1995, 253) and leads to de-carburisation of cast iron. In a way, the process is the opposite of the one described in chapter II.2 as carburisation (Fig.6).

(a)

Fig.28 (a) : large amorphous mass of iron found on 10.7.96, during the excavation in the section east of the West trench.

(b)

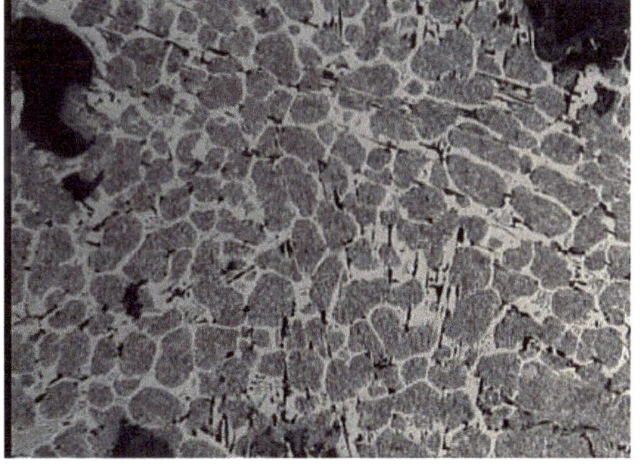

Fig.28 (b) optical microscopy of MES 127 in low magnification x66, displaying white cast iron structure.

Both processes aim towards a final product with a desirable amount of carbon and thus a good quality steel ready to be worked by the smith. The metallography and the distribution of the phases on MES126 suggests that this is a mass of cast iron where re-heating took place (Kostoglou and Navasaitis 2006).

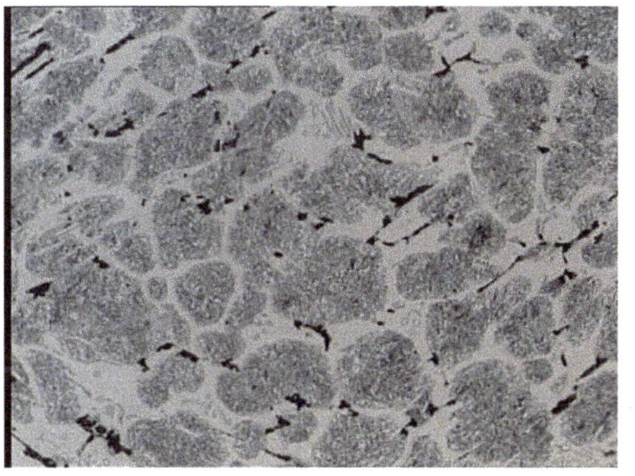

(c)

Fig. 28 (c) Optical microscopy of MES 127 in magnification x132 showing islands of pearlite with austenite around the grain boundaries. The black areas are pores resulting from the air entrapped in the iron

This re-heating expelled excessive carbon in the form of gaseous carbon monoxide. That is the reason why the flakes of graphite have little holes around and they have also turned to needles of cementite. Sample MES 139 also resulted from a similar process.

V. 2ii. Messemvria-Zone. Metallography of artefacts.

74 artefacts were recorded, representing almost all the iron artefacts found in the site so far. 22 have been analysed optically and 7 in the SEM-EDAX microanalyser. Table 13 depicts both the typology and metallography of the artefacts studied.

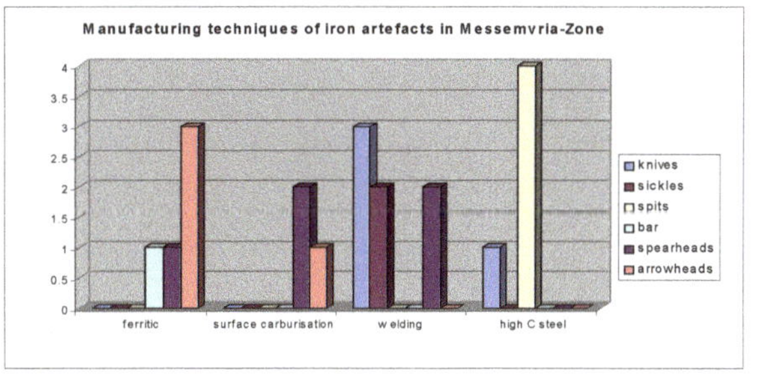

Table 13. Manufacturing techniques of iron artefacts in Messemvria-Zone (based on table 15 in the appendix).

Fig.28 (d). MES126

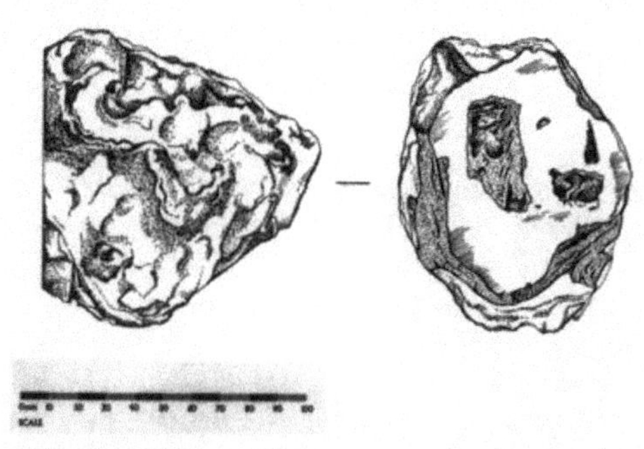

Fig. 28(e). MES 126, found n 28.6.96 in the West Trench. Mass of iron. Weight 2.724gr. 5th century BC. The black areas in the middle are mineralised charcoal fragments

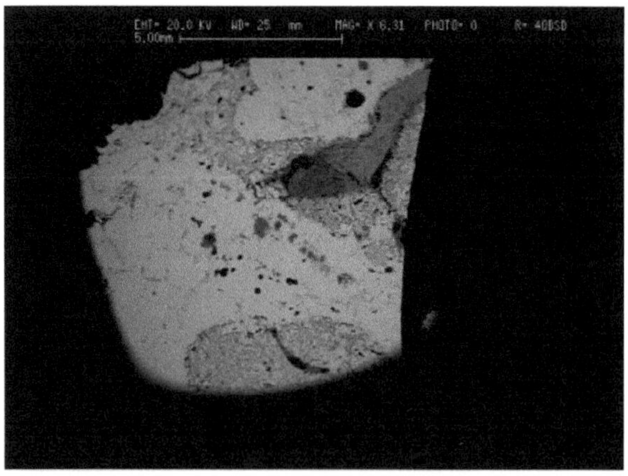

Fig28 (f). SEM photograph of MES 126 showing the distribution of different phases within the sample.

Fig.28 (g). Optical microscopy x66 (right) showing two phases due to different graphite (black) and cementite (white) both in pearlitic matrix.

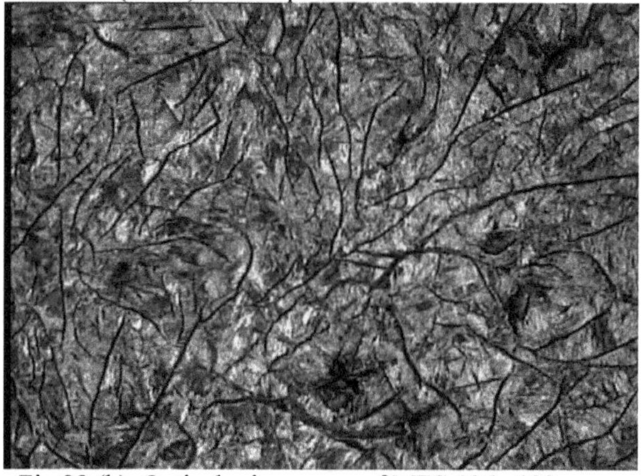

Fig.28 (h). Optical microscopy of MES 126 x132 (right) displaying long flakes of graphite in pearlitic matrix.

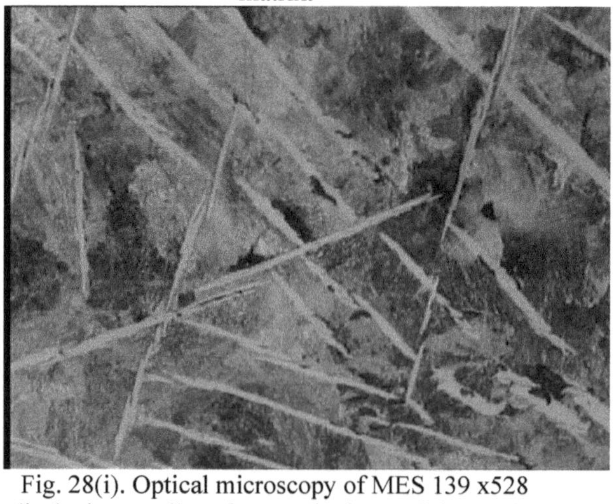

Fig. 28(i). Optical microscopy of MES 139 x528 displaying needles of cementite in pearlitic matrix.

.

Ferritic (PLATE 5 in the appendix).
(Samples: arrowheads MES 1, 2, 29; spearhead MES 23; straight knife MES 54; currency bar MES 21).

Wrought iron, with a lot of slag inclusions, has been used for the fabrication of arrowheads MES 1, 2, 29 (plate 5 in the appendix). The ferrite grains are large and slightly deformed, suggesting cold working to give the shape of a pyramid to these 4th century BC arrowheads. Spearhead MES 23 is also made by wrought iron. Ferrite grains and slag inclusions in this sample are elongated in the direction of forging. The structure of this blade is very soft, that is why the blade has been damaged (deformed point) during use. The fact that it could be easily repaired, attributes to the act of the dedication of the blade to the God Apollo a more symbolic meaning.

Fig. 29. Photomicrograph in magnification x66 of spearhead MES23 displaying ferrite structure.

The currency bar MES 21 has been found with the iron spearheads and the spits in the sanctuary of Apollo during the 1989 excavation, in the same context with pottery and artefacts of the late 6th century BC. This is a unique find, to my knowledge. Iron bars are not a common find in Greek or Mediterranean excavations. The absence of evidence rather than evidence itself prompted Pleiner (1969) and Tylecote (1987) to suggest that wrought iron was circulated in the ancient Greek world in the form of iron sickles and/or spits.

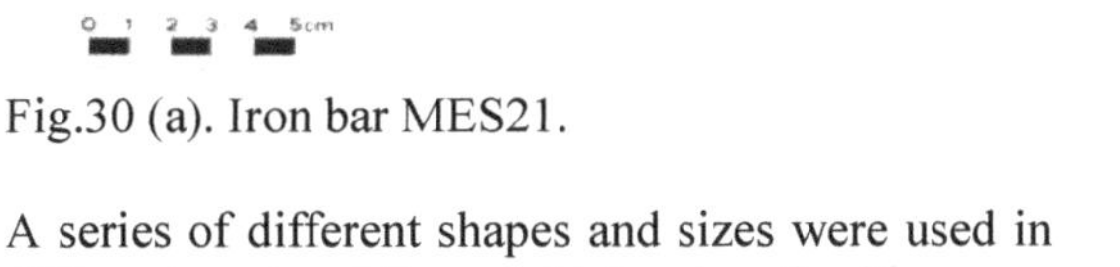

Fig.30 (a). Iron bar MES21.

A series of different shapes and sizes were used in different parts of the world to give the bloomery iron a marketable form. Thus, there are bars or flattened balls or plates used in Europe, Asia and Africa (Rostoker and Bronson 1990, 96). Most of them are again in use in the Medieval Ages and the usual quality control takes the form of a deep incision to expose the metal in the interior of the bar.

Most of the examples of iron bars found in Europe so far are mainly steel bars dated after the 13th century AD. They are much larger since they are the discard product of medieval water-powered bloomeries (Tylecote 1987, 254). Recent finds include the iron bars from Snorup-Sweden (Host-Madsen and Buchwald 1999) and from Saone-France (Dieudonne-Glad *et al.* 2001), but most of the found artefacts are Medieval in date. There are also the sword-type bars of pre-Roman age in southwestern England and the spindle-shaped bar from Carthage (Rostoker and Bronson 1990).

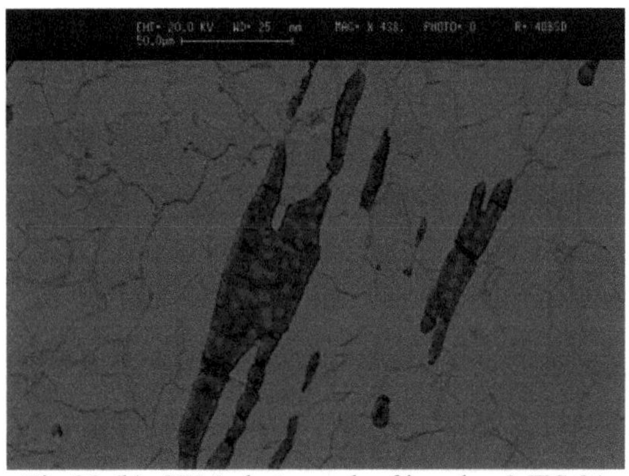

Fig.30 (b). SEM photograph of iron bar MES 21 displaying large ferrite grains and slag inclusions.

On the other hand, there is some evidence about the weight and the size of Roman blooms but there is no evidence about the kind of currency iron bars in the Mediterranean during the first millennium BC (Tylecote 1987, 255).

No evidence about standardization in the circulation of iron in terms of shape, weight, and carbon content exists so far. Quality control of such products could be assumed as early as the end of the 6th century BC. This is, for example, evident in the quality control of the bronze alloys mentioned in an inscription of the 4th century BC (Varoufakis 1979, note 13). Risberg has recently included in one of her publications artefacts described as iron bars found in Kommos, Crete (Risberg 1997, 188). Since no photograph or other details are included in the publications, the identification of the artefacts remains uncertain.

The Messemvria iron bar is 0.26m long, 0.45m wide, 0.017m tall and it weights 615.6gr (Fig. 30a). A cross section was removed from the left edge of the bar and was examined under the metallographic microscope and the Scanning Electron Microscope with EDAX micro-analyser (Kostoglou 2003). The section showed a uniform distribution of large, amorphous ferrite grains with a lot of slag inclusions but no welding lines (Fig.2b), indicating that the bar was forged from a single bloom. Ferrite grains are slightly deformed on the edges due to hammering, in order to give a bar shape to the bloom. Slag inclusions are elongated in the direction of hammering.

The chemical analyses of the slag inclusions revealed a three-phase slag with globular wustite (97.46% iron oxide), plates of fayalite (an iron silicate rich in manganese oxide 2.09%) and interstitial glass (iron-silicon-aluminium silicate). Manganese is a safe trace element for the Messemvria samples suggesting that the currency bar MES 21 is made locally from the smelting of manganese ferrous ores located in proximity to the site (Kostoglou 2003). The shape (elongated and flattened in order to be handled easier), the weight (almost that of a single bloom) and the context (votive, in the same group with the spits) suggest that this is not a simple iron bar, but most likely a currency bar or an ingot.

Surface Carburisation (PLATE 6 in the appendix)

(Samples: spearheads MES 7, 25; arrowhead MES 62; and MES 53 falx)

Surface carburisation of wrought iron is evident on the samples removed from the edges of spearhead MES 7 and arrowhead MES 62. It is also evident on the edge of medium carbon steel in MES 25 spearhead and MES 53 falx. A low piece of wrought iron or of medium carbon steel is heated up in charcoal to raise the carbon content to an aimed point that could be considered empirically as good quality steel (see chapter II.2).

SEM analyses in the slag inclusions of samples MES 25, 62 and 53 showed an interesting differentiation in chemical composition. Spearhead MES 25 has three-phase slag inclusions with minor amounts (less than 1%) of manganese oxide. Arrowhead MES 62 is rich in manganese oxide in the wustite phase of the slag inclusion (see discussion of the smelting slag). The sample from falx MES 53 shows a high concentration of the manganese oxide in the silicate phase of the slag (4.58%). Manganese in small amounts is considered as a beneficiary element in iron since it can eliminate hot shortness and increase hardenability. It seems that the smith was able to acknowledge this and use it accordingly.

Fig.31. Photomicrograph x132 showing edge carburisation on spearhead MES25.

Considering the typology of the artefacts and their chemistry, a local origin for the falx MES 53, the arrowhead MES 62, the spit MES 33 and the currency bar MES 21 could be suggested. The spearheads MES 23 and MES 25 are more likely of foreign origin.

Welding (PLATE 7 in the appendix).

Welding high carbon steel to low carbon steel is evident in the microstructure of MES 10, 22, 17 straight knives. Evidence for the same manufacturing technique exists in the microstructure of sickle MES 44. There, medium carbon steel was forged to low carbon steel. Slag inclusions are elongated in the direction of forging. Welding was not successful as the number of slag inclusions and the low diffusion of carbon in iron suggest.

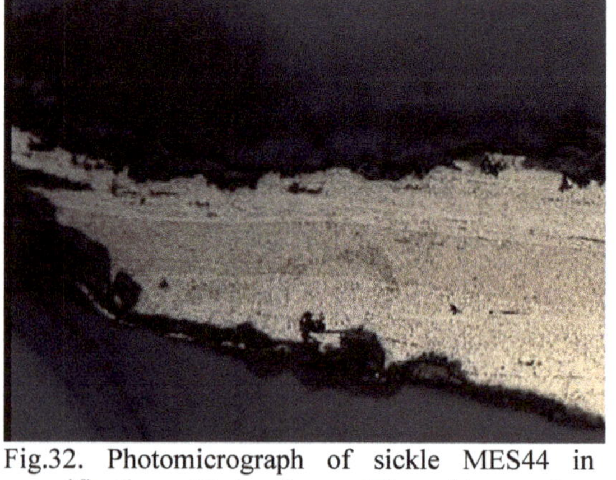

Fig.32. Photomicrograph of sickle MES44 in magnification x66 showing welding of low carbon steel (white) and copper rich iron (middle sheet).

This might be due to the copper and phosphorus amounts present in the slag (fig.32 and SEM table in the appendix) left in the artefact from the wrought iron used in the first place. This is a unique artefactm to my knowledge. Welding high carbon steel to medium carbon steel is evident in

MES 65 curved and MES 50 straight knife. Welding and twisting of two high carbon steel sheets is evident in MES 32, 33 spits. SEM analyses in MES 33 showed that the slag inclusions are rich in manganese oxide suggesting that the spit was manufactured locally.

A total of 6 spits (chapter IV) were also studied. All the samples examined displayed steel microstructures and the best example is depicted in Fig. 33. The spits analysed were made by welding and then twisting low carbon steel to high carbon steel. In order to check if this was the case with all spits, I sampled another four from the ancient settlement of Avdera, a Greek colony on the Thracian coast (Fig.1). The Avdera spits were found in a domestic context during the excavation of the Classical settlement. They are made not from steel as the Messemvria ones but from ferrite iron with a lot of slag inclusions.

This evidence, although contradictory, is very important. The Messemvria-Zone spits seem to be of special value and use because they are found in a ritual context and they were made of steel. The Avdera spits found in a domestic context were made of ferrite iron because their use –roasting- did not require any of the properties of the steel.

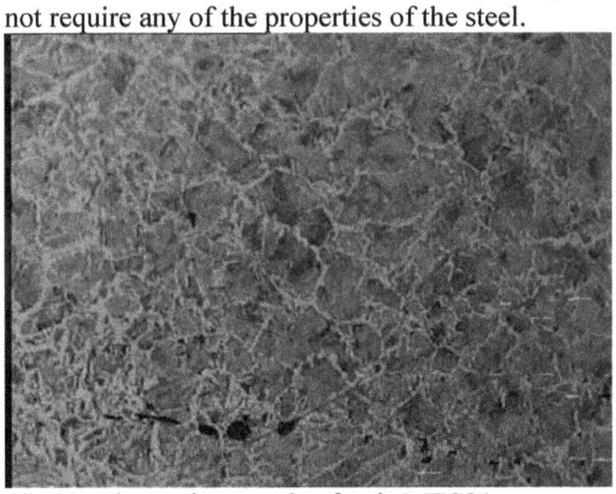

Fig.33. Photomicrograph of spit MES31 x264 displaying hypper-eutectoid steel structure with pearlite (grey) and cementite around the grain boundaries (white).

These new finds suggest a clear differentiation in terms of quality and use of spits. Further research is required to establish whether the shape, weight and carbon content of iron and steel currency bars were standardized. On the basis of the above evidence it may be concluded that the iron bar, probably forged from single blooms, was the marketable form, an ingot maybe, for the circulation of ferrite iron in ancient Greece at the end of the Archaic period (6th century BC). The spits, more easily carburised to steel due to their long and thin shape, were used as steel currency bars or ingots.

High carbon steel (plate 8 in the appendix).

High carbon steel with hypereutectoid composition was used for the manufacturing of tools that reacquire extra strength and hardness. With the exception of the welded structures in Messemvria-Zone spits/bars, only one other sample revealed high carbon composition (see chapter II.2). This is the sample from knife MES9 (Fig.34a and 34b) that displays spherodised cementite in pearlitic matrix. Current research showed that such structures are almost impossible to reach with carburisation and usually result from the decarburisation of cast iron because carburisation of wrought iron can hardly achieve such uniform distribution of carbon and in such high amounts (Godfrey *et al.* 2001). If this is the case, the knife MES9 dated to the 5th century provides an indirect indication for the use of cast iron in Messemvria-Zone.

Fig.34 (a). Photomicrograph of knife MES9 in low magnification x66 showing uniform distribution of high carbon structure.

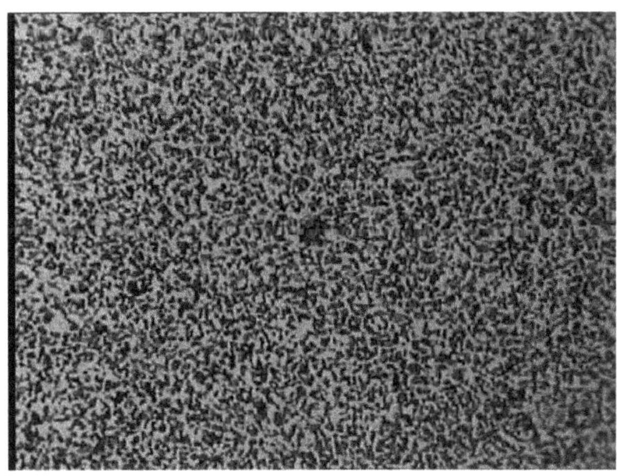

Fig.34 (b). Photomicrograph of knife MES9 in higher magnification x264 showing uniform distribution of spaerodised cementite (white) and pearlite (black).

V. 2iii. Conclusions: iron technology in Messemvria-Zone.

Three groups of smelting slag have identified in the material under study. Groups 1 and 2 correspond to the smelting of manganese/ferrous ores that are located in proximity to the town (circa 5 km north). These ores were deliberately smelted and date between the 6th and the 4th century BC. Manganese is considered as beneficial element; its presence in iron increases the hardenability of the finished product. Manganese has been also detected in the slag inclusions of artefacts providing a tracer element for the differentiation between local and foreign products. Thus, manganese-rich iron has been used in the manufacture of falces, arrows, and spits. On the contrary, the spearheads found in the sanctuary of Apollo were probably made elsewhere and it could be suggested that they are votives of not-local people.

Group 3 gives evidence for the smelting of iron pyrites rich in copper. Such ores are located in close proximity to the site at a site called Kirke. Copper-rich iron has been used in the manufacture of sickle MES44, indicating a rare archaeological example of the use of such ores.

The main product of smelting operations in Messemvria was bloomery iron, which was later forged to a marketable shape, such as the iron bar MES 21. On a later stage bloomery iron was shaped in long, thin rods, such as spits, which could be carburised more efficiently. Since artefacts, such as iron bar and steel spits, are found in the same context, it is suggested that the iron bar was the marketable form for the circulation of wrought iron. The spits, which carburised to steel more easily due to their long and thin shape, were the form for the circulation of steel. Both shapes were in use during the same times. The only parallel find, to the knowledge of the author, is the Snorup bars, Jutland, Denmark (Host-Madsen and Buchwald 1999). Two kinds of bars were found in association with bloomeries from slag-pit furnaces. The first group consisted of the so-called 'slender' bars of steel and the second of 'massive' bars of wrought iron. According to the analytical data, the authors suggested that while the massive bars of wrought iron were a local product, the steel slender bars were imported probably from Norway. All bars are dated between 330 and 560 AD and they could be used as a direct parallel for the argument stated above. Smelting furnaces were probably located in the northeast corner of the enclosed settlement. Slag was found in the same context with handmade pottery in layers of the 5th century BC. These are the destruction layers corresponding to the formation of a platform, where the town of the 5th-4th century was built, according to the Hippodamian system. The archaeological evidence strongly suggests that most of the industrial activity took place before 5th century BC. Therefore, a connection of the industrial activities and the local iron objects in the Apollo sanctuary should be suggested. Continuity in smelting of manganese ferrous ores from the end of 6th century (bar, spit) to the 4th century (arrowhead, falx) is evident. The fact that most of the slag is found along with handmade pottery, the so-called-Thracian, suggests a connection with the indigenous population.

The availability of raw materials, such as different ores and fuel, in Messemvria-Zone did not dictate the technological choices made. In the bloomery process different ores were used (e.g. manganeseferous/cupriferous) as the analyses suggest. The smelter had a good empirical knowledge of the properties of the ores and an ability to work the material in such a way to overcome potential difficulties and successfully produce tools and weapons. The smelter(s) was also able to operate the furnace in different ways in order to produce wrought iron and/or cast iron. Cast iron was decarburised and there are indications that hich-carbon, uniform, steel edges were probably produced by this method. These are advantages in the technology that –although they were not excluded theoretically- they were not evident in antiquity. The Messemvria-Zone smelter(s) and smiths are masters of their art. Their achievements can compare only with modern metallurgy. The fact that the layers of the ancient town of Messemvria-Zone (see chapter IV) were found almost undisturbed made possible the discovery of unique finds such as the iron and steel currency bars, the cast iron fragments and many tools and weapons.

V. 3i. Avdera (5th century B. C. to 2nd century AD). Metallography and SEM-EDAX analyses of industrial waste.

A total of 41 fragments (almost half of the total existing number) of industrial waste (total weight 9.355gr) have been collected and recorded for this study. 22 samples were analysed under the metallographic microscope and 9 were examined by SEM-EDAX. Most of them come from the area of trench Γ in the Classical enclosure of the city (see chapter IV, distribution map) and they bellong to the Roman times, approximately between the 2nd century BC and the 1st century AD. Most of them (AV 116, 114, 122, 123, AV 129, 133a, 130b, 103a) are small and medium size amorphous fragments of slag. Two (AV 108, 128) are shaped slag. There are also four samples of metallic iron (AV 141, 135b, 135a, 136) one sample from a bloom (AV 112) and few samples coming form fragments of vitrified ceramic material (AV 124, 126, 130a, 131, 134, 135a, 138). In the present discussion, only the very characteristic samples of each category are included.

Smelting slag (PLATE 9 in the appendix).

The samples AV 114, 116, 122, 129 are most probably smelting slags on the base of their metallography (three phase slag with small amounts of iron oxide or iron prills) and chemical composition (Fig.35). The AV 123, 128 are smelting slags, probably furnace bottom since they are round shaped. They contain iron oxide, iron prills, and entrapped ore (for slag typology see chapter II.2)

Fig.35. SEM photograph of smelting slag AVD114 showing needles of fayalite (light grey), iron oxide (white) and glassy matrix (dark grey).

Generally, the Avdera smelting slags have moderate amount of silica (8-30% in area analyses) and lime (2-13% in the area analyses). The alumina content is considerably high, more than 10% in most of the slags, and there are minor amounts of sodium and potassium. The amounts of oxides, manganese and magnesium vary between approximately 2-4% in the fayalitic phase of the slag. The phosphorus content in the interstitial glass is also between 2% and 4%. Mg, Mn and P in the studied samples seem to change proportionally (tables 13a and 13b in the appendix).

These three elements along with the other soil components, namely silica and alumina point to the smelting of iron oxides. This conclusion corresponds to the analyses of the entrapped ore in AV 123, as well (tables 13a and 13b in the appendix). Oxides have been used extensively in antiquity as a primary source for iron. They contain low iron quantities because of the advanced weathering process they have been through, but they have been often chosen for smelting. Their porosity and self-fluxing properties, due to gangue material, along with the abundance of their sources made these ores the bulk quantity of iron ores used diachronically.

Sample AV 103 is intriguing. Under the microscope it displays the three typical phases of a bloomery slag, namely, wustite, fayalite and interstitial glass. Wustite is mainly iron oxide (94.23%) rich in aluminium oxide (1.48%). Fayalite is an iron silicate rich in alumina (2.39%), lime (2.78%) and sulphur (3.58%). The interstitial glass is a lime-alumina silicate. The prills display a very high arsenic (45.91%) and sulphur (49.03%) content in spot analyses. The arsenic content in the area analyses is 1.18%, the sulphur 7.45% and there are also minor amounts of zinc oxide (1.08%) and high amounts of barium oxide. The last two are not present in the prills.

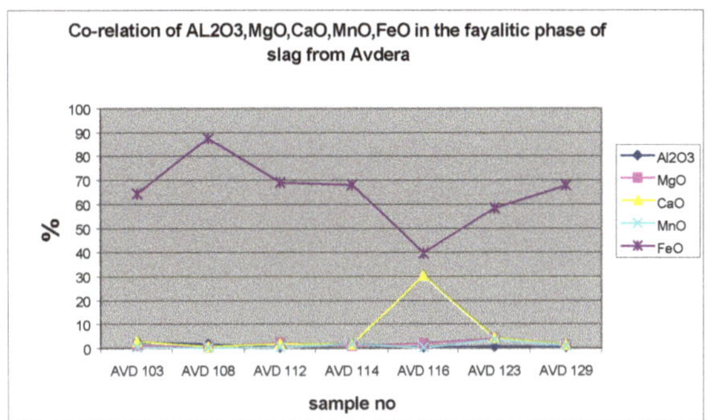

Table 14. Co-relation of major elements in slag from Avdera (based on tables 13a and 13b in the appendix).

The metallic prills indicate a very good control of the temperature during the smelting operation. Thus is not probable that the sample AV 103 is an accidental product formed in the furnace or that it comes from an unsuccessful smelting operation. Occasional fluxing with lime (CaO) also took place, as it is indicated in table 14. As can be noticed in sample116 table 14, high lime component results to smaller amounts of iron lost in the slag. The presence of arsenic, zinc, sulphur and barium needs an explanation. Arsenic and copper are associated with the mixed sulphide mineralisation of the Rodhope Mountains (see section II.2). The closest location of this kind of ores is at Kimmeria (Fig. 3 and 10).

Smithing slag (plate 10 in the appendix).
Smithing slag is represented with a lot of small and amorphous fragments that display mainly wustite under the microscope. Sometimes their macroscopic characteristics indicate tapping as in the case of AVD129.

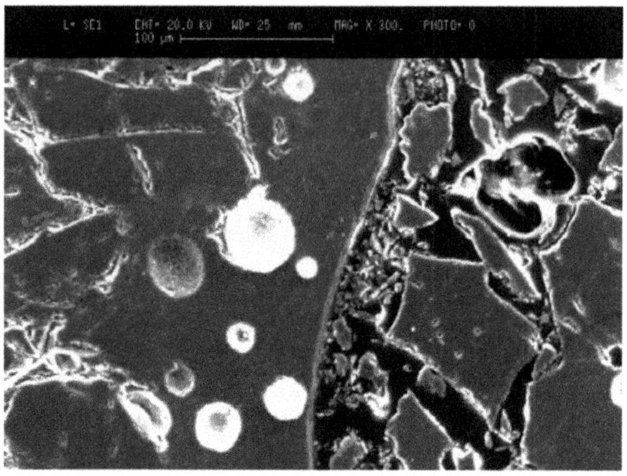

Fig.36. SEM photograph of furnace lining AVD130 showing titanium rich prills of iron and copper in silicate matrix.

Furnace lining (PLATE 11 in he appendix).

Samples AVD124, 135, 138, and 130 (tables 13a and 13b) are small fragments of furnace lining (Fig.36). They are all highly vitrified ceramics, mainly silica and alumina with fewer amounts of lime and iron oxide.

Metallic iron (PLATE 12 in the appendix)

(Samples AVD 112, and 141, display a low carbon bloom surrounded by three-phase slag).

V. 3ii. Metallography of artefacts from Avdera

163 artefacts were collected and recorded for this study. This number includes only some of the nails found because the total number of iron nails found so far exceeds 400. Not all artefacts where available for sectioning because they are currently on display in the Avdera archaeological museum. Advanced corrosion was a problem during sectioning in almost all finds studied. Out of almost 70 artefacts appropriate for study with the methodology of this work, 25 where examined optically and 3 under SEM- EDAX micro-analyser.

Ferritic iron (PLATE 13 in the appendix).

(Samples: curved knife AVD 38, strigil AVD 1, spit AV 50.)

Fig.37. Photomicrograph of strigil AVD1 showing ferritic grains slightly deformed on the point due to cold-hammering.

Cold working, meaning hammering wrought iron in low temperatures, has been used to shape tools that did not reacquire strength or hardness such as the curry comp AVD 1 (Fig.37) or spit AVD 50, which display ferritic structures. All artefacts contain a lot of slag inclusions.

Surface Carburisation (PLATE 14 in the appendix)

(Samples: straight knife AVD 36, sickle AVD 23, spit AVD 44, and axe AVD 39).

Carburisation is either done on the edge of wrought iron as in AVD 39, or on the edge of low carbon steel as in samples AVD 36, AVD 43 and AVD 44. In all the above samples carburisation was never complete, namely no hard steel edges were produced, indicating a low degree of efficiency in steel manufacturing techniques.

Welding (PLATE 15 in the appendix)

(Samples: curry comp AVD 5, straight knives AVD 48, 26, 18, and 53).

A number of different artefacts showed evidence of welding, suggesting that this was a common technique in Avdera. Usually, a sheet of low carbon steel was forged, while hot, with a sheet of wrought iron in a way that the steel was either placed on the core, or on the edge(s). The joining, however, was not done successfully as indicated by the many slag inclusions left behind (Fig. 38 and 39). The high phosphorus content (see SEM table 13a and13b) in AVD 18 should cause considerable difficulties during forging and it might be responsible for the final braking of the knife into two parts.

Fig. 38. Photomicrograph x66 of saw AVD26 showing welding between a wrought iron (upper part) and low carbon iron (lower part) with a lot of slag stringers between.

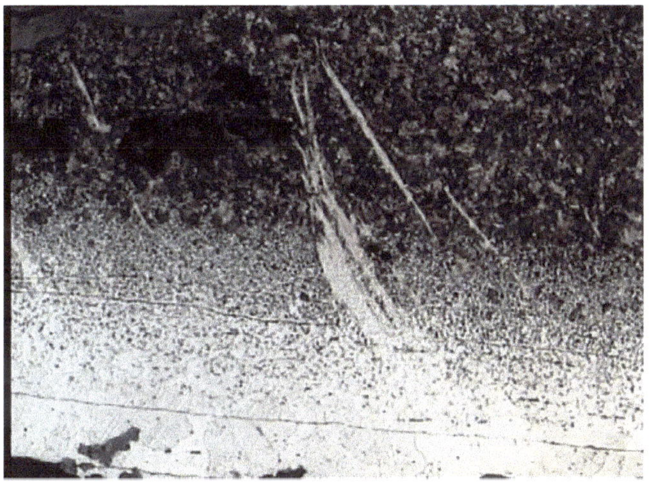

Fig. 39. Photomicrograph x66 of knife AVD53 showing welding between wrought iron (lower part) and high medium carbon steel (upper part). The gradual diffusion of carbon into the wrought iron can be seen along the welding lines.

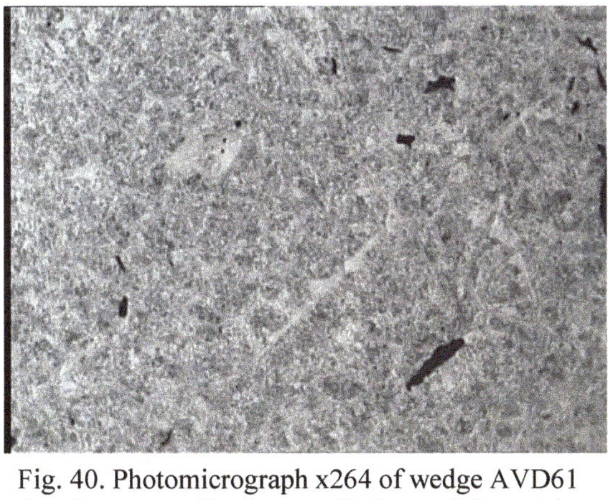

Fig. 40. Photomicrograph x264 of wedge AVD61 showing spaerodised cementite in pearlitic matrix.

High carbon steel (PLATE 16 in the appendix)

(Samples: straight knife AVD 38; axe AVD 30, wedges AVD 49, and 61).

High carbon steel with hypereutectic composition has been used for the manufacturing of tools that require extra strength and hardness such as wedges and axes (Fig.39).

It is interesting though that while the axe AVD 30 has the same typology with the axe AVD 39 (plate 16 in the appendix) they both display different structures and thus properties. The AVD 39 axe is made from soft wrought iron and showed some evidence for carburisation on the edge. Nevertheless, the edge of AVD 39 is not a hard one.

On the other hand, the axe AVD 30 is made by very hard steel (fig). SEM-EDAX analyses in slag inclusions of AVD 30 showed a very high phosphorus component and unusually high for Avdera samples (see table 13a and 13b) titanium oxide (4.22%). High titanium and phosphorus oxides were also detected in AVD 108 and 130 (see analyses of industrial waste). The presence of titanium in those two samples in the specific phases with the slag composition was associated with the composition of clay used for furnace lining. On the basis of its metallography (high carbon steel) and its chemistry (titanium rich) the axe AVD 30 might have originated in mainland Macedonia where there is evidence for smelting titanium-rich magnetite sands (Photos 1987).

V. 3iii. Conclusions: iron technology in Avdera.

The Avdera slags are characterised by low silica and lime components. Alumina is generally high and the oxides of magnesium, manganese and phosphorus change proportionately, suggesting the smelting of oxide ores. Smelting sulfide ores is also evident. Charcoal was the main source of fuel. The product was low carbon bloom.

The samples analysed and their distribution (see chapter IV) suggest that smithing and probably small-scale smelting workshops were located in the area out of the west gate of the Classical enclosure during Roman times. During the year 1991 only, excavations in the area brought to light many small fragments of slag of different types with a total weight of 1383gr and more than 320 iron nails, 2 knives, 2 doors' equipment, and 1 wedge. Smithing furnaces used both for iron and bronze artefacts indicating a close interaction between the two technologies in terms of their material, installations, energy sources and people who performed them. The Avdera smiths were not as qualified as the

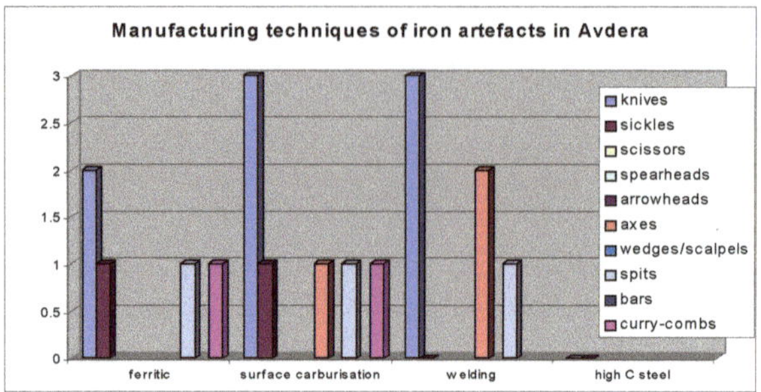

Table 15. Graph showing manufacturing techniques of iron artefacts in Avdera (based on table 16 in the appendix).

smiths at Kalyva or Messemvria as it is evident by the bad quality of the steel artefacts found so far. They were not able to deal with the difficulties of phosphorus-rich iron either (see AVD 18).

Provenance of ores is an interesting issue in the case of the city-state of Avdera. Both bog ores and mix sulphides are located outside of the territory (*choral*) of the city-state. This should cause considerable difficulties in acquiring raw material in early times, when indigenous tribes were controlling the mining areas. During the Roman times, however, this was not a problem since the geographical and political boundaries changed under the central Roman administration.

V. 4i. Kalyva – Kastro (2nd and 3rd centuries AD).

Metallography and SEM-EDAX analyses of industrial waste.

The industrial waste form Kalyva -Kastro consists of thirty five samples (total weight = 9 Kg) of which thirty were sectioned and examined optically; seventeenth of them were analysed with SEM-EDAX. Most of the samples were small (4-6cm) amorphous, dark grey displaying only minor porosity. The analytical results are presented as oxides and in their normalised form (tables 14a and 14b in the appendix).

Smelting slag (PLATE 17 in the appendix).

Most slag samples (KL73, KL82, KL91, KL95, KL99b, KL98, KL87, KL70, KL102a) fall in this, often difficult to describe, category of smithing/smelting (?) slag on account of their high iron oxide or fayalite component. On the basis of their metallography and chemical composition, samples KL73, 82, 87, 89, 91, 98, 99a, 102a, are more likely coming from oxide ore smelting. Under the metallographic microscope and in low magnification (KL 73, 87, 89, 91,98) they exhibit a highly porous three- phase slag with a lot of metallic inclusions and prills.

SEM-EDAX chemical analyses showed that these samples have a high iron component as it could be expected from the formation of wustite and the presence of metallic inclusions (smelting slag in Fig. 41). They (table 16) also have low silica and lime composition but high potash (K_2Na) and soda (Na_2O).

Potassium and sodium are part of the charcoal and can attribute minor amounts to the slag. These elements (K and Na), if present, can function as melting point depressants (Rostoker and Bronson 1990, 92). Thus, the slag has a low melting point (approximately at 1250° C), which practically means loss of iron. As it is noticed in table 16 (KL99A 99B) minimum amounts of K and Al result to major loss of iron in the slag (>70%). Iron at this stage is reduced again to metal and enters the slag in the form of amorphous or round prills, as in Fig. 42. This kind of reaction can take place in small bowl furnaces (see chapter II.2), where there is no sufficient draft. The produced bloom can have the form of a small furnace bottom such as the bloom KL 75 (Fig.42), which is approximately 0.15 in diameter and weights1664gr.

Smelting slag KL 87 (Fig.41) has a very high P content (5.70% in area analyses and 14.90% in interstitial glass), which is rich in copper oxide (1%). Sample KL 95 comes probably from a smelting slag as well and it is also rich in copper oxide (1.01-3.41%), arsenic oxide (1.13%) and phosphorus (2.87%). KL 95, and KL 87 point to the smelting of the iron oxides of the area which are always associated with manganese, zinc, and copper deposits because they are part of the mixed sulphide mineralisation (Epitropou and Bitsios 1987). One possible source of the ore used at Kalyva-Kastro could be located in the mining area of Potamaki approximately 5kilometrs NW of Kalyva-Kastro (Favas 1993). Iron oxides, also rich in titanium, result from the weathering of the iron pyrite. This could explain the minor amount of titanium oxide present in KL 79, 74 and 85, where titanium-rich iron prills were detected (tables 14a and 14b).

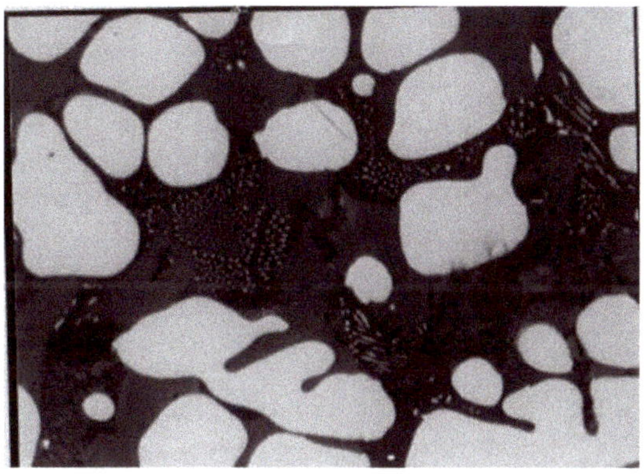

Fig.41. Photomicrograph x200 of smelting slag KL87 showing wustite (white), fayalite (light grey) and glass (dark grey).

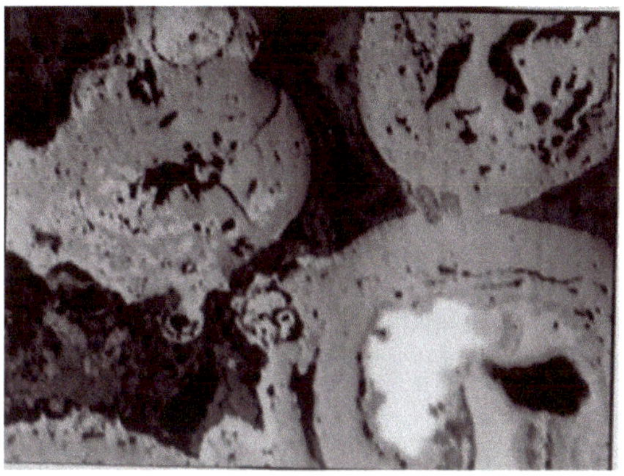

Fig.42. Photomicrograph of bloom KL75 showing metallic iron (white) and weathered iron oxide (light grey). The black areas are holes.

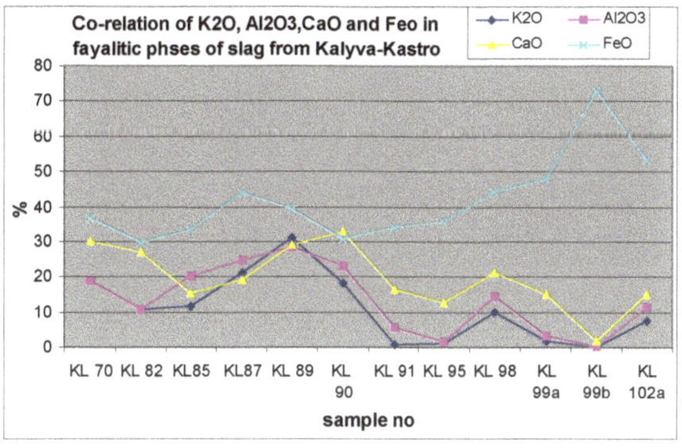

Table 16. Graph showing the co-relation between major elements in slag from Kalyva-Kastro (based on tables 14a and 14b in the appendix).

If iron pyrites where smelted in Kalyva-Kastro in the 2nd century AD, then installations for roasting should be found. Roasting of iron pyrites was needed to reduce the size, to help the formation of pores on the surface and to eliminate the sulphur content, which evaporates during the heating. It is a complicated time-, effort, fuel- and flux-consuming act. The smelting of iron pyrites demands, except for roasting, the addition of fluxes, rich in silica and lime, to help the formation of a fluid slag and the reduction of the metal. The analytical data do not support such an operation.

Smithing slag (PLATE 18 in the appendix)

Most of the small amorphous fragments are smithing slags with a chemical composition of, mainly, iron oxide.

Furnace lining (PLATE 19 in the appendix)

Finally, the two samples KL 74 (Fig.43), and KL 85 are mainly silica, lime, alumina, iron oxide and they are rich in titanium oxide. Some fine dendrites of iron oxide (wustite) and few metallic prills are evident close to the external and highly vitrified surface. Both are highly vitrified ceramics and they probably come from furnace lining as their optical examination and chemical composition indicates. Titanium should be one of the elements in the clay. The formation of prills suggests very high working temperatures.

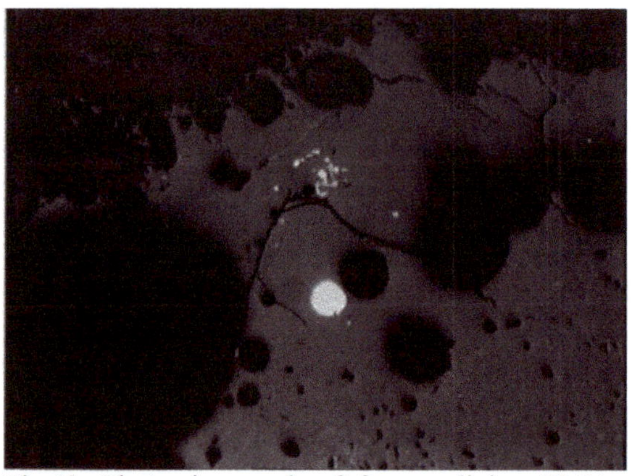

Fig.43. Photomicrograph of furnace lining KL74 showing highly vitrified ceramic (grey) with iron prills (white).

V. 4ii. Metallography of artefacts.

The analytical data suggest that steel making techniques where well-known in Kalyva –Kastro during the 2nd and the 3rd centuries AD. A variety of different techniques have been used in the manufacturing of steel tools and weapons.

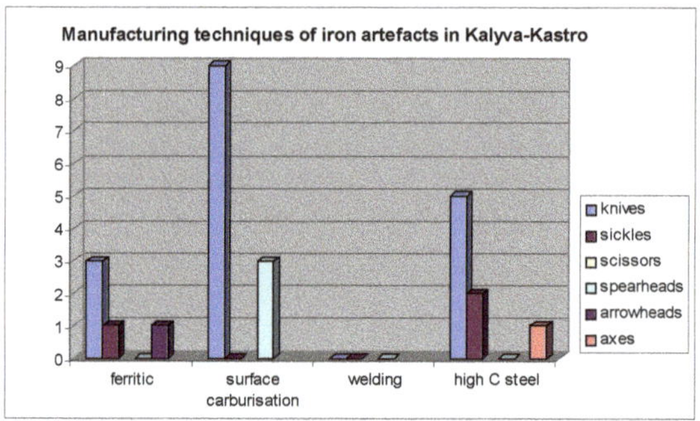

Table 17. Manufacturing techniques of iron artefacts from Kalyva-Kastro (based on table 17 in the appendix).

Ferritic iron (PLATE 17 in the appendix)

Ferritic structures are evident in samples from the scissors KL 59, the straight knife blade KL22 and the back of the curved knife blade KL 23. Ferrite grains in the above samples are large and polygonal, slightly transformed due to forging. Slag inclusions are also present. The low hardness of the samples (119 Hv in KL 22, and 152 Hv in KL 23) is in accordance with their structure. The blades KL 59, KL 22, and KL 23 were made by forging sheets of wrought iron at room temperature. In the case of scissors KL 59, P-rich iron should be the source for the metal.

The curved blade KL 23 has a very soft structure that is not appropriate for its use, which requires strength. It is thus probable that this blade has been discarded or deposited before its fabrication was complete, namely before carburisation took place. This conclusion is also supported by a high incidence of curved knives at the site show evidence of surface carburisation.

Surface Carburisation (PLATE 18 in the appendix)

It is evident in the samples of: straight blades KL 62 and KL 20, curved blades KL 65 and KL23 edge, large triangular blade KL 67, and socket of spearhead KL 17.

A single sheet of wrought iron (pure ferrite with large grains as in the core of the v shaped sample in Fig. 44) has been first heated in charcoal for a couple of hours to rise the carbon content up to the eutectic point. This is evident by the lamellar pearlite structure in the middle. Heat treatment followed on the edge. The re-heating and forging caused the absorption of more carbon and thus the formation of spaerodised pearlite (black) and broken cementite (white) along the surface of the edge. Surface carburisation of wrought iron was a well-known and generally applied technique at Kalyva-Kastro. It has been used to give a steel edge to a variety of tools (straight and curved knives) and weapons (spearheads) requiring one or two good, thrusting edge(s).

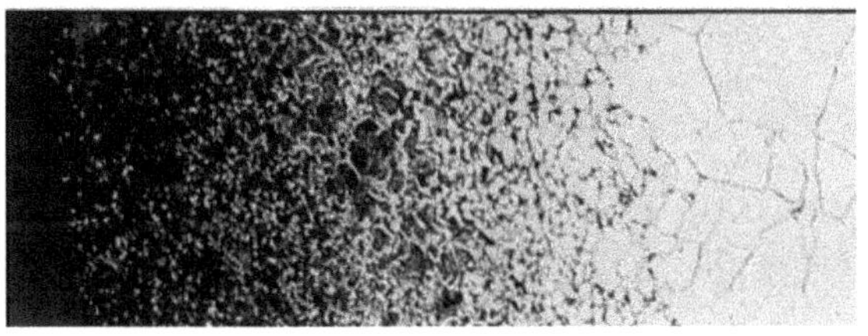

Fig. 44. Microphotopraph of knife KL62 x200 showing carburisation of wrought iron.

The carburisation is either applied on the right or on the left edge (Fig.44), and some times on both (plate 21 in the appendix). The different cutting edges (left versus right) might be related to the use of knifes by right-handed and left-handed people respectively (personal communication with anthropologist Dr. N. Agelarakis, Adelphi University, NY). This fact indicates that the blacksmith at Kalyva-Kastro produced knifes that were ordered for specific people and for specific uses, suggesting a high level of specialization in iron production.

Welding (plate 19 in the appendix)

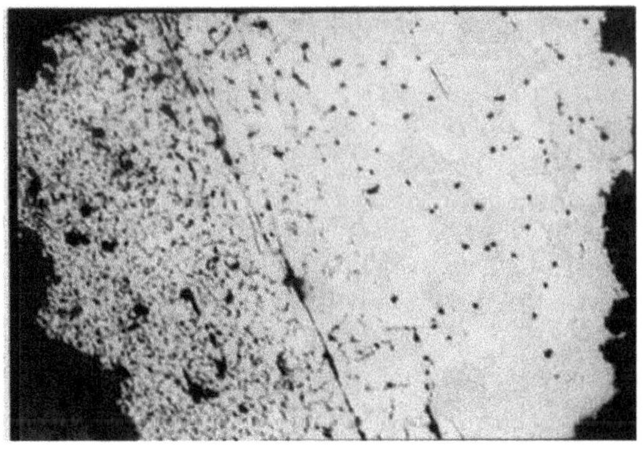

Fig. 45. Photomicrograph x50 of knife KL12 showing welding of wrought iron (right) to low carbon steel (left).

Evident in samples from: curved blade KL 12, sickle KL 4, straight knife KL 48, jointers KL 30 and KL 31. Forge welding of sheets of steel to wrought iron was not a common practice. A sheet of steel is welded to wrought iron (see KL 12) or two sheets of low and medium carbon steel are forged together and then folded and hammered again (see KL 4). The wrought iron used in both cases is not of very good quality, since it has a lot of slag inclusions which affect its properties and its resistance to corrosion. High carbon steel has been used only in the case of jointer KL 31 because its use as a building material required extra strength.

High carbon steel (PLATE 20a in the appendix)

High carbon steel has been used for the fabrication of heavy-duty tools and some weapons. Both the spearhead KL 60 and the billhook KL 64 are made from single pieces of uniform steel with compositions very close to the eutectic (0.8%). These structures are appropriate for equipment where strength and resistance to de-formability are appropriate.

The structures of the axe KL55 sickle KL33, and straight knives KL37, KL39, KL46 are tempered martensite. High carbon steel has been first fast cooled, probably by quenching into water or oil. The resulted needle-like martensite crystals required further treatment to reduce brittleness. Thus, the blades have been heated again in charcoal up to a temperature less than 723° C where re-crystallization occurs and hammered while hot.

Sickle KL 33 has a very high titanium oxide content (6.31-7.71%) in the slag inclusions. Titanium oxide was present in minor amounts (less than 2%) in the slag (KL 79) and also in the furnace lining fragments (KL74 and KL 85) and it was probably associated with the smelt of the iron pyrites of the area. Titanium was used as a safe tracer element for fingerprinting the provenance of ore in the case of iron artefacts from mainland Macedonia (Photos 1987). There, titanium was associated with the smelt of titanium-rich magnetite sands located in the coastline of Macedonia. The smelt of iron sands is not testified by the Kalyva industrial waste samples. However, because the iron purites in th earea are rich in titanium oxide, it could be safely suggested that sickle KL33 was probably made locally. The high amount of phosphorus in the same sickle (1.58-1.90%) probably caused considerable difficulty during hot forging and it might have been responsible for the final brake up of the blade. The broken sickle was found stored in a pithos along with other iron objects and it could be suggested that it was part of a group of iron artefacts to be used for re-smelting in a smiths workshop.

Finally the curved knives KL 14, KL 41 are also made of very hard steel with hypereutectic composition (carbon content close to 1.2%). This metallographic structure, as in the case of Messemvria-Zone, can be produced either by carburisation wrought iron or decarburisation of cast iron.

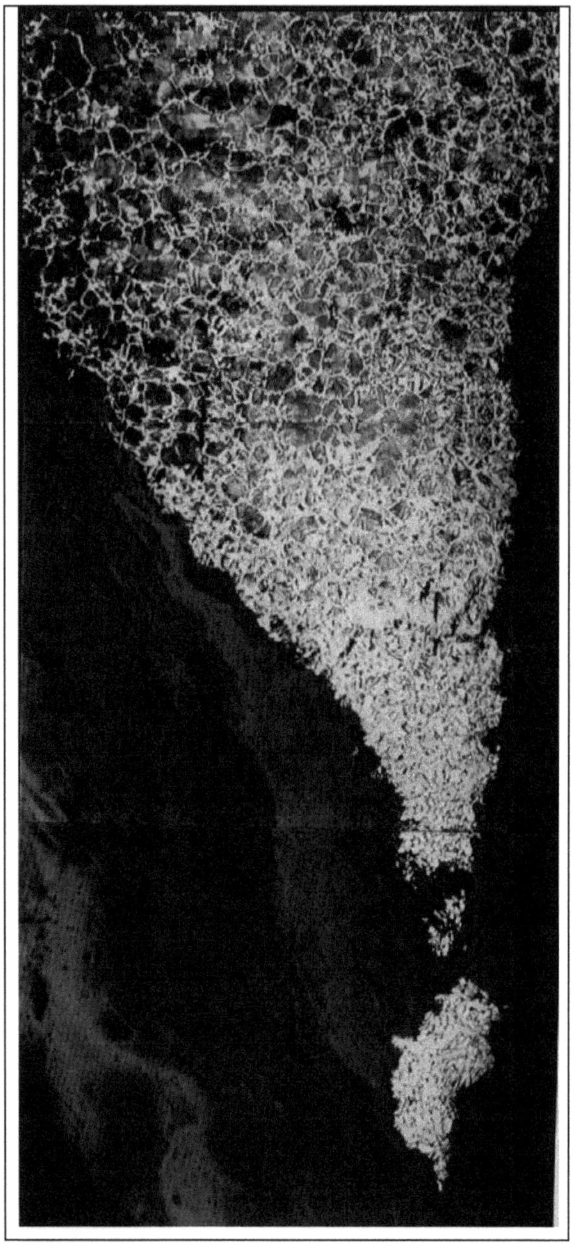

Fig.46. Photomicrograph of curved blade KL 14 x50. The v shaped section displays uniform distribution of hypereutectic composition of steel with pearlite and cementite around the grain boundaries. The point of the v shaped section is de-carburised.

Hv on F+P=205
Core=F, 180

Similar is the straight blade KL41.

V. 4iii. Conclusions: iron technology in Kalyva-Kastro.

Iron was produced locally by smelting iron oxides and later (2nd century AD) iron pyrites. On the basis of their deposition, distribution and metallography, iron artefacts and industrial waste indicate the area of a workshop within the castle. Small-scale smelting and smithing activities are located within the same area. Most of the industrial waste has been found in area B-2 (see chapter IV) where a small workshop was probably located. The analytical data suggest the smelting of bog ore in the early phases probably in small bowl furnaces. The smelting of iron pyrites is evident in the 2nd century AD. No roasting furnaces or other structures within the castle were found close to the mine. Roasting probably took place elsewhere. Charcoal was the primary fuel, as it is evident form charcoal inclusions in the slag. The final product was the bloom (KL 71) which was then forged and manufactured to produce everyday tools.

Iron ore sources are located in close proximity to the fort, probably in the site called Potamaki. The raw material comes to the fort in the form of ore ready for smelting. Smelting, bloom smithing and manufacturing of artefacts takes place within the fort. Charcoal was the main fuel used for smelting and smithing. The smiths had a good knowledge of manufacturing and repairing iron and steel blades, able to deal with the difficulties of phosphoric iron. Steel making techniques where applied to the artefacts without any typological distinction.

A high degree of specialization in the manufacture of tools is suggested by the analytical data. A close relationship between the tools and the individuals who used them is evident. The knife is a strictly personal tool/equipment and specific care is taken during its manufacture. The under study technology, according to the archaeological data, probably reflects the technological level of the indigenous population, since most of the artefacts are agricultural or domestic tools. This conclusion may be seems controversial since we described the Thracians mainly as famous warriors and it seems that this is the area where iron had to play an important role both in technological and cultural terms (see chapter VI, interpretation).

V. 4. Discussion: the metallurgy of iron in Aegean Thrace.

The analytical data presented in this chapter in combination with the archaeological data of chapter IV, support the conclusion that different small-scale household or community-based workshops, practising different local technologies were operating in the region. The individual characteristics, the development and the end of each technological tradition follow the life of each settlement. Each local technological tradition is characterized by continuity in terms of ores, fuel, and manufacturing techniques used. The existing local traditions are variations within the metallurgy of iron and create a diverse picture on regional scale. The emerging picture contradicts the one described in the conventional historical framework of the region (chapter I) and in the documentary evidence (chapter III) that report major discontinuities in the local developments.

From the river Nestos to the river Evros and from the Rodhope Mountains to the Aegean Sea (Fig. 3 and 10) different local technological traditions emerge. These can be summarised as follows:

Messemvria-Zone
Smelting took place in shaft furnaces where manganese/ferrous or copper-rich iron ores were reduced to metallic iron with the aid of charcoal and occasionally possibly lime as flux. All primary sources of ores and fuel were located in proximity to the site. The bloomery furnace under different operations could produce both wrought and cast irons, the latter most probably is accidental.

Edged tools and weapons were also produced locally and they can be distinguished from the imported examples by the constant amount of manganese present in the metallic iron and the slag inclusions in the artefacts. All known techniques for manufacturing steel tools and weapons are evident (table 13) and all were successfully applied.

Iron was traded in small elongated bars forged probably from a single steel bloom which was circulated in the form of long, thin bars such as spits.

Avdera
No iron finds and/or industrial waste are found in Avdera from the 7th to the 5th centuries BC. The first iron artefacts are found in the settlement and in the cemetery of the Classical colony. The first evidence for smelting and smithing iron in the site bellongs to the Roman period. Access to primary resources was complicated because both ores and fuel deposits were located outside the territory of the city-state of Avdera. This problem was solved only after the re-organisation of the territorial boundaries and under the administration control of the Romans. The majority of the artefacts and all the slag found in Avdera bellong to the Roman times.

Most of the artefacts excibited either surface carburisation or welding of low carbon to high carbon steel. Neither of these two techniques was applied very successfully, with a few exceptions. Only the wedges were made of high quality steel, but there is no evidence for their local origin.

Kalyva-Kastro
Despite the fact that fort is situated in one of the richest metalliferous zones of Greece there is evidence only for the smelting of iron oxides in bowl furnaces. The Kalyva-Kastro metalworkers were efficient since they were able to produce enough metallic iron from relatively pure iron oxide by using appropriate fluxes.

The artefacts were probably made locally and they mainly include implements used in agro-pastoral activities. Smiths should be highly skilful as the manufacture of artefacts suggests. In Kalyva-Kastro surface carburisation of wrought iron or heat treatment of high carbon steel produced steeled tools and weapons. There is also evidence for the use of phosphorus rich iron. None of the artefacts sampled showed a welded structure.

Other technological traditions in North Greece, as discussed in chapter IV, include:

- The smelting of manganese/ferrous ores in Symvolon Mountains.
- The use of radioactive magnetite sands in Vrontou Mountains.
- The smelting of titanium-rich magnetite sand on the island of Thassos.

The archaeological and analytical evidence examined in this thesis (chapters IV and V) suggest that mining in North Greece took place in small, dispersed mines that were located close to the settlements. Roman occupation of the region brought no major changes in this model of mining exploitation. There is no archaeological evidence to suggest imperial control of the mines. The assumption suggested by the documentary evidence (chapter III), that indigenous people were controlling the mines of Aegean Thrace, is also supported by the archaeological evidence.

The study of the excavated material provides further evidence towards the above interpretation. The fact that the household or community-based workshop remains on the centre of the production from 6th BC to Roman times, leads to the conclusion that the application of colonial/imperial strategies to Aegean Thrace did not change old and well-established technological traditions. Local traditions continued to operate within the settlements by using the same raw materials and smelting techniques for centuries.

The analytical data suggest that different technological choices were made in the production of iron and steel in the three sites examined in this work. The people involved in iron and steel production made a series of choices in terms of raw materials, energy sources, attempted products, manufacturing techniques to name some of them. These choices have contributed to the creation of a unique iron technology in each site.

Each local technology is characterised by the use of particular raw materials, as mentioned in the beginning of this section, and by the application of specific manufacturing techniques. The study of the manufacturing techniques in each site can provide a better insight in the understanding of local traditions. Tables 18 and 19 show the frequency and distribution of the four main manufacturing techniques in the sites under study. The artefacts under study are knives (table 18) and the total of edged tools and weapons (knives, sickles, axes, wedges, and spearheads in table 19). As it has been explained in chapter II.2 the study of steel manufacturing techniques should include edged finds since they are the ones needing the properties of steel. Because of this limitation, iron finds such as strigils are excluded from these two diagrams. Comparing the manufacture of knives in table 18 with the manufacture of all edged finds in table 19, no major differences are noticed. This mean that, methodologically speaking, to treat edged tools and weapons as a total as in table 19 and not as different categories as in table 18, provides a better statistical example for the study of techniques. In other words, it means that the typological classification of the archaeological finds is irrelevant to the study of their technology since what matters is the use of the finds (i.e. heavy duty versus light duty tools) and not their type. This suggestion does not over-rule the correlation between function and style, but it aims to give another dimension to it.

What is apparent from these two tables is that manufacturing techniques appear to be very diverse on a regional scale and constant on a local scale. For example, in table 19, it can be seen that:

- ferritic edged tools and weapons are absent from the record of Messemvria-Zone;
- high carbon steel edges are not found at Avdera; and
- welded structures are not in use at Kalyva-Kastro.

Table 18.Based on tables 15,16,17 in the appendix.

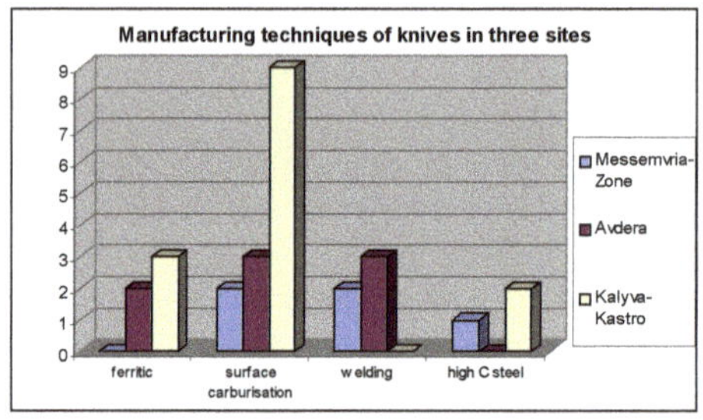

Table 19 based on tables 15,16, 17 in the appendix

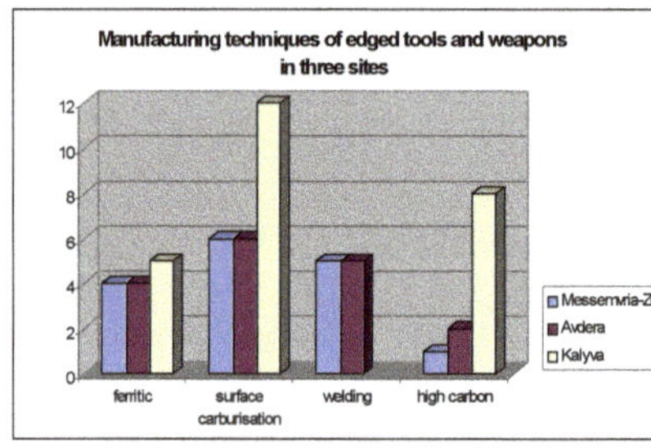

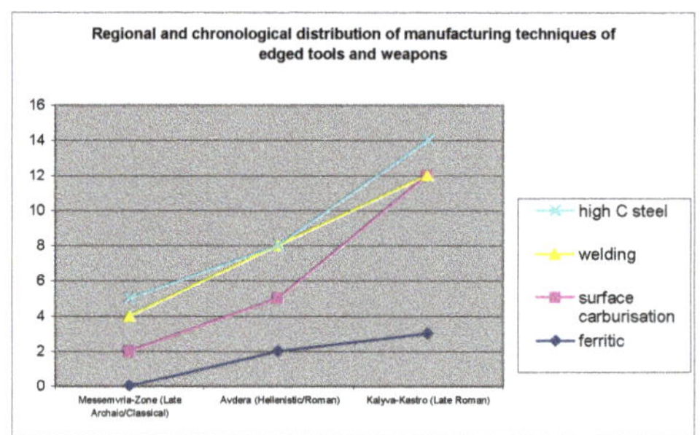

Table 20. Based on 15, 16, and 17 in the appendix

On the other, hand a variety of techniques is used at all three sites. The evidence presented in tables 18 and 19 suggests a pattern where individual traditions with preferences for the use or the exclusion of some techniques exist on local scale and lack of standardisation is apparent on regional scale. The deliberate use of specific raw materials such as the manganese/ferrous and copper-rich iron ores in Messemvria-Zone, or the iron oxides in Kalyva-Kastro along with the preference for some manufacturing techniques (indicating a series of technological choices) characterise each local technology. Skilled metalworkers and smiths existed in Messemvria-Zone (chapter V.2) and in Kalyva-Kastro (chapter V.4) where the archaeological evidence indicates cultural continuity with the indigenous population of the region (chapter IV). The situation seems to be different in Avdera (chapters V.3 and IV), where iron is not locally produced and commonly used during the early phases of the colony. The manufacturing techniques under study appear in the archaeological record as early as the end of the 6th century BC. There is no evidence to suggest that new techniques were introduced to the iron technology of the region during the Roman times. Table 20 depicts the evidence from all three sites and illustrates this conclusion. No innovation or invention in terms of new ores, fuel, and techniques of manufacture in Roman times has been noted so far. On the contrary, advanced techniques of production, such as the use of manganese rich iron ores in Messemvria-Zone, did not find their way into the new system. However, new territories such as in the case of Avdera were opened for exploitation under the new socio-political system. It can be concluded that both the effects of Greek and Roman cultures to the others, the indigenous, as well as the idea of the large innovations introduced to barbaric people should be re-addressed by the archaeologists. In the case of iron technology in Aegean Thrace Romans probably did not introduce new techniques but they used skills, experience and materials that already existed.

Chapter VI. Synthesis and conclusions.

The general discussion following, attempts a collective interpretation of all the data presented in this thesis. The technology of iron production in a regional level is discussed in the light of the archaeological context to extract information about the production, distribution and consumption of iron in Aegean Thrace from Classical to Roman periods. In this context, technological continuity or discontinuity is evaluated, along with other changes noted in the archaeological record in order to see if changes in technology are part of an entirely new socio-economic structure or not, and if technological choices are culturally oriented. The indigenous involvement in iron metallurgy and the role of iron in the relationship between the indigenous population on the one hand, and the Greek colonists or the Romans on the other is also assessed in order to establish transformations in local identities from Classical to Roman times.

This thesis followed step by step the theoretical and methodological model suggested in chapter II. The final synthesis of all evidence discussed in this thesis follows that model as well. The main points are summarized as follows:

1. Interpreting material culture ↔ analysing material culture (technology in its own terms)
2. Defining continuity and change in material culture ↔ defining continuity (=stability) and change (=innovation) in technology.
3. Explaining change in material culture ↔ explaining correlations between changes in one form of material (i.e. settlement patterns and/or burial practices) and technological changes.
4. Whose technology are we looking at? Or, issues of identity ↔ Is new technology part of an entirely new socio-cultural structure?

Chapter I showed that historical conventions and colonial biases dictate current interpretations of material culture in Aegean Thrace and they appear inadequate in accommodating cultural diversity and local developments. Theories of 'Hellenisation' or 'Romanisation' on the one hand, and culture historical approaches on the other hand, have over-emphasized discontinuities in material culture (chapter II.1) and used ancient written sources (chapter III) as indicators of ancient realities and identities. The analysis of both the archaeological evidence (chapter IV) and analytical data (chapter V) strongly questions the validity of such ideas in the light of current theoretical concerns about the study of material culture, cultural identity and post-colonial theory (chapter II.1). Iron technology and the use of its products (everyday iron tools, locally produced with the same empirical technique, used and re-used for long periods), carry less of an opportunity for displaying social ideology in a symbolic manner than might happen with other categories of material such as pottery. In this context it could be argued that iron artefacts carry a more explicit social significance that could be traced archaeologically by the study of distribution and consumption patterns. On the contrary, iron technology, as a cognitive system, carries an implicit knowledge that is shared between the producers of a small community. The combined study of both the explicit and implicit aspects (artefacts, technology, and cultural attitudes), of iron production and consumption offers a better insight in the way that technologies are embedded in culturally distinct societies.

Starting with the conclusion from the analytical chapter, attention should be drawn to the high variability and diversity of the data. It was not possible to establish standardization of iron production in terms either of chemical composition of the industrial waste or of the application of steel manufacturing techniques. This is the case with the material from both the Classical and the Roman periods. Although, the successful and consistent application of expert knowledge is evident (with the exception of Avdera), no correlation between types of artefacts and manufacturing techniques can be traced. The smelting by the bloomery process of any possible iron ore is evident: oxides, mixed sulfides, manganese/ferrous ores. Steel manufacturing techniques, from surface carburisation and welding to de-carburisation of high carbon steel are applied to any tool or weapon (a crude distinction should be made between heavy and light duty implements).

The intriguing diversity noticed in the analytical studies from a strictly metallurgical point of view, could be attributed to the following reasons:

1. time span (samples come from different smelting/smithing operations) and they represent either different periods or experimentation within the same period;
2. nature of process-since it is an empirical and highly risky process that does not produce the same results twice;
3. different raw materials used (ores, fuel, fluxes);
4. different techniques used.

The above conclusions could not be that attractive either to an archaeologist of the Greco-Roman world. Firstly because, Greeks and Romans are considered highly specialised craftsmen (chapters II and III). Secondly, because it is understood that steel making techniques are well known and widely used in Roman times (chapter II.2). However, analytical work has demonstrated that to find a qualified smith to produce and repair a steel blade is not as common as it was originally thought, even in Late Roman periods (Tylecote and Gilmour 1986). The lack of statistically large sample of analytical work on iron from Greece in particular (chapters II.2, IV.1, and V) makes any comparisons difficult.

The diversity of the analytical data could also signify the diversity and lack of standardization in iron production as noted in chapter V. Such diversity would not be expected in a system with centralised control of the production. This inevitably leads to the conclusion that smaller scale production activities operating within the local societies were the common practice. There is an imbalanced impression of Roman mining (Edmonson 1989) in which a few large-mines and production centres usually under imperial control (see chapter IV) over-shadow smaller scale production. These smaller mines and production centres within smaller settlements gain more in social significance if technology is studied on a local/regional scale.

The characteristics of iron production in Aegean Thrace (chapter V) from Classical to Roman times can be summarised in the following three points:

- locality: the locus of the metallurgical activity is defined by the settlement that includes the same technological tradition;
- diversity on regional scale (dispersed local traditions, lack of standardisation);
- the household or community-based workshop remains at the center of the production.

Those three characteristics remain the same even when major historical events (such as the Roman annexation) occur in the area. In the case of Aegean Thrace the dynamics of small scale mining/smelting east of the river Nestos could be compared with large-scale mining in the west area, between the rivers Strymon and Nestos. In the neibouring region in Philippi, mines were under Imperial control and Philippi was the centre of Roman administration (chapter IV).

In studying the impact of the Roman rule in a very small region between Strymon and Evros, part of the same province (see chapter I) different forms and transformations of social processes are noticed. This diversity strongly questions the conventional history of the area, and in particular the assumed technological superiority of some cultures (chapters I and II.1).

The analytical (chapter V) and archaeological evidence (chapter IV) showed that advantages/innovations exist in the area in the context of indigenous developments as early as the end of the 6th century BC, long before the first evidence for Roman presence in the region. An advanced knowledge of iron and steel manufacturing techniques is evident. The important socio-economic role of iron can be seen in the votive iron and steel currency bars as early as the end of the 6th century BC. Its highly symbolic meaning is emphasised by the iron rings found in Avdera and Messemvria in elite tombs. The modern metallurgist could point the production of cast iron (chapter V) as the greatest innovation of all, if purposfull production could be demonstrated. Nevertheless, as in chapter II.2, our understanding of technology has no common ground with the ancient one. Iron production, in the given society, has techniques that could share knowledge, raw materials, and probably the same operators. Initially, probably only few, such as the metalworkers and their groups, could appreciate the advantages of the new material. The lack of products from the intermediate stages of production, such as slag, masses of iron, and blooms in ritual or burial contexts (chapter IV) suggests that only the final object and its use was of an economic or symbolic significance.

This brings the discussion to the second and third steps of the theoretical model as these were summarised earlier in the section. Does any change in technology occur? How could this be related and explained in co-relation with other changes in material culture?

The technology of iron in Aegean Thrace appears to be more conservative in changes, and not only because of the availability of raw materials and/or techniques. Anthropologically oriented studies (i.e. Lemmonier 1992; Herbert 1983) showed that people are very selective in their technological choices and they do not always chose the best available material. Technological choices are culturally oriented and depend on factors such as accumulated knowledge, evident need, and cultural/social acceptability (for a more extended discussion see chapter II). This was demonstrated very vividly by the example of Messemvria-Zone.

In the case of Aegean Thrace, no technological innovation or invention in connection with the Greek colonial or the Roman imperial context has been documented so far. Major changes in the social organisation have not been documented either. Even had they been they are not reflected on the material under study. The over-emphasised discontinuity in the archaeological record seems to be the result of world system theory and colonialist approaches to material culture (see chapters II.2 and IV). A closer inspection of the archaeological record suggests the opposite: the dramatic transition from Classical to Roman times seems to be an artificial construction in this region.

First, no major changes in the settlement patterns have occurred. The habitation of the fortified settlements of the Classical and Hellenistic periods continue, as has been shown to happen in the Messemvria-Zone and at Kalyva-Kastro. As already noted the mines remain under the exploitation and probably the management of indigenous people. Most of the industrial waste related to iron production is also found in the context of indigenous material culture.

The archaeological data suggest a rural economy based on trade along the coast, and a mixed economy based on agriculture and pasture, warfare and mining as seasonal occupations. The village (*kome*) is at the centre of this model from Classical to Roman times as archaeological data suggest (chapter IV). Not many permanent settlements occur in the upland and lowland zones. Survey data showed that the countryside of the Greco-Roman world it was no different (sea also chapter III). The effect of the Greek colonisation and of the Roman presence is limited. Continuity in settlement patterns is also evident (chapter IV). Absence of features such as provisional cities, agricultural plains to support urban population, planned urban layout, infrastructure, domestic and public architecture, elite display or emulation of Rome, suggest the limited impact of the Roman rule on Aegean Thrace. Even the foundation of new towns along via Egnatia such as Traianoupolis, Andrianoupolis, Plotinopolis had no impact on the local settlements; it is also interesting that no free Roman merchants settled in the region (Loukopoulou 1997), suggesting that the Roman presence was limited to military units. Local populations (both of Greek and Thracian origin) continued living in the colonies of the coastline and in the fortified settlements of the hinterland (chapter IV).

The new road network was based on the pre-existing one (chapter I, Fig. 5). It remained under the control of the indigenous villages. The importance of the location of Thrace (routes and military passages to the Balkans and the Central Europe) has been also transformed during the centuries since the attention shifted from the maritime trade roads in Classical times to the inland road system in the Roman period (chapters I and IV). Burial practices display continuity with the Iron Age practice of cremation covered by soil or stone burial mounts. Burial mounds are located both in present day Bulgaria (Dimitrova 1987) and in Aegean Thrace (Triantafyllos 1987), along the main communication roads. The archaeological record supports the existence of people with strong family and community identities that were negociated through burial and religious rituals in the city's cemeteries and temples. In these local communities we have to imagine the metalworkers mastering their art in house-based workshops and transferring their empirical knowledge to pupils of the next generation assuring the continuation of their tradition. The study of iron made it possible to touch upon transformations of identities noticed in the archaeological record. By studying the transformations in iron technology it was possible to demonstrate how indigenous cultural identity in Aegean Thrace moved from its ethnic origin in Classical times (colonial/Greek versus indigenous/Thracian) towards a more social/political identity in Roman times (indigenous/of Greek or Thracian origin versus Roman).

The documentary-iconography evidence (chapter III) in the light of the discussion of the archaeological evidence (chapter IV) appears misleading. The idea of a Thracian in the mind of a Greek or a Roman writer does not reflect reality. Indigenous people of Thracian origin were living within the colonies (i.e. Avdera) or they were using (or appropriating) Greek material culture extensively (i.e., Messemvria-Zone). The boundaries and the population of Aegean Thrace did change over the centuries. The 'Thracian' with the long heavy coat, boots and tattoos that probably the first colonists met in 8th –6th centuries BC did not stay the same over the years. Over the next centuries close contact, either by peaceful co-existence or war, brought the two populations together. By the time Romans came to Aegean

Thrace and the province of Thracia was created everyone was called a 'Thrachian', despite his/her origin. Identity was build around the socio-political agenda of the Roman Empire.

To conclude, I would like to suggest that during the early periods, namely Classical and Hellenistic, identity in Aegean Thrace had a stronger ethnic element expressed clearly in the documentary-iconography evidence but less so in the material culture and the related technologies. The stronger dichotomy between colonial/imperial and indigenous elements, expressed as coast versus hinterland relationships, is artificial. The archaeological and analytical data suggest the existence of local, continuous technological traditions within the sites clearly seen in the production and consumption of pottery and iron as well as in burial practices and settlement patterns. During the Roman Period, an interesting change seems to have occurred: both colonial and indigenous elements came closer to each other and in contrast to the Romans. Identity under the Roman Imperial control was based on socio-political criteria. Everybody living within the province of Roman Thracia was a 'Thrachian' regardless of Greek or indigenous origin.

Conclusions

At a context specific level this work has sought to (a) introduce Aegean Thrace in the map of archaeological and archaeometallurgical studies focussing on its industrial past, and (b) propose a new reading of the archaeological, textual and analytical data towards the study of social transformations in Aegean Thrace. The scope of this thesis was to (a) use science-based archaeology in an archaeological context testing its ability to provide answers to specific archaeological problems, and (b) to create a holistic theoretical/methodological model for the social study of technology on local/regional scale. Documentary, archaeological and analytical evidence agrees that mining took place in small-dispersed mining areas usually with polymetalic mineralisation. Different ore deposits provided the raw material. There is evidence for smelting of manganese-ferous ores, copper rich iron ores and iron oxides. Iron was produced locally at each site by smelting in bloomery furnaces. The bloomery process produced mainly wrought iron that was further carburised to steel. There is also evidence for the production of cast iron, using the bloomery process, in Messemvria-Zone, but this is most probably accidental. Charcoal was used for fuel in all the sites examined.

In all sites implements such as billhooks, axes, falces and sickles, were found suggesting rural activities (pruning trees and preparing stony ground for cultivation of vineyards and olives) were not uncommon. Steel making techniques, with the exception of Avdera, were well known and practised in the region from the 5th century BC. Steel was manufactured mainly by either carburising wrought iron or by forging high carbon bloom. Locality and diversity on a regional scale and continuity of technological traditions on a local scale are supported by both the archaeological and analytical results. In an area smaller that 80 km by 50 km numerous variations of iron technology can be found based on the smelting of different ores (sands, oxides, pyrites, manganese or copper rich iron deposits). No apparent innovation or invention in terms of new ores, fuel, techniques of manufacture, is found in connection with Greek colonial or Roman imperial context.

The iron and steel currency bars support that iron had a high socio-economic value at least since the end of the 6th century BC. The symbolic meaning of iron is traced in the study of iron rings that provide the only direct connection between iron and ideology. There is no metallurgical evidence to suggest that other stages of the production were ritualised as the ethnographic record suggests. All the evidence examined supports continuity that is evident in settlement patterns, burial practices, hand-made pottery and in iron technology. Technological continuity suggests also cultural continuity because technologies tend to be conservative and they are tightly linked to local cultural traditions. That is why the study of technology can provide insights to the study of identity. In the case of Aegean Thrace, the study of iron has been proved useful in interpreting transformations of societies and identities in a multi-cultural context. The cultural identity of the population living in Aegean Thrace was transformed from an ethnic-based onto one based on socio-political factors during Roman times. The aim of this work is to illustrate how the study of iron technology can be used as a methodological tool in understanding locality and indigenous agency within a colonial/imperial setting. Small-scale activities such as local iron technologies make clear the existence of very diverse, local societies, which might go unnoticed with the use of a more traditional and unified approach to cultural processes. I would like to end by proposing that both the impact of Greek and Roman cultures to the others, on the indigenous, as well as the idea of the major innovations introduced to less advanced cultures should be re-addressed by the archaeologists. In the case of iron technology in Aegean Thrace, Greeks and Romans did not introduce new techniques but they used skills, experience and materials that already existed.

GLOSSARY (based on Shackelfold 1992, Scott 1991, and Tylecote 1962)

Alloy
Metal composed of more than one element.

Annealing
A process of heat-treatment carried out on a metal or alloy, usually to soften the material to allow further deformation.

Annealing twin
In FCC metals, a process of re-crystallisation (often of worked and annealed metals) in which a mirror plane in the crystal growth results in two parallel straight lines appearing across the grain when the metal is etched.

Austenite
Face centred cubic (γ) phase of iron or steel.

Bainite
A term more commonly applied in the past to the decomposition of martensite in steels by tempering at about 200-350 degrees Celsius.

Binary diagram
Two-component phase diagram.

Bloom
A roughly finished metallic product; specifically, the spongy mass of iron produced in a bloomery furnace in which the iron is reduced *in situ* and is not molten during reduction. The crude iron bloom must be extensively worked at red heat to consolidate the iron and remove excess slag and charcoal.

Body centred cubic (BCC)
Common atomic arrangement with one atom located at the centre of the cube and eight atoms on the corners of the cube.

Brittle
Lacking in deformability.

Carbon steel.
Ferrous alloy with nominal impurity level and carbon as the primary compositional variable.

Carburisation
The process of increasing the carbon content of the surface layers of a metal (often wrought iron) by heating the metal bellow its melting point with carbonaceous matter such as wood charcoal.

Case-hardening
A process consisting of one or more heat treatments for producing a hard surface layer on metals as in carburisation. In the case of steel the carbon content of the surface can be increased by heating in a medium containing carbon followed by heat treatment.

Cast Iron
Ferrous alloy with 2-wt % carbon

Casting
Material processing technique involving pouring of molten metal into a mould, followed by solidification of the liquid.

Cementation
The term has several meanings. Cementation of gold alloys with salt in cupel may remove silver leaving pure gold behind. Carbonaceous material may be used to cover the wrought iron, which is then heated. Carbon can diffuse into the structure creating a low-carbon steel surface by cementation.

Cementite
The hard brittle component of iron-carbon alloys, containing about 6.6% carbons corresponding to the phase Fe3C and crystallizing in the orthorhombic system. It is soluble in molten iron, but the solubility decreases in austenite to about 0.9% at the eutectoid temperature. Pearlite, the eutectoid of ferrite and cementite, is the most common component containing cementite in ancient and historic steels.

Ceramic
Non metallic inorganic material.

Clay
Fine-grained soil, composed chiefly of hydrous aluminosilicate minerals.

Cold-working
The plastic deformations of a metal at temperature low enough to cause permanent strain hardening. The treatment usually consists of rolling, hammering, or drawing at room temperatures when the hardness and tensile strength are increased with the amount of cold-work, but the ductility and impact strength are reduced.

Dendritic
Shaped like the branches of a tree. Dendrites are common in cast alloys and may look like an intersecting snowflake pattern.

Diffusion
The migration of one alloy or metal into another. Inter-diffusion also occurs with the secondary metal

migrating into the first. Usually heat is required for this process to occur.

Dislocation
Defects in the crystal structure of the metal that allow movement of planes or atoms within the lattice. Edge and screw dislocations are two common types.

Ductility
Deformability.

Eutectic composition
Composition associated with the minimum temperature at which a binary system is fully melted.

Eutectic reaction
The transformation of a liquid to two solid phases upon cooling.

Eutectic temperature
Minimum temperature at which a binary system is fully melted. For the iron-carbon alloy this is 723 C.

Eutectoid
Decomposition form a solid phase into two finely dispersed solid phases creates a structure called a eutectoid. The eutectoid point is fixed in binary alloys and in morphology may resemble the eutectic.

Face-centred cubic (FCC)
Common atomic arrangement in the crystal of metals with one atom located at the centre of each surface of the cube.

Fayalite
The most common component of ancient slags, often occurring as slag stringers in wrought iron. Fayalite is an iron silicate, 2FeO. SiO2 which melts at about 989 degrees Celsius, crystallizes in the orthorhombic system and usually takes the form of broken elongated grey laths in silicate based slags.

Ferrite
Ferrous alloy based on the bcc structure of pure iron at room temperature.

Ferrous alloy
Metal alloy composed of predominately iron.

Glass
No crystalline solid, unless otherwise noted, with a chemical composition comparable to a crystalline ceramic.

Grain
Individual crystallite in a polycrystalline structure.

Grain boundary
Region of mismatch between two adjacent grains in a polycrystalline structure.

Graphite
An allotropic form of carbon, occurring in the trigonal system as grey, soft, lustrous plates. It is the form of carbon found in steels and cast irons and usually occurs in grey cast irons as thin flakes or nodules.

Hardness
Resistance of a material to indentation.

Heat treatment
Temperature versus time history necessary to generate a desired microstructure.

Hot-working
Deformation of the metal or alloy above the temperature necessary for plastic deformation of the metal.

Hv
Hardness on the Vickers or Diamond Pyramid Number (DPN) scale.

Hypereutectic
Composition greater than of the eutectic.

Hypoeutectic
Composition less than that of the eutectic.

Impurity
A chemically foreign constituent, sometimes present due to raw material sources and, sometimes, purposefully added.

Interstitial
A small element that may occupy lattice spaces without causing too great a distortion of the original lattice structure. An example is carbon in iron. Carbon is a small element and can insert into the cubic iron lattice.

Iron
An element with atomic number 26, atomic weight 55.85, mp 1528 degrees Celsius, specific gravity 7.87. A heavy whitish metal and one of the most abundant and widely distributed. Iron was known early as meteoric iron, which usually contains some nickel. Extraction by the bloomery process was common by the early centuries B.C., and this continued until the invention of the blast furnace. Later methods of extraction in the Western world were able to melt iron and, by the 13th century A.D., make cast iron. In China, iron (with carbon) could already be molten during the time of Christ

and cast iron was produced much earlier than in the West.

Magnetite
Important ferrous compound (Fe_3O_4) with magnetic behaviour.

Martensite
Often used only for the hard, needle like component of quenched steels, but more generally, any needle like, hard transformation product of the quenched alloy. The most common in ancient materials is martensite in low-carbon steels, or martensite in beta-quenched bronzes.

Matrix
The portion of a composite material in which a reinforcing, dispersed phase is embedded.

Matte
A liquid or solid phase of sulphides usually FeS and Cu_2O.

Melting point
Temperature at which a solid to liquid transformation occurs upon heating.

Oxidation
Reaction of metal with atmospheric oxygen.

Oxide
Compound between an elemental metal and oxygen.

Pearlite
The fine mixture of ferrite and cementite found in steels. The eutectoid, pearlite, will be complete when the carbon content reaches 0.8%. In most ancient steels, a mixture of ferrite and pearlite is common.

Phase
A homogenous chemical composition and uniform material, describing one component in a metallic system.

Prill
In the extraction of copper from primitive smelts, the metal is produced as small droplets or particles in a slaggy matrix. These small metallic particles are called prills and were often extracted by breaking up the smelted product and sorting the metal. In crucible processes, prills are small droplets of metal adhering to the crucible lining. In the ancient iron making prills are small droplets of pure metallic iron often noted on the smelting slags.

Pottery
Burned clay ware ceramic.

Property
Observable characteristic of a material.

Quenching
The act of quickly cooling a metal or alloy by plunging into cold water or oil.

Quenched structures
Usually non-equilibrium structures or phases that have been made metastable in an alloy by quenching in water or oil. The most common quenched products are martensite in steels and martensite in high-tin bronzes. Quenching may also be used to suppress ordering reactions, especially in gold alloys and some ancient texts refer to this practice to avoid embrittlement.

Slag
A glassy phase or mixture of phases usually to be found in ancient or historic wrought iron or steel. The slag is an important by-product of the smelting metals. It may be incorporated in copper or iron alloys as a result of incomplete separation or incomplete melting during extraction, as in the bloomery process.

Slag stringers
Small pieces of slag that have become incorporated into the metal and are then strung out as small elongated ribbons as a result of working the metal to shape it.

Smelting
Chemical reaction between the ore and the fuel resulting to the production of metal.

Smithing
Forging hot or cold the metal or the bloom.

Steel
A malleable alloy of iron and carbon that contains about 0.1-1.9% carbon. The carbon is present as cementite, usually as a component of pearlite. Low-carbon steels contain from 0.09% carbon to 0.2% and are soft. Medium carbon steels contain 0.2 - 0.4 % carbon and high carbon steels more than 0.4%.

Tempered martensite
An $a+Fe_3C$ microstructure produced by heating the most brittle martensite phase.

Tempering
Usually applied to steels in which some of the hardness is removed by heating at moderate temperatures, from 450 to 650 degrees Celsius, depending on the type of tempering required. In ancient metals, tempering, or self-tempering of martensite would have been carried out to reduce the brittleness of the fully quenched steel cutting edges of swords.

TTT diagrams
Time-temperature-transformation diagrams. Most useful in considering the nature of the quenched microstructure to be found in steels and high tin bronzes.

Vickers
A hardness scale that uses a diamond having an angle of 136 degrees between the faces. The loading can be easily used for micro hardness measurements. It is one of the most useful hardness testing scales.

Weld
A term used to describe a joint made between two metals made by the heating and joining the separate parts with no solder applied. Ancient welds were often made in precious metals, such as gold and silver and in the joining of iron components, especially in the fabrication of wrought iron.

Widmanstatten
A type of structure that forms when a new solid phase is produced from a parent solid phase as plates or laths along certain crystallographic planes of the original crystals. The structure is associated with many meteoric irons, and with changes upon cooling in worked iron and copper alloys. Most commonly found as an incidental feature of wrought low-carbon steels.

Wrought
The process of hammering or deforming a metal or alloy, as opposed to casting.

Wrought iron
Iron that has been produced from the bloomery process and has been consolidated by hammering and annealing into wrought product. Wrought iron usually contains slag stringers that have been elongated and flattened in the process of working from the bloom.

Bibliography

Alcock, S. 1996. *Greacia capta. The landscapes of Roman Greece.* London: Routledge.

Aldington, R. (ed.) 1959. *New Larousse encyclopedia of mythology*. London: The Hamlyn Publishing Group limited.

Alestam, T. 1990. *Bird migration*. Cambridge: Cambridge University Press.

Andersen-Daamgaard H.W.H., S. Houby-Nielsen, and A. Rathje (eds) 1997. *Urbanisation in the Mediterranean in the 9th to 6th centuries BC.* Danish Studies in Classical Archaeology, Acta Hyperborea 7. Copenhagen: Museum Tusculanum Press

Andreou, S., M. Fotiadis, and Kotsakis K. 1996. Review of Aegean Prehistory V: the Neolithic and Bronze Age of Northern Greece, *American Journal of Archaeology, 100: 537-597.*

Appadurai, A. 1986. Toward an anthropology of things, in A. Appadurai (ed.), *The social life of things*: 3-63. Cambridge: Cambridge University Presss.

Arafat, K. and C. Morgan. 1994. Athens, Etruria and the Heuneburgh: mutual misconceptions in the study of Greek-barbarian relations, in I. Morris (ed.), *Classical Greece: ancient histories and modern archaeologies*: 108-134. Cambridge: Cambridge University Press.

Archibald, Z.H. 1983. Greek Imports: some aspects of the Hellenic impact on Thrace, in A.G. Poulter (ed.), *Ancient Bulgaria, International Symposium on the Ancient History and Archaeology of Bulgaria*: 304-332. Nottingham.

Aufrecht, W.E. (ed.) 1997. *Urbanism in antiquity from Mesopotamia to Crete. .* .Sheffield: Sheffield Academic Press.

Avramea, A., 1994. Thrace in the Roman Period, in vol. *Thrace*: 135-146. Thessalonica: General Secretariat of East Macedonia and Thrace.

Azank, R. 1991. Ancient Warriors, *Art Forum 29*: 691-671.

Bachmann, H.G. 1982. *The identification of slags from archaeological sites*. London: Occasional publication no 6, Institute of Archaeology, UCL.

Bailey, D.W. 1998. Bulgarian archaeology: ideology, sociopolitics and the exotic, in L. Meskell (ed.) *Archaeology under fire*: 87-110. London: Routledge.

Bakalakis, G. 1959. *Proanaskafikes ereunes stn Thraki*. Thessalonica: Kyriakidis Press.

Bakirtzis, C., 1994. Byzantine Thrace (AD 330 1453), in vol. *Thrace*: 151-210. Thessalonica: General Secretariat of East Macedonia and Thrace.

Barret, J. 1997. Romanisation: a critical comment, in D.J. Mattingly (ed.) *Dialogues in Roman imperialism: power, discourse and discrepant experience in the Roman Empire*: 51-65. Portsmouth, Rhode Island.

Bartel, B. 1989. Acculturation and ethnicity in Roman Moesia Superior, in T.C. Chamnpion (ed.) *Center and periphery*: 173-185. Cambridge: Cambridge University Press.

Blakely-Westover, S. 2000. *Smelting and sacrifice: comparative analysis of Greek and Near Eastern cult sites from the Late Bronze Age through the Classical period,* in S Young *et al.* (eds) *Metals in Antiquity*: 86-90. BAR Brittish Series 792.

Bassiakos, Y., M. Kostoglou, and S. Pournis. 1999. Radioactive ancient metallurgical remains in Macedonia and Thrace, Greece: technological implications, in *Iron working in Europe*, International Conference Proceedings, Jutland, Denmark.

Belllon, P. 1553. *Les observations de plusieurs singularites et choses memorables trouvees en Grece, Asie. Inde, Egypte, Arabie et autres pays etranges*. Paris.

Bernstein, H., T. Hewitt, and A. Thomas 1992. Capitalism and the expansion of Europe, in T. Allen and A. Thomas (eds) *Poverty and development int the 1990s*: 168-184. Oxford.

Best, J. 1969. *Thracian peltasts and their influence on Greek warfare*. Cronigen: Wolters-Noordhoff.

Bintliff, J. 1997. Regional survey, demography, and the rise of complex societies in the ancient Aegean: core-pariphery, neo-Malthusian, and other interpretive methods. *Journal of Field Archaeology, 24(1): 1-37.*

Boardman, J. 1980. *The Greeks overseas: their early colonies and trade*. London: Thames and Hudson.

Bradley, R. 1981. Economic growth and social change: two examples from prehistoric Europe, in B.G. Sheridan (ed.) *Economic archaeology*: 231-237. BAR Brittish Series.

Bradley, R. 1985. *Consumption, change and the archaeological record.* University of Edinburgh, Department of Archaeology, Occasional Paper.

Brown, K.S. 1994. Seeing stars: character and identity in the landscapes of modern Macedonia. *Antiquity, 68: 784-796.*

Cary, M. 1949. *The geographical background of Greek and Roman history.* Oxford: Clarendon Press.

Casson, S. 1968. *Macedonia, Thrace, and Illyria. Their relations to Greece from the earliest times down to the time of Philip son of Antypas.* Groningen: Bonma's Boekhnis.

Cavanagh, W.G. 1992. Surveys, cities and synoecism, in R. A. Wallace-Hadrill (ed.) *City and country in the ancient world*: 97-118. London: Routledge.

Champion, T.C. (ed.) 1989. *Center and periphery: comparative studies in archaeology.* Cambridge: Cambridge University Press.

Chang, C. 1993. Pastoral transhumance in the southern Balkans as a social ideology: ethnoarchaeological research in northern Greece, *American Anthropologist, 95(3): 687-703.*

Cherry, J.F. 1987. Power in space: archaeological and geographical studies of the state, in J.M. Wagstaff (ed.) *Power in Space*: 146-172. Oxford: Basil Blackwell.

Chiotis, E., C. Koukouzas, and G. Papadimitriou 1996. Old mining and metallurgical activities in Agistron-Serres, Macedonia, in *2nd Symposium of the Hellenic Archaeometry Society*. Thessalonica.

Clogg, R., 1992. *A concise history of Greece.* Cambridge: Cambridge University Press.

Coles, J.M. and A.J. Lawson 1986. *European wetlands in prehistory.* Oxford: Oxford Clarendon Press.

Conolly, P., 1981. *Greece and Rome at war.* London: Macdonald.

Conofagos, C. and G. Papadimitriou 1981. *La metallurgie du fer et de l'acier en Grece, pendant la period clasique,*

Courchin, L. 1991. *Roman Spain: conquest and assimilation.* London: Routledge.

Craddock, P.T. and N.D. Meeks 1987. Iron in ancient copper, *Archaeometry, 29(2): 187-204.*

Craddock, P.T. 1989. The scientific investigation of early mining and metallurgy, in in J. Henderson (ed.) *Scientific analysis in archaeology*: 178-212. Oxford: Oxford University Committee for Archaeology.

Craddock, P.T. 1995. *Early metal mining and production.* Edinburgh: Edinburgh University Press.

Davis, O. 1935. *Roman mines in Europe.* Oxford: Claredon Press.

Davis-Kimball, J. 1997. Warrior women of the Eurasian Steppes. *Archaeology 50: 44-48.*

Delopoulos, G. 1994. *Geographical description of Thrace*, in vol. *Thrace*: 351-374. Thessalonica: General Secretariat of East Macedonia and Thrace.

Diaz-Andreou, M. and T. Champion, 1996. *Nationalism and archaeology in Europe.* London: Universtiy College London Press.

Diaz-Andreu, M., Constructing identities through culture: the past in the forging of Europe, in P. Graves-Brown, S. Jones, and C. Gamble (eds) *Cultural Identity and archaeology: the construction of European communities*: 48-61. London: Routledge.

Dieudonne-Glad N., Parisot J., Dupont E., Ronchail F, Gauttier W., and D. Wissler, 2001. Metallography of five flat iron bars with socket from the river Saone (France), *Journal of the Historical Metallurgy Society, 35 (2): 67-73.*

Dietler, M., 1997. The Iron Age in Mediterranean France: colonial encounters, entanglements, and transformations, *Journal of World Prehistory, 11(3): 269-357.*

Dietler, M. 1998. Consumption, agency, and cultural entanglement: theoretical implications of a Mediterranean colonial encounter, in J.G. Cusick (ed.) *Studies in culture contact*: 288-311. Illinois. Southern Illinois Universtity Press.

Dobres, M. A. and J. Robb (eds) 2000. *Agency in Archaeology*. London: Routledge
Drougou, S. 1991. *Hellenistic pottery from Macedonia.* Thessalonica: Aristotle University.

Earle, T. 1997. *How chiefs came to power: the political economy in Prehistory.* California: Stanford University Press.

Edmondson, J. C. 1989. Mining in the later Roman Empire and beyond: continuity or disruption, *Journal of Roman Studies, LXXIX: 84-102.*

Epitropou, N. and D. Mpitzios 1987. O oriktos ploutos tis Thrakis, *Thrakika Chronika 1987: 239-243.*

Euaggelou, E. 1976. *Ekthesis axiologiseos ton stichion tis geochimikis ereunis is tin periochin 'Profitis Ilias' Aisimis, Alexandroupoleos.* Athens: Institute of Geological and Mineralogical Exploitation (IGME).

Eustratiou, N. 1987. Ethnoarchaeologikes ereunes stous oreinous oikismous tis Rodopis, *The archaeological work in Macedonia and Thrace 1987: 479-485.* Thessalonica.

Faklaris, P. 1994. Aegae: determining the site of the first capital of the Macedonians. *American Journal of Archaeology 98: 609-616.*

Gassman, G. 2001. Actual excavations at an early Celtic smelting settlement in the Swabian Mountains, Southern Germany, in *International Conference Abstract Publication: The introduction of iron in Eurasia.* 4th -8th October 2001, Uppsala, Sweden.

Godfrey, E., M. van-Nie, and G. McDonnell 2001. Early evidence for the use of ultrahigh carbon steel in Europe, in *International Conference Abstract Publication: The introduction of iron in Eurasia.* 4th -8th October 2001, Uppsala, Sweden.

Gotzev, A. 1997. Characteristics of the settlement system during the early Iron Age in ancient Thrace, in Andersen H D, Houby-Nielsen S, Rathjie A (eds) *Danish Studies in Classical Archaeology, Acta Hyperboreia 7.* Copenghagen: Museum Tusculaneum Press, University of Copenghagen.

Granados, B. 1990. Warriors Dance. *Southwest Art 20(June 1990): 116-126.*

Graves-Brown, P., S. Jones, and C. Camble (eds) 1996. *Cultural Identity and Archaeology.* London: Routledge.

Greene, K. 1993. The study of Roman technology: some theoretical constraints, in E. Scott (ed.) *Theoretical Roman Archaeology: first conference proccedings: 39-48.* Aldershot.

Greene, K. 1993. The study of Roman technology: theoretical constraints, in E. Scott (ed.) *Theoretical Roman Archaeology: first conference proccedings:* Aldershot.

Haarer, P. 2001. Obelloi and iron in Archaic Greece. Unpublished Ph.D. thesis, Department of Archaeology , University of Oxford.

Hadrill, A.W. 1992. City and country in the ancient world: Introduction, in W. Hadrill and Rich (eds) *City amd country in the ancient world*: ix-xviii. London: Routledge.

Hall, J.M. 1995. Approaches to ethicity in the Early Iron Age Greece, in N. Spencer (ed.) *Time tradition and society in Greek archaeology*: 6-17, London: Routledge.

Hall, J. 1997. *Ethnic identity in Greek antiquity.* Cambridge: Cambridge University Press.

Halleux, R. 1974. *Le probleme des metaux dans la science antique.* Paris: Les Bellles Lettres.

Hamilakis, Y. and E. Yialouri 1996. Antiquities as symbolic capital in modern Greek society. *Antiquity 70: 117-129.*

Hammond, N.G., 1972. *A history of Macedonia.* Oxford: Claredon Press.

Hammond, N.G., 1981. *Atlas of Greek and Roman world in Antiquity.* Park Ridge, NJ: Noyes Press.

Hauptman, A., E. Pernicka, and G.A.W. G.A. (eds) 1989. Old World Archaeometallurgy. Vol. Bhft 7. Bochum: Der Anschnitt.

Healy, J., 1978. *Mining and metallurgy in the Greek and Roman world.* London: Thames and Hudson.

Hedeager, L. 1987. Empire, frontier and the barbarian hinterland: Rome and Northern Europe from AD1-400, in T. C. Champion (ed.) *Centre and periphery in the ancient world*: 125-140. Cambridge: Cambridge University Press.

Hedeager, L. 1987. Empire, frontier and the barbarian hinderland: Rome and Northern Europe from AD 1-400, in I. Hodder *(ed.) Archaeology as long-term history*: 125-140. Cambridge: Cambridge Univeristy Press.

Herbert, E. 1993. *Iron, gender, and power: rituals of transformation in African societies.* Indiana University Press.

Hingley, R. 1996. The 'legacy' of Rome: the rise, decline and fall of the theory of Romanisation, in N. Webster and J. Cooper (eds) *Roman Imperialsm: post-colonial perspetives*: 35-48. Leicester: Archaeology Monographs.

Hodder, I. 1982. *Symbols in Action*. Cambridge: Cambridge University Press.

Hodder, I. 1986, *Reading the past: current approaches to interpetation in archaeology*. Cambridge: Cambridge University Press.

Hodder, I. 1987. The contextual analyses of symbolic meanings, in I. Hdder (ed.) *The archaeology of contextual meanings*; 1-10. Cambridge: Cambridge University Press.

Hodder, I., 1991. Archaeological theory in contemporary European societies: the emergence of competing traditions, in I. Hodder (ed.) *Archaeological Theory in Europe the last three decades*: 1-24. London: Routledge.

Hoddinott, R.F. 1975. *Bulgaria in Antiquity: an arcaeological introduction*. London: London and Tonbridge.

Hoddinott, R.F. 1971. *The Thracians*. London: Thames and Hudson.

Horne, L. 1982. Fuel for the metal worker. *Expedition, Anthropology (25, 1): 6-13.*

Host-Madsen, L. and V.F. Buchwald 1999. The characterization and provenancing of ore, slag and iron from Iron Age settlement at Snorup, *Journal of Historical Metallurgy Society **33**(2): 57-67.*

Hupchick, D.P. 1996. *A conscise historical atlas of Eastern Europe*. New York: St. Martin's Press.

IGME, 1973. *Geological map of Greece. Athens:* Institute of Geology and Mineral Exploration.

Isaac, B.H. 1986. *The Greek settlements in Thrace until the Macedonian conquest*. Leiden: Brill.

Issac, B.H. 1992. *The limits of empire: the Roman army in the East*. Oxford: Clarendon Press.

Johnston, R. 1998. *Approaches to the perception of landscape. Archaeological Dialogues 5(1): 54-68.*

Jones, J. 1984. Ancient Athenian silver mines, dressing floors and smelting sites. *Journal of the Historical Metallurgy Society 18: 65-82.*

Jones, S. and P. Graves-Brown 1996. Introduction: archaeology and cultural identity in Europe, in Jones, S. and P. Graves-Brown (eds) *Cultural Identity and Archaeology*: 1-24. London: Routledge.

Jones, S. 1997. *The archaeology of ethnicity*. London: Routledge.

Jonson, L. 1992. *Birds of Europe with North Africa and the Middle East*. Princeton: Princeton University Press.

Juleff, G. 1996. An ancient wind-powered iron smelting technology in Sri-Lanka. *Nature 379: 60-63.*

Killick, D. and R. Gordon 1987. Microstructure of puddling slags from Fontley, England and Roxbury, Connecticut, USA, *Journal of the Historical Mettalurgy Society 21(1): 28-36.*

Killick, D., van der Merwe, N. J., Gordon, R. B, and D. Grebenart 1988. Reassessment of the evidence for early metallurgy in Niger, West Africa, *Journal of Archaeological Science 15: 367-394.*

Killick, D. 1991. The relevance of recent African iron smelting practice to reconstruction of prehistoric smelting technology, in P. Glumac (ed.) *Recent Trends in Archaeometallurgical Research*: 47-54. Philadelphia: MASCA, Papers in Science and Archaeology.

Knapp, A.B. 1992. Archaeology and Annales: time, space and change, in A. B. Knapp, A.B. (ed.) *Archaeology, annales and ethnohistory*: 1-21. Cambridge: Cambridge University Press.

Knapp, A.B. 2000. Archaeology, science-based archaeology and the Mediterranean Bronze Age metals trade, *European Journal of Archaeology 3(1): 31-55.*

Knapp, A.B. 1996. Archaeology without gravity: postmodernism and the past, *Journal of Archaeologicall Method and Theory 3(2): 127-157.*

Knapp, A.B., V.C. Pigott, and E.W. Herbert (eds) 1998. *Social approaches to an industrial past: the archaeology and anthropology of mining*. London: Routledge.

Knapp, A.B. and M. Given 1995. A report on the Sydney-Cyprus Archaeological Survey Project (SCYSOP) - third season (1995): 295-366. Report to the Department of Antiquities, Cyprus.

Kostoglou, M. 2009. Material culture and materials science: studying social choices in ancient iron production. *Archaeometry* (forthcoming)

Kostoglou, M. 2008. Towards an intergradeted approach for the study of ancient Greek technology. In *Fakorelis Y, N. Zacharias and N. Polikreti (eds) Proceddings of the 4th IOnternational Symposium of the Hellenic Archaeometry Society, British*

Archaeological Reports International Series, 459-467.

Kostoglou, M. and Navasaitis J. 2006. Cast Iron in Ancient Greece: myth or fact? *Mediterranean Archaeology and Archaeometry, vol. 06 No 2, 53-56.*

Kostoglou, M. 2003. Iron and steel currency bars in ancient Greece. *Mediterranean Archaeology and Archaeometry, vol. 03, No 1, 63-68..*

Kostoglou, M. 1996. *Excavation in Messemvria-Zone*. Excavation diary, courtecy of the 19th Ephoreia of Prehistoric and Classical Antiquities of Thrace. Komotini.

Kotsakis, K. 1991. The powerful past: theoretical trends in Greek archaeology, in I. Hodder (ed.) *Archaeological theory in Europe, the last three decades*: 65-90. London: Routledge.

Kotsakis, K. 1998. The past is ours: images of Greek Macedonia, in L. Meskell (ed.) *Archaeology under fire*: 44-67. London: Routledge.

Koukouli-Chryssanthaki, C. 1988. Excavation at Avdera, in *The historical, archaeological and laographical research in Thrace*: 39-74. Thessalonica: Institute of Balkan Studies.

Kriastiansen, K. 1998. Chiefdoms, states and systems of social evolution, in K. Kristiansen (ed.) *Social transformations in archaeology*: 243-267. London: Routledge.

Kwok, K. 1995. Keris, the power of metal, *Oriental Art 41(Summer): 66-67.*

Lazaridis, D. 1971. *Avdera kai Dikaia*. Athina.

Leake, W.M. 1841. *Travelers in North Greece*.

Lechman, H. and A. Streinberg, 1979. The history of technology: an anthropological point of view, in G. Bugliarello and D.B. Donner (eds) *The history of philisophy of technology*: 135-162. Urbana: University of Illinois Press.

Lemonier, P. 1986. The study of material culture today: toward an anthropology of technical systems, *Journal of Anthropological Archaeology 5: 147- 186.*

Lemonier, P. (ed.) 1992. *Elements for an anthropology of technology.* Anthropologiacl papers vol 88. Michigan: Museum of Anthropology, Universityof Michigan.

Lemonier, P. 1993. *Technological choices: transformation in material cultures since the Neolithic*. London: Routledge.

Lightfoot, K.G. 1995. Culture contact studies: redefining the relationship between prehisoric and historic archaeology, *American Antiquity 60 (2): 199-217.*

Love, J.A. 1989. *Eagles*. London: Whittet Books.

McGovern, P.E. 1995. Science in archaeology: a review, *American Journal of Archaeology 99: 79-142.*

Mack I., McDonnel G., Murphy S., Andrewa P., and K. Wardley 2000. Liquid steel in Anglo-Saxon England, *Journal of Historical Metallurgy Society 34 (2): 87-96.*

Mackay, D.B. 1997. A view from the outskirts. Realignments from modern to postmodern in the archaeological study of urbanism, in H. D. Andersen *et al. Urbanisation in the Mediterranean in the 9th to 6th centuries BC:* 279-285. Copenhagen: Museum Tusculanum Press.

Maratos, G. and V. Andronopoulos 1966. *Mineral wealth of Thrace and Macedonia (in Greek).* Athens: Institute of Geological and Mineralogical Exploitation (IGME).

Maratos, G. and V. Andronopoulos 1972. *O oryktos ploutos Thrakis-Macedonias.* Xanthi: Institute of Geological and Mineralogical Exploitation (IGME).

Mattingly, D.J. (ed.) Dialogues in Roman imperialism: power, discource, and discrepant experience in the Roman Empire. *Journal of Roman Archaeology, Supplement Series.* Portsmouth: Rhode Island.

McCovern, P.E. and M.D. Notis (eds) 1989. *Cross-craft and cross-cultural interactions in ceramics,* ceramics and civilisation. Westerville, OH: American Ceramic Society.

Melfos, V.I. 1995. *Ereuna ton vasikon kai eugenon metallon stin perirodopiki zoni tis Thrakis.* Unpublished Ph.D. thesis Department of Geology, Aristotle University, Thessalonica.

Meskell, L. (ed.) 1998. *Archaeology under fire: nationalism, politics and heritage in the Eastern Mediterranean and Middle East.* London: Routledge.

Merkel J and K Barrett 2000. The adventitious production of iron in the smelting of copper revisited: metallographic evidence against a tempting model, *Journal of the Historical Metallurgy Society 34 (2): 59-66.*

Meyer, J.H. 1971. *Car-borne radiometric survey Macedonia and Thrace* , I.A.E.A.: Vienna.

Miller, D. 1994. *Modernity, an ethnographic approach: Dualism and mass consumption in Trinidad (Explorations in Anthropology)*. Oxford: Oxford/Providence.

Miller, D. 1995. *Worlds apart, modernity through the prism of the local*. London: Routledge.

Millett, M. 1990. *Romanisation: historical issues and archaeological interpretation*, in T. Black and M. Millett (eds) *The early Roman Empire in the West*: 35-41. Oxford.

Moore, H. 1987. Problems in the analyses of social change: an example from the Marakwet, in I. Hodder (ed.) *Archaeology as long-term history*: 85-104. Cambridge: Cambridge University Press.

Morris, I. 1992. The early polis as city and state, in R.A. Wallace-Hadrill (ed.) *City and country in the ancient world*: 25-58. London: Routledge.

Morris, I. 1992. *Death and ritual in ancient Greece*. Cambridge: Cambridge University Press.

Morris, I. 1994. Archaeologies of Greece, in I. Morris (ed.) *Classical Greece: ancient histories and modern archaeologies*: 8-47. Cambridge: Cambridge University Press.

Morris, I. 2000. *Archaeology as cultural history: wards and things in Iron Age Greece*. Oxford: Blackwell Publishers ltd.

Morton, G. and J. Wingrove 1966. Construction of bloomery slags-Part 1: Roman, *Journal of the Iron and Steel Institute 207: 1556-1564.*

Morton, G. and J. Wingrove 1972. Construction of bloomery slags-Pat II: Medieval, *Journal of the Iron and Steel Institute 201: 478-488.*

Muhly, J.D. 1997. Recent works in archaeometallurgy, *American Journal of Archaeology 101(October 1997): 771-3.*

Nash, P. 1987. Imperial expansion under the Roman Republic, in M. Rowlands (ed.) *Centre and periphery in the ancient world*: 87-103. Cambridge: Cambridge University Press.

Nesbitt, R.W. *et al.* 1988. The geological setting of base metal mineralization in the Rodhope region, Northern Greece, in P.M. Herzig (ed.) *Base Metal Sulphide Deposits*: 499-514. Berlin: Springer-Verlag.

Nijboer, A.J. 1995. Iron production and iron votive offerings at Borgo Le Ferrieres/Satricum (8th to 6th centuries BC), *Palaeohistoria, Acta et Communicationes Instituti Bio-Archaeologici Universitatis Groninganae 35/36: 89-105.*

Nijboer, A.J. 1997. The role of craftsmen in the urbanisation process of central Italy (8th to 6th centuries BC), in H.W.H. Daamgaard-Andersen, S. Houby-Nielsen, and A. Rathje (eds) *Urbanisation in the Mediterranean in the 9th to 6th centuries BC*: 383-401. Copenhagen: Museum Tusculaneum Press.

Olsen, H. 1974. *Car-borne scintillometer survey of central and east Macedonia and Thrace, U.N.D.P.-I.A.E.A.-G.A.E.C.*

Pantos, P. 1980. The circulation of the coins of Thasos, Neapolis and Philippoi in Greek Thrace, in *First Local Conference Proccedings: Kavala and its Region*. Thessalonica.

Papadakis, A. 1975. The black sands of Loutra Eleutheron near Kavala, Greece, *Science Annals, Faculty of Physics and Mathematics, University of Thessalonica 15: 331-388.* Thessalonica.

Papadakis, A. and K. Michailidis 1977. Polyframboidal pyrite formations in Greek beaches, *Science Annals, Faculty of Physics and Mathematics, University of Thessalonica 17: 353-364.* Thessalonica.

Papadimitriou, G., C. Garagunis, and J. Kordatos 1996. Examination of the iron shels of wooden poles used for the foundation of an ancient pier on the Strymon River near the ancient city of Amphipolis, *in Proccedings of the Second Symposium of the Hellenic Archaeometrical Society*. Thessalonica.

Papadopoulos, J.K. 1997. Phantom Euboians, *Journal of Mediterranean Archaeology 10(2): 191-219.*

Papageorgiou, S. *et al. 1996.* Diereunisi tis schesis metaxi metavolis tis paraktias topografias kai tis istorias tis katikisis sti Macedonia kai Thraki, in *Proccedings of the Second Symposium of the Hellenic Archaeometrical Society.* Thessalonica.

Papanikolaou, N. 1964. *Geofysiki anagnorisis peiochis metalliou Kirkis/Alexandroupolis, periochis metalliou King Arthur.* Xanthi. Institute of Geological and Mineralogical Exploitation (IGME).

Papastamataki, A. 1975. *The exploitation of the mineral wealth in the ancient Greece: Research on the content of the ancient Greek metallurgical slags in gold and other metals (in Greek)*. Athens: Institute of Geological and Mineralogical Exploitation (IGME).

Papastamataki, A. 1985. *Oi Skories tis archaias Ellinikis Metallourgias*. Athens: Institute of Geological and Minerological exploitation (IGME).

Papastamataki, A. 1986. *A study of ancient slags and their contribution to the revealing of ancient metallurgical technology: geological and geophysical research*. Athens: Institute of Geological and Mineralogical Exploitation (IGME).

Papastamataki A, Orfanos B., Triantafyllos D, and Ch. Karadima 1996. Ena agnosto metallio sth Maroneia Komotinis, abstract in the *3rd Symposium of Archaeometry*. Athens.

Papoulia, V. 1988. Phaseis entaxeos ton archaion phylon tis chersonisou tou Aimou ston Elliniko kosmo, in *I Istoriki, archaeologiki kai laographiki ereuna sti Thraki*: 203-214. Thessaliniki.

Papoulia, V. 1994. Ancient Thrace as a historical unity, in *vol. Thrace*: 13-34. Thessalonica: General Secretarial of East Macedonia and Thrace.

Pentazos, E. 1971. Archaeologikai ereunai en Thraki, *PAE: 86-118, plates 104-146.* Athens.

Phavas, N. 1982. *Ekthesi epii tis geologikis kai koitasmatologikis anagnoriseos stin metamorphomeni zoni Kimmerion Xanthis.* Xanthi: Institute of Geological and Mineralogical Exploitation (IGME).

Phavas, N. 1990. *Chemical analyses from Potamaki, Drama*. Xanthi: Institute of Geological and Mineralogical Exploitation (IGME).

Phavas, N. 1993. *Ekthesi koitasmatologikis anagnorisis Farassinou Dramas*. Xanthi: Institute of Geological and Mineralogical Exploitation (IGME).

Photos, E. 1987. *Early extractive metallurgy in north Greece: a unified approach to regional archaeometallurgy.* Unpublished Ph. D. thesis, Institute of Archaeology, University College London.

Pigott, V.C. 1996. The study of ancient metallurgiccal technology: a review, *Asian Perspectives 35(1): 89-97.*

Pigott, V.C. 1996. Near Eastern archaeometallurgy: modern research and future directions, in *The study of the ancient Near East in the twenty-first century.* Eisenbrauns.

Pleiner, R. 1969. *Iron working in ancient Greece*. Praha: National Technical Museum.

Pleiner, R. 2000. *Iron in Archaeology. ThEuropeaN Bloomery smelters.* Praha.

Poulios, B. 1988. The mines of Mount Pangaion, *Archaeologikon Deltion 36, B2 Chronika.* Athens.

Psilovikos, A. and G. E. Syrides 1992. The archaic city of Avdera, geomorphological investigation, in *2nd International Symposium of Thracian Studies.* Alexandroupolis.

Randsborg, K. 1991. *The first millennium A.D. in Europe and the Mediterranean: an archaeological essay.* Cambridge: Cambridge University Press.

Ray, K. 1982. Material metaphor, social interaction and historical reconstruction, in I. Hodder (ed.) *The archaeology of contextual meaning.* Cambridge: Cambridge Univerity Press.

Rees, S. E. 1979. Agricultural implements in Prehistoric and Roman Britain, BAR British Series 69 (i).

Render, J. 1987. Natural draft furnaces, *Archeomaterials 2: 47-58.*

Rich, J. and A. Wallace-Hadrill (eds) 1992. *City and country in the ancient world.* London: Routledge.

Ripoll-Perello, E. 1982. *Ampurias: description of the runes and the Monographic Museum.* Barcelona.

Risberg, C. 1994. Evidence of metal working in early Greek sanctuaries, in Ch. Risberg *et al.* (eds) 4th and 5th international workshops: *Trade and production in premonetary Greece: production and the craftsman.* Jonsered, Sweden: Paul Astroms Forlag.

Robinson, D. 1931. The Robinson collection, *Corpus Vasorum Antiquorum, fas. vol. III, I, fas 2.* Cambridge, MA: Harvard University Press.

Robinson, H. 1975. *The armour of Imperial Rome*.

London: Ards and Simon Press.

Rosen, S.A. 1997a. Craft specialization and the rise of secondary urbanism: a view from the Southern Levant, in H. W. H. *Andersen et al. Urbanisation in the Mediterranean in the 9th to 6th centuries BC*: 82-91. Copenhagen: Museum Tusculaneum Press.

Rosen, S.A. 1997b. Craft specialization and the rise of seondary urbanism: a view from the souhern Levand, in N.A.M. Aufrecht and S. W. Gauley (eds) *Urbanism in antiquity from Mesopotamia to Crete*: 82-91. Sheffield Academic Press.

Rostoker, W. and B. Bronson 1990. *Pre-industrial iron. Its technology and ethnology*. Philadelphia, Pennsylvania: Archaeomaterials Monograph No1.

Rowlands, M. (ed.) 1987. *Centre and periphery in ancient world*. Cambridge: Cambridge University Press.

Rowlands, M. 1987. Centre and periphery: a review of a concept, in M. Rowlands (ed.) *Centre and periphery in the ancient world*: 1-11. Cambridge: Cambridge University Press.

Rowlands, M. 1995. Inconsistent temporalities in a nation--space, in D. Miller (ed.) *Worlds apart, modernity through the prism of the local*: 23-33. London.

Rowlands, M. 1998. Objectivity and subjectivity in archaeology, in R. M. K. Kristiansen (ed.) *Social transformations in archaeology*: 29-39. London: Routledge.

Rowlands, M. and R. M.K. Kristiansen 1998. Introduction, in R.M. K. Kristiansen (ed.) *Social Transformations in archaeology, global and local perspectives:* 1-26. London: Routledge.

Sakelariou, M. 1980. *Les Proto-Grecs*. Athina: Ekdotike Athinon.

Salapata, G. 1997. *Hero warriors from Corinth and Lakonia, Hesperia 66(April/June 1997): 245-260.*

Samsaris, D.K. 1980. *O exellinismos tis Thrakis kata tin archaeotita*. Thessalonica: Kyriakidis Press.
Samsaris, D. 1987. Les mines et la metallurgie de fer et de cuivre dans la province romaine de Macedoine, *Klio 69(1): 152-162.*

Sanmarti-Greco, E. *et al.* 1995. Amphores grecques et trafics commerciaux en Metiterranee occidentale au IV s. av. J.-C, *Etudes massalieres 4: 31-47.*

Schiffer, M.B. and A.R. Miller 1999. *The material life of human beings*. London: Routledge.

Scott, E. 1993. *Theoretical Roman Archaeology, first conference proceedings*. Avebury.

Shackelford J. F. 1992. Introduction to material science for engineers. New Jersey: Prentice Hall.

Shalganova, T. and A. Gotzev 1995. Problems of research in the Early Iron Age, in P.I. Bailey D. W. Panayotov (eds) *Prehistoric Bulgaria*: 327-342. Madisson, Wisconsin: Prehistory Prees.

Shanks, M. and C. Tilley 1987. *Social theory and archaeology*. Cambridge: Blackwell Polity.

Shanks, M. and C. Tilley 1989. Archaeology into the 1990s, *Norwegian Archaeological Review 22(1): 1-21.*

Shanks, M. 1996. *Classical archaeology of Greece: experiences of the discipline*. London: Routledge.

Shefton, B.B. 1994. *Massalia and colonisation in the North-Western Mediterranean*, in D.A.F. Tsetslehladze (ed.) *The archaeology of Greek colonisation*: 61-86. Oxford: Oxford University Press.

Sheridan, A. and G. Bailey (eds) 1981. *Economic archaeology, towards an integration of ecological and social approaches.* . BAR International Series, Vol. 96.

Sherratt, A. 1992. What archaeologists can learn from annalistes? A.B. Knapp (ed.) *Archaeology, Annales, and Ethnohistory*: 135-142. Cambridge: Cambridge University Press.

Sherratt, S. 1994. Commerce, Iron and Ideology: metallurgical innovation in the 12th-11th century Cyprus, in Symposium on *Cyprus in the 11th century BC*. Nicosia, Cyprus: Anastasios G. Leventis Foundation.

Scmidt, P.C., and S. T. Childs 1995. Ancient African iron production, *American Scientis 83(Nov.-Dec.): 524-533.*

Snodgrass, A.M. 1967. A*rms and armour of the Greeks*. London: Thames and Hudson.

Snodgrass, A. 1980. *The dark age of Greece*. Cambridge: Cambridge University Press.

Snodgrass, A. 1987. *An archaeology of Greece.* London: University of California Press.

Snodgrass, A.M. 1992. Archaeology and the study of the Greek city, in Rich, J. and A. Wallace-Hadrill (eds) *City and country in the ancient world*: 1-24. London: Routledge.

Spencer, N. 1995. Multi-dimmensional group definition in the landscape of rural Greece, in N. Spencer (ed.) *Time, tradition and society in Greek archaeology*: 28-42. London: Routledge.

Spencer, N. 1995. *Time, tradition and society in Greek Archaeology*. London: Routledge.

Stavropodis, J. and S. Pournis 1971. *Car-borne scintillometer survey in Macedonia and Thrace (in Greek)*. Athens: Greek Atomic Energy Commission.

Stavropodis, J. 1983. Radioactive metallurgical slags in Macedonia and Thrace (In Greek), *Mineral Wealth 24: 41-46.*

Stoss-Gale, Z.A. *et al.* 1998. Lead Isotope data from the isotrace labouratory, Oxford: archaeometry data base 5, ores from Bulgaria, *Archaeometry 40(1): 217-226.*

Svolopoulos, K. 1994. Western Thrace in the body of the Greek State, in vol. *Thrace*: 267-286. Thessalonica: General Secretariat of East Macedonia and Thrace.

Syrides, G. 1996. The Holocene stratigraphy and palaeoenvironments in the area of ancient Avdera (Greece), in the *2nd Symposium of Hellenic Archaeometrical Society: Archaeometrical and Archaeological Research in Macedonia and Thrace*. Thessalonica: Aristotle University.

Talbert, R. 1985. *Atlas of Classical history*. London & Sydney: Groom Helm.

Taylor, T. 1985. Bronze Age Thrace reviewed, *Antiquity 59: 129-132.*

Taylor, T. 1987. Flying stags: icons and power in Thracian art, in I. Hodder (ed.) *The archaeology of contextual meanings*: 117- 132. Cambridge: Cambridge University Press.

Tholander, E. and S. Blomgren 1985. On the classification of ancient slags by microscopic examination, *ISKOS 5: 415-425.*

Thomas , J. S. 1996. *Time, culture and identity: an interpretive archaeology*. London: Routledge.

Thomas, N. 1991. *Entangled objects : excange, material culture, and colonialism in the pacific*. Cambridge, MA: Harvard university Press.

Tilley, C. 1981. Economy and society, in B.G. Sheridan (ed.) *Economic Archaeology, towards an integration of ecological and social approaches*: 131-148. BAR International Series.

Triantafyllos, D. 1972. Episimansi Archaeotiton, *Archaeologiko Deltio 27: 536-547, plates 472-488.* Athens.

Triantafyllos, D. 1973. Episimansi Archaeotiton. *Deltio 28, Chronika): 463-464.*

Triantafyllos, D. 1983. Archaiko nekrotafeio sti Dytiki Thraki, in *Estratto dell' Annuario Della Squola Archaeologica Di Atene e Delle Mission Italiani in Oriente*. Athens.

Triantafyllos, D. 1984. Anaskafi sta Rizia Evrou, *PAE 1984: 95-106, plates 91-93.* Athens.

Triantafyllos, D. 1990. Anaskafi tymvon sta nekrotafeia tis archaias Strymis, in *To Archaeologiko Ergo sti Macedonia kai Thraki*. Thessalonica.

Triantafyllos, D. 1988. Excavation at Kalyva, in *To Archaeologiko Ergo sti Macedonia kai Thraki*. Thessalonica.

Triantafyllos, D. 1994. Ancient Thrace, in *vol. Thrace*: 35-97. Thessalonica: Generat Secretariat of East Macedonia and Thrace.

Triantafyllos, D. and D. Terzopoulou 1996. Tafikoi tymvoi tis Thrakis, in *To Archaeologiko Ergo sti Macedonia kai Thraki to 1988*. Thessalonica.

Trigger, B. G. 1989. *A history of archaeological thought*. Cambridge: Cambridge University Press.

Trigger, B.G. 1991. Post-processual developments in Anglo-American archaeology, *Norwegian Archaeological Review, 24(2): 65-75.*

Tsaimou, K.G. and G.D. Papadimitriou 1996. A Greek inscription of Roman times referring to metallurgical activities of Greeks in the Black Sea, in *2nd Symposium of the Hellenic Archaeometric Society*. Thessalonica.

Tsatsopoulou, T. 1988-1996. Messemvria-Zone: excavation diaries. 19th Ephoreia of Prehistoric and Classical Antiquities of Thrace. Komotini.

Tsatsopoulou, T. 1997. *The excavation at Mesemvria-Zone during the last decade*, in *The Archaeological Work in Macedonia and Thrace*. Thessalonica.

Tsimpidis-Pentazos, E. 1971. Archaeologikes ereunes sti Thraki, *PAE: 86- 118, plates 104-146.* Athens.

Tylecote, R. F. 1962. *Metallurgy in Archaeology.* London: Edward Arnold.

Tylecote, R.F. 1976. *A history of metallurgy.* London: The Metals Society.

Tylecote, R.F. 1981. Iron sands from the Black Sea, *Anatolian Studies 3: 137-140.*

Tylecote, R. and Merkel, J. F. 1985. Experimental Smelting techniques: Achievements and Future, in P.C.M. Hughes (ed.) *Furnaces and Smelting Technology in Antiquity.* London: British Museum.

Tylecote, R.F. and B.J. Gilmour 1986. *The metallography of early ferrous edge tools and weapons.* BAR British Series.

van der Merwe, N.A., 1988. Science and magic in African technology: traditional iron smelting in Malawi, in R. Maddin (ed.) *The beginning of the use of metals and alloys*: 245-260. Cambridge, MA: MIT Press.

van-Andel, T.H. and C.N. Runnel 1987. *Beyond the Acropolis: a rural Greek past.* Stanford.

van-Dommelen, P. 1993. Roman peasants and rural organisation in central Italy, in E. Scott (ed.) *Theoretical Roman archaeology*: first conference procedings: 167-186. Aldershot.

van-Dommelen, P. 1997. Some reflections on urbanisation in a colonial context: west central Sardinia in the 7th to 5th centuries BC, in H. Andersen *et al.* (ed.) *Urbanisation in the Mediterranean in the 9th to 6th centuries BC*: 243-278. Acta Hyperboreia 7. Copenhagen: Museum Tusculanum Press.

van-Dommelen, P. 1997. Colonial constructs: colonialism and archaeology in the Mediterranean, *World Archaeology 28:* 305-323.

van-Dommelen, P. 1998. *On colonial grounds: a comparative study of colonialism and rural settlement in first millennium BC west-central Sardinia.* Leiden: Archaeological Studies, Leiden University.

Varoufakis, G. 1979. *Chimiki kai metallourgiki ereuna gyro apo 19 siderenia podia tripodon tis geometrikis epoxis.* Athens: Ethniko Metsovio Polytechnio.

Vaughan, S.J. 1995. Observations from an Aegean perspective on the relationship of archaeology and science as reflected in the study of ancient pottery, in I. Morris (ed.) *Klados: essays in honour of J N Coldstream.* London: University of London.

Vavelidis, M., *et al.* 1996. *Skapti Ili: ena archaeo metallio chrisou stin topothesia Mandra Kari, Palias Kavalas.* Thessalonica.

Vavritsas, E. 1966. Archaeologikes ereunes sti Mesimvria, *PAE: 59-69.* Athens.

Vavritsas, A. 1967. Anaskafi Messemvrias, *PAE: 89-95, plates 67-73.* Athens.

Vavritsas, A. 1968. Anaskafi Messemvrias, *PAE: 80-83, plates 67-73.* Athens.

Vavritsas, A. 1969. Anaskafi Messemvrias, *PAE 1969: 60-69, plates 71-91.* Athens.

Vavritsas, A. 1970. Anaskafi Messemvrias, *PAE: 66-75, plates 87-116.* Athens.

Vavritsas, A. 1971. Anaskafi Messemvrias, *PAE: 119-123, plates 147-160.* Athens.

Vavritsas, A. 1973. *Anaskafi Messemvrias, PAE: 70-82, plates 78-106.* Athens.

Vavritsas, A. 1975. Anaskafi Messemvrias, *PAE: 141-145, plates 113 - 124.* Athens.

Vavritsas, A. 1976. Anaskafi Messemvrias, *PAE: 142-146, plates 107-111.* Athens.

Vavritsas, A. 1977. Anaskafi Messemvrias, *PAE: 136-139, plates 77-85.* Athens.

Vavritsas, A. 1978. Anaskafi Messemvrias, *PAE: 94-98, plates 77-80.* Athens.

Vavritsas, A. 1979. Anaskafi Messemvrias, *PAE: 108-113, plates 73-80.* Athens.

Vavritsas, A. 1980a. Anaskafi Messemvrias, *PAE: 5-7, plates 2-12.* Athens.

Vavritsas, A. 1980b. Anaskafi Messemvrias, *PAE: 20-26, plates 20-33.* Athens.

Vavritsas, A. 1981. Anaskafi Messemvrias, *PAE: 1-6, plates 1-20.* Athens.

Velkov, V. 1980. *Roman cities in Bulgaria: collected studies.* Amsterdam: Hakkert Press.

Veranis, N. 1989. *Gold-bearing mineralizations in the Serbomacedonian zone:*

application of 'panning' for their localisation and selection of exploration targets, Thessalonica: Institute of Geological and Mineralogical Exploitation (IGME).

Voudouris, P. 1993. *Mineralogische, mikrothermometrische und geochemische Untersuchoungen an epithermalen Au-Ag-Gangmineralisationen bei Kassiteres/Sape (Nordostgriechenland).* Universität Hamburg.

Wagner, D. B. 1999. The earliest use of iron in China, in S. Young *et al.* (eds) *Metals in Antiquity: 1-9.* BAR International Series 792.

Wagner, D.B. 1985. *Dabieshan. Traditional Chinese iron production techniques practised in southern Hennan in the Twentieth Century.* Scandinavian Institute of Asian Studies Monograph Series, vol. 52. Curzon Press.

Wagner, G.A. and G. Wesgerber (eds) 1988. *Antike Edel- und Buntmetallgewinnung auf Thasos.* Der Anschnitt: Bochum.

Wagner, G.A., *et al.* 1981. Ancient gold mines on Thasos, *Naturwissenschaften, 68 (8): 263.*

Wagstaff, J. M. (ed.) 1987. *Landscape and culture: geographical and archaeological perspectives.* Oxford: Basil Blackwell.

Ward-Perkins, J.B. 1991. Cities of ancient Greece and Italy: planning in Classical antiquity, in G.R. Collins (ed.) *Planning and cities.* New York: George Braziller.

Webster, J. 1996. Roman Imperialism and the 'post-imperial' age, in J. Webster and N. Cooper (eds) *Roman Imperialism: post-colonial perspectives*: 1-18. Leicester.

Webster, J. and N. Cooper 1996. *Roman Imperialism: post-colonial perspectives.* Leicester Archaeology Monographs vol. no 3. Leicester.

White, K.D. 1967. *Agricultural implements of the Roman world.* Cambridge: Cambridge University Press.

Whithouse, R.D. and J.B. Wilkins 1989. Greeks and Natives in South-East Italy: approaches to the archaeological evidence, in T.C. Champion (ed.) *Centre and periphery*: 102-141. Cambridge: Cambridge University Press.

Whitley, J. 2001. *The archaeology of ancient Greece.* Cambridge: Cambridge University Press.

Wingrove, J. 1970. Identification of iron oxides. *Journal of the Iron and Steel Institute 208(1): 258-264.*

Woolf, G. 1990. World system analysis and the Roman Empire, *Journal of Roman Archaeology 12(2): 223-234.*

Woolf S. J. 1996. Nationalism in Europe: a reader. London: Routledge.

Xidakis, G., J.V. Diamantis, and R.G. Marinos 1988. Building stones and geological environment in three ancient cities of the Aegean Thrace, Greece, in *The engineering geology of ancient works, monuments and historical sites: preservation and protection. Athens: Greek National Group of IAEE.*

Young, R. 1990. *White mythologies. Writing history and the west.* London.

APPENDIX A

Table 1. Chronological table

Period.	Events.
1200-800 BC	Dispersion of Thracian tribes
800-650 BC	Foundation of Greek of colonies in Thrace.
650-550 BC	Foundation of "commercial colonies" (emporia).
491 BC and 480 BC	Persian campaign in Thrace.
c.480 BC	Foundation of the Odrycian State.
352 and 342-340 BC	Phillip II captures Avdera and Maroneia.
306 BC	Lyssimachos became " King of Thrace".
146 BC	Thracian Peninsula under Roman possession. Avdera and Maroneia declared free
A.D 46	Emperor Claudius abolished the privileges of Thracian Princes and made Thrace a Roman Province.

Table 2. List of Spaces lived in Thrace

Plants	Animals	Birds
Alnus Incana	Ursus arctos	Tetrao urogalus
Anemone nemorosa	Capreolus capreolus	Bonasa bonasia
Betula Pentula balcanica	Canis lupus	Aquila chrysaetos
Campanoula moesiana	Vulpes vulpes	Picoides tridactylos
Carex fava	Sus scrofa	Scolopax rusticola
Centaurea scabiosa	Sciurus vulgaris	Ficadula parva
Cherophyllum bulbosum		Corvus corax
Cicerbita alpina	Picus canus	
Listera cordata		Loxia curvirostra
Melampyium sylvaticum		Turdus merula
Radiola linoides		Cuculus canorus
Selaginella helvetica		Dentrocopus major
Silene flavescens		Dryocopus martius
Stachys officinalis		Caprimulgus-europaeus
Syringa vulgaris		
Valeriana montana		Columba oenas

Table 3. List of vases picturing scenes related to the Thracians.

Vase	Origin	Decoration	Reference
1.Oinochoe Tampa 86. 41	Attic Black Figure, Etruria, c. 510 BC	Side A: Horsemen, Thracian cloak, Corinthian helmet	(http://www.perseus.tufts.edu, 7.4.1999)
2. Neck amphora Wurzburg L 186	Attic Black figure, Etruria, c. 519 BC	Side B: Horseman dressed in Thracian cloak	Simon 1975, 105
3. Hydria Harvard 1969.14	Attic Black figure, c.540-530 BC Unknown region	Shoulder: horseman in Thracian cloak-zeira.	Pedley 1971

4. Cylix Boston 03.844 	Attic Red figure c. 520-510 BC Unknown region	Inside: Horseman in chlamys wearing Thracian cap and boots.	(http://www.perseus.tufts.edu, 7.4.1999)
5. Cylix Munich 2620	Attic Red figure, Etruria, c. 510-500 BC Unknown region	Tondo: Horseman in chiton and petassos wearing Thracian cloak and boots.	Simon and Hirmer 1976, 99-100.
6. Cylix Boston 10.196	Attic Red figure, c. 520-500 BC Unknown region	Inside: archer in trousers wearing soft cap. Side B: horseman in Thracian cloak	Caskey and Beazley no 71, plates 5.1 and 6.
7. Albertson University of Chicago 1967.115.243	Attic red figure, c. 510-500, Unknown region	Sides A and B: peltasts wearing Thracian cap and holding peltae	ARV2, 157. No 87
8. Cup Louvre G448	Attic red figure, c. 460-450 B.C., Unknown region	Side A: youths, one of them dressed in Thracian boots.	ARV2, 880 no 5
9. Cylix Harvard 1959.219 	Attic Red figure, c. 470-460 B.C., unknown region.	Tondo: Peltast holding long spear and pelta, Thracian coat, cap and boots.	Robinson CVA, 2, plate III. I
10. Cylix Boston 10.195	Attic Red figure, c. 490-470 B.C. unknown region.	Side A: youth with Thracian boots and cap.	ARV 2, 402, no 15.
11. Column Crater Berlin V. I. 3172	Attic Red figure, c. 440 B.C., from Sicily	Side A: Orpheus and men in Thracian long coat, holding javelins and wearing skin cap.	ARV 2, 1103. 1,1683
12. Bell crater Toledo1967.154	Attic Red figure, c. 440-430 B.C., unknown region.	Side A: Menelaos and Helen; he wears Thracian helmet.	CVA, USA fasc.17, 28, plate 43.
13. Fragment University of Chicago 1967.115.60	Attic red figure, c. 465-460 B.C. from Attic.	Decoration: warrior wearing Thracian helmet.	ARV2, 607, no 92.
14. Bell crater London 1961.7-10.1	Attic Red figure, c. 470-460BC unknown region.	Side A: warriors, the older wears Thracian helmet.	ARV2, 592.33 bis1, 660
15. Lekythos Cleveland 28.660	Attic Red figure, c. 480-470 BC	Thracian helmet on top of a stool.	ARV2, 648, no 37
16. Bell crater Chicago1922. 2197	Attic Red figure, c. 450 B.C., from Campania	Side B: Thracian helmet	ARV2, 610, no 21
17. Calyx crater Boston 59.178	Attic Red figure, c. 470-460 B.C. unknown region.	Side A: combat, a Greek warrior attacks a Trojan who wears a Thracian helmet.	ARV2, 590, no.11
18. Lekythos Boston 13. 202	Attic red Figure, c. 460 B.C., unknown region	Death of Orpheus; the Thracian woman wears chiton and fawn skin boots.	ARV2, 1002, no 11
19. Stamnos Louvre G 416	Attic Red figure, c. 470-450 B.C., from Campania.	Orpheus' death, Thracian women in Greek costume with	CVA, France 4, Louvre 3, p. 10; Robertson 1992, 173-175

		tattoos, sickle.	
20. Hydria Boston 90.156 	Attic Red figure, c. 460 B.C., unknown region.	The death of Orpheus by Thracian women in Greek costume holding implements; Orpheus wears fawn skin boots and diadem; Thracian man wearing skin cap is watching.	(http://www.perseus.tufts.edu, 7.4.1999)
21. Bell crater Harvard 1960.343	Attic red figure, c. 440-430 B.C., unknown region.	Side B: the death of Orpheus by Thracian women.	ARV2, 1042, 2., and Robinson Collection III. I. Plate XLVII.
Vase	**Ware, Region, and Date**	**Decoration (Key words)**	**Reference**
22. Hydria-kalpis Harvard 1960.342	Attic red figure, c. 440-430 B.C. , from Attic.	Domestic scene with loom. Thracian nurse (slave) wearing belted tunic similar to the cloak.	CVA, Robinson 2, 31-32, plate 43.

Table 4: selected list of reliefs depicting scenes related to the Thracians.

Description	**Origin**	**Date**	**Reference**
1. Thracians with romphaeae	• Trajans'column in Rome.	Roman	
2. Thracian Hero 	• Komotini.		
3. Votive-Artemis-Bendis wearing boots, mantle and holding long spear London 2155	• Attica from sanctuary	C. 350-325 B.C.	LIMC 3.1
4. Funerary Slave girl (has been interpreted as Thracian) Athens NM 4496	• Attica	C. 350-325 B.C	Catalogue of the National Archaeological Museum, Athens.

Table 5. References to ancient literature used with the text in chapter III.

References on:	**Literature**
Thracian geography	• Herodotus *Histories* 4.99.1, V.3, V.17 • Thucydides *Histories* 2.99.3, 2.97.1-2 • Strabo *Geography* 7, fragments 9, 10, 14, 16 (general), 33 (Nestos is he boundary between Macedonia and Thrace as fixed by Philip and his son Alexander), 7, fragment 47 (topography of the Thracian coastline in Maroneia, Evros navigable up to Cypsela, in 7.3.13 he also says that the Romans moved their military equipment through the river), 9.5.16 and 10.3.17 (parts of Thrace added to Macedonia, change in names), 7 fragment 43 (topography of the Thracian coast; Nestos River floods often)
Thracians as foreign or barbaric people	• Herodotus *Histories* 5.23.2 (Greeks and foreign), 7.185.1 (the Greeks of Thrace) • Aristotle *Politics* 1324b (non-Hellenic nation) • Thucydides *Histories* 7.29.4 (the bloodiest of the barbarians) • Strabo *Geography* 7.7.1

	• Flavius Josephus *Jewish Antiquities* 13.379
Thracians against Greeks and vice versa	• Herodotus *Histories* 1.168.1 (against Avdera), 9.75.1 (against Athenians in Amphipolis) • Xenophon *Anabasis* 1.1.9, 1.3.4, 2.6.2, 2.6.5, 6.4.2, *Hellenica* 3.2.10 (wall around 12 cities in Chersonese) • Thucydides *Histories* 4.102.1 (against the Athenians in Amphipolis with Brasidas)
Thracians Macedonians and vice versa	• Thucydides *Histories* 2.95.1-2 (Sitalkes against Perdiccas) • Strabo *Geography* 7.3.8 (Alexander invaded the country of the Triballoi as far as the river Ister) • Diodorus *Historical Library* 16.3.7
Thracians vs Romans	• Strabo *Geography* 6.4.2, 7.5.6, 7.5.11
Thracian tribes against other Thracian tribes	• Xenophon *Anabasis* 7.5.13 • Thucydides *Histories* 4.101.5 (Odryssae against Tribaloi)
Thracian customs	• Herodotus *Histories* 1.168.1, 2.167.1, V.6 (honour in war), V.308, • Xenophon *Hellenica* 3.2.5 (races in the honour of the dead) • Strabo *Geography* 7.3.3, 4
Thracian women (maenads or slaves)	• Pseudo-Apollodorus, *Library* vol 1.19 and vol. 2.87 • Pausanias *Description of Greece* 9.30.5 and 7.5.8
Tribal organisation and kingship	• Herodotus *Histories* 7.111.1 (Satrae free and excellent warriors) • Xenophon *Kyropaedia* 1.1.4, *Hellenica* 5.2.17 (Thracians under no king) • Thucydides *Histories* 2.29.2-3, 5.6.2 • Strabo *Geography* 7, fragment 47 (Thrace has 22 tribes, it has been devastated but can afford fifteen thousand cavalry and two hundred thousand infantry).
Tribal territories	• Xenophon *Anabasis* 7.5.13 (stone boundaries in the coast of Black Sea) • Strabo *Geography* 7.3.2, 7.5.1, 7 fragment 47, 12.3.3 • Strabo *Geography* 7.7.4 (the seaboard inhabited by the Greeks and Thracians occupy all the region from beyond the river Strymon to the Pontus and the Haemus Mountains-some add the area between Strymon and Nestos to the Macedonians).
Population movements, dislocations caused by migrations, wars, or administrative policies, changes in names and boundaries	• Plutarch *Pericles* 11.5 (a thousand settlers sent to Thrace to settle with the Bisaltae) • Strabo *Geography* 7.3.10 (Aelius Catus as governor of Macedonia, transplanted fifty thousand people from the far side of the river Ister into Thrace, some of them Getae, but all called now Moesi), 7.3.11 (Celti intermingled with the Thracians and Illyrians), 7.3.13 (constant migration inn both sides of the river Ister, Daci, Getae and Moesi migrations), 7.5.6 (the population of Thrace has been reduced in number because of war between each other and against the Macedonians and the Romans), 7 fragment 11 (synoikismos and settlers in Olyntos), 12.4.4 (difficult to mark boundaries between tribal territories)
Gift exchange and/or diplomacy	• Xenophon *Anabasis* 7.3.16 • Thucydides *Histories* 2.29.5, 2.97.4 • Strabo *Geography* 7.3.8
Thracian religion (Orpheus)	• Euripides *Rhesus* 920 • Strabo *Geography* 10.3.16 • Pausanias *Description of Greece 6.20.18*, 9.17.7, 9.30.4-12
Language	• Euripides *Rhesus* 290 • Strabo *Geography* 7.5.7
Clothing	• Xenophon *Hellenica* 2.4.11, *Anadasis* 7.4.4,
Eating and drinking habits	• Xenophon *Anabasis* 7.3.22-23 • Strabo *Geography* 7.3.3
Thracian flaura and fauna	• Herodotus *Histories* 5.23.2 (wood for ship building) • Xenophon *Anabasis* 7.2.1 (barley and wheat)

	• Xenophon *Anabasis* 7.5.12 (millet) • Xenophon *On Hunting* 11.1 (lions, leopards, lynxes, panthers and bears,)
Music and dance	• Xenophon *Anabasis* 6.1.5-6, • Strabo *Geography* 7.5.7, 10.3.6, 10.3.17, 7 fragment 40
Thracians as specialised forces	• Herodotus *Histories* 5.14.2, Xenophon *Anabasis* 6.3.11 (guides) • Xenophon *Hellenica* 1.3.10, *Anabasis* 1.5.13 (horsemen) • Herodotus *Histories III.75.1* and IV, 94, Xenophon *Hellenica* 3.2.2, *Anabasis* 2.2.7, 6.2.16, 6.3.6, 7.4.4, Thucydides *Histories* 1.98.1-2, 2.29.5, 2.79.4, 3.94.4, 3.97, 4.28.4, 4.123.4, Strabo *Geography* 12.4.4 (Thracian peltasts); Plato Laws 7.348b,e, 7.349c, 7.350a, Isokrates *Panegyricos* 4.115, 141, 144, 8.118, 15.111, Diodorus Sicelus *Historical Library* 14.91.3, 15.44.3, 16.24.2 (peltasts in the Greek army introduced by Iphicrates in the 4th century BC). • Thucydides *Histories* 4.129.2, 5.6.2 and 4, 7.27.2 (two expensive to held mercenaries-one drachma a day), Xenophon *Anabasis* 2.2.7, Pausanias *Description of Greece* 1.23.3, Strabo Geography 10.4.10 (mercenaries, usually as foot soldiers) • Thucydides *Histories* 2.96.2 and 7.27.1, Xenophon *Kyropaedia* 6.2.10 (swordsmen 'machairophoroi' from the hill Thracians, in Mount Rhodope) • Euripides *Rhesus* 290 (archers and other light armed troops • Diodorus Sicelus *Historical library* 17.17.4 (scouts in Alexander's campaign to Asia) • Cicero *Philippics* 7.17 (the armor of a Thracian
Romphaiae	• Xenophon *Kyropaedia* 6.2.10
Thracian involvement in mining	• Herodotus *Histories* 5.23.2 (Silver mines in Thrace), 7.110-112 (the Thracian tribes of Pierians, Odomanti and Satrae have gold and silver mines in the Mount Pangaion), 6.46.3 (tax in gold talents from the 'Dug forest' to Athens), 9.75.1 (control over the gold mines on the Thracian coast opposite Thassos causes war). • Xenophon Ways and Means 4.14 (a thousand slaves in the silver mines of Laurium under the management of a Thracian) • Herodotus Histories VII.112, Thucydides *Histories* 1.100.2-3, 4.105.1, Euripides *Rhesus* 920 and 970 , Xenophon *Hellenica* 5.2.17, Demosthenes *On the Chersonese* 45, Strabo *Geography*14.5.28, Plutarch *Cimon* 4.1,(gold and silver mines in Mount Pangaion) • (bout gold and silver in Mount Pangaion) • Demosthenes *Against Pantaenetus* 4 (Maroneia, small district in the mining area of Laurium) • Strabo *Geography* fragment 45, 14.5.28 (The Thracian tribe Sintoi specialised in mining and connected to Sintiens who colonised Lemnos, the most ancient center of metallurgy), 7.7.4 (Philip of Macedon organized very large revenues from the mines and other natural resources from the area between Strymon and Nestos) • Titus Livius XLV, 29, 10-11 (mines of iron and copper exploited by the indigenous villages which remained free but hey had to pay to the Romans half of what they paid to the Macedonian king while roman colonists exploited the mines around the Philippi) • Plutarch *Cimon* 4.1 (he refers to Thucydides mines) • Diodorus *Historical Library* 11.70.1 (quarrel between Athenians and Thasians about the gold and silver mines in Mount Pangaion • Notitia Dignitatum (Thrace one of the armories of the Empire)

Table 6. Messemvria-Zone: list of industrial waste.

Sample no	Excav. No	Context	Date	No of fragm	Description	Weight
MES 101	ΑΠΚ28	20.6.96 Pottery kiln south of the NE tower of the enclosed settlement		1	Large amorphous fragment of tap slag	1855gr
MES 105	ΑΠΚ75	1987 East 1, section 2 North of the area π		Many	Small amorphous	620gr
MES 106	ΑΠΚ46	17.9.87 East 1, section B Depth 1.10m		1	Medium size amorphous	128gr
MES 111	ΑΠΚ45	24.9.87 East 1, section B, north of T7		3	Small amorphous	911gr
MES 118		8.11.88 Trench north of the enclosed settlement North road, depth 0.40m		Many	Small amorphous	362m
MES 119		8.11.88 Trench north of the enclosed settlement North road, depth 0.60m		2		1258gr
MES 123		16.9.87 East 1, section B, depth 1.10m		6		1664m
MES 124	ΑΠΚ88	8.9.87 East 1, section E, depth 1.40m		1	Large shaped slag	134gr
MES 125		17.11.87 Area A east of the wall of the enclosed settlement, square Δ, depth 1.25m		1	Slag	
MES 127	ΑΠΚ 208	10.7.96 ΔT4, section A North of T2 (D-E)(4-6) Depth 0.80m				766gr
MES 129	ΑΠΚ95	28.6.96 ΔT4, trench north of the enclosed settlement				243gr
MES 130	ΑΠΚ 234	12.7.96 ΔT4, section B A vertical road Depth 1.70m (Y4/95)		1	Slag similar to MES143	87.7gr
MES 132	ΑΠΚ 457	6.8.96 ΔT4, section B , section E/95 area B, depth 1.15m (Y2)		2		553gr
MES 133		17.6.96 Monumental burial Surface find			Large amorphous mass of iron	1437gr
MES	ΑΠΚ	31.7.96				212gr

134	379	ΔT2, section B 1st parallel road, depth 0.50m				
MES 135	ΑΠΚ 443	2.8.96 ΔT2, 1st parallel road, east of section A/95		2		
MES 136		17.6.96 north of the NE tower of the enclosed settlement		Many	Shaped	6000gr
MES 137	ΑΠΚ 132	4.7.96 ΔT4, section Δ, North of T2 (O-A)(0-2) Depth 0.80m				146.8gr
MES 139	ΑΠΚ 267	23.7.96 A vertical road Depth 1m (Y3/96)				610gr
MES 143	ΑΠΚ 196	4.7.96 ΔT4, section B A vertical road Depth 1.70 (Y4/95)				89.6gr
MES 145	ΑΠΚ 109	3.7.95 ΔT4, section Δ 1st parallel road Depth 1.70 (Y4/95)			Large shaped	540gr
MES 146	ΑΠΚ 192	17.7.95 New Tower			Large shaped	304gr
MES 147	ΑΠΚ 154	6.7.95 ΔT4, section A, North of T6 (B-G)(16-18) depth 0.70m(Y2)			Large shaped	195gr
MES 148	ΑΠΚ 200	19.7.95 New Tower SW corner of T6-T7 3μ (Ψπ)		2	Large shaped	371gr
MES 149	ΑΠΚ49	26.6.95 West Gate		3	Large shaped	1325gr
MES 150	ΑΠΚ 181	14.7.95 New Tower West of the west city wall				230gr

Table 7b.Messemvria-Zone: list of artefacts.

Sample no	Excav. no	Context	N of frag-ments	Description	Measurements
MES1		1.11.69 Enclosed settlement	1	Arrowhead (pyramid in shape)	0.09m
MES2	ΑΓΚ853	1.11.69 Enclosed settlement	1	Arrowhead (from catapult)	0.075m
MES3	ΑΓΚ852	1.11.69 Enclosed settlement	1	Arrowhead (pyramid in shape)	0.04m
MES4	ΑΓΚ798	27.11.67 Enclosed settlement	2	Straight knife	
MES5	ΑΓΚ751	1.11.69 Enclosed settlement	1	Key (laconic type)	0.10m
MES6	ΑΓΚ920	7.10.70 Enclosed settlement	1	Large straight knife blade	0.37m
MES7	ΑΓΚ788	23.11.67 Enclosed settlement	1	Spearhead	0.17m
MES8	ΑΓΚ762	7.12.71 Enclosed settlement	1	Arrowhead (pyramid)	0.055m
MES9	ΑΓΚ840	1966 Enclosed settlement	1	Straight knife blade	0.13m
MES10	ΑΓΚ839	1969 Enclosed settlement	1	Spearhead	0.14m
MES11	ΑΓΚ858	1969 Enclosed settlement	1	Spearhead	0.14m
MES12	ΑΓΚ750	19.12.71 Enclosed settlement	1	Spit	0.17m
MES13	ΑΓΚ4603	4.8.83 Trench north of the enclosed settlement, area north of the wall 3, Depth 1.75m	1	Arrowhead (pyramid in shape)	0.05m
MES13b	ΑΓΚ5247	3.8.87 East 1, square Γ, depth 0.60m	1	Arrowhead (pyramid in shape)	0.045m
MES14	ΑΓΚ5238	12.8.87 East 1, square Γ, area ι, depth 0.60m	1	Spearhead	0.21m
MES15	ΑΓΚ4640	17.8.84 Area ι, east of T2	1	Arrowhead (pyramid in shape)	0.06m
MES16	ΑΠΚ365	3.8.89 AT2, square Δ (Θ–Ι)(12–14) 1.80μ (Υ8β)	1	Dagger	0.13m
MES17	ΑΠΚ329	31.7.89 AT2, square Δ North of T4 1.80μ (Υ8β)	1	Straight knife blade	0.11m
MES18	ΑΠΚ371	1.8.89 AT2, square Δ North of T4 1.80μ (Υ8β)	1	Chain ring	0.06m

MES19	AΠK202	25.7.89 AT2, square H (18-20)(E–ΣT) 1.80μ (Y10)	1	Nail	0.065m
MES20	AΠK296	21.7.89 AT2, square Δ North of T4 1.80μ (Y8β)	1	Key, small laconic type (from box)	0.02m
MES21 MES 21a	AΠK351	31.7.89 AT2, square Δ North of T4 1.80μ (Y8β)	1	Spit Currency bar	0.26m 0.26x0.045x 0.017m 615.6 gr
MES22	AΠK233	2.8.89 AT2, square Δ North of T4 (18–20)(H–Θ) 1.80 μ (Y8β)	1	Spearhead	0.35m
MES23	AΠK235	3.8.89 AT2, square Δ North of T4 (12–14)Θ 1.80 μ (Y8β)	1	Spearhead	0.19m
MES24	AΠK166	31.7.89 AT2, square H North of T4 (0–2)(I–Θ) 1.90μ(Ψ10)	1	Spearhead	0.13m
MES25	AΠK426	31.7.89 AT2, square Δ North of T4 1.80 μ (Y8β)	1	Spearhead	0.18m
MES26	AΠK179	31.7.89 AT2, square Δ North of T4 1.80 μ (Y8β)	1	Spearhead	0.15m
MES27	AΓK11054	1.8.89 AT2, square Δ North of T4 1.80 μ (Y8β)	1	Nail	0.014m
MES28	AΠK220	10.7.89 AT2, square Δ North of T4 1.55μ (Y8β)	1	Arrowhead (leaf shaped)	0.047m
MES29	AΠK227	27.7.89 AT2, square Δ North of T4 1.80μ (Y8β)	1	Arrowhead (leaf shaped)	0.077m
MES31	AΠK343	31.7.89 AT2, square Δ North of T4 1.80μ (Y8β)	3	Spit	a.0.19m b.0.15m c.0.16m

MES30	ΑΠΚ309	27.7.89 AT2, square Δ North of T4 1.70μ (Y8β)	1	Sheet	0.06x0.015m
MES32	ΑΠΚ345	31.7.89 AT2, square Δ North of T4 1.80μ (Y8β)	1	Spit	0.30m
MES33	ΑΠΚ375	2.8.89 AT2, square Δ North of T4 1.80μ (Y8β)	1	Spit	0.50m
MES34	ΑΓΚ11056	25.7.89 AT2, square Δ North of T4 1.80μ (Y8β)	1	Spit	0.26mm
MES35	ΑΠΚ349	1.8.89 AT2, square Δ North of T4 1.80μ (Y8β)	4	Spit	a.0.18m b.0.28m c.0.20m d.0.18m
MES36	ΑΓΚ9054	20.7.90 Sofoglou Sakir field Section 12, west of feature (B) Depth 1.30m	Many Fragme nts of 1	Curry comp	
MES37	ΑΠΚ137	20.7.90 Sofoglou Sakir field Section 12, west of feature (B) Depth 0.30m	Many Fragme nts of 1	Striggil	
MES38	ΑΠΚ33	13.7.90 Sofoglou Sakir field Section 12, west of feature (B) Depth 1.10m	1	π shape clamp	0.22m
MES39	ΑΓΚ9045	25.7.91 ΔT3,square B (H–Θ)(14–16) west of T17 depth 0.50m (Y2)	1	Straight knife blade	0.05m
MES40	ΑΠΚ21	5.8.91 ΔT3,square B between walls T4.T6,T8	1	Straight knife blade	0.054m
MES41	ΑΠΚ41	1.7.92 ΔT3,square Γ/91 area Γ (ΔE,8- 10)1.85m (Y21)	1	Spit	0.18m
MES42	ΑΠΚ42	1.7.92 ΔT3,square Γ/91 area Γ (ΔE,8–10)1.85m	1	Plough	0.07m

		(Y21)			
MES43	ΑΠΚ102	16.7.92 ΔΤ3,square H (AB,14-15) 1.15m (Y23)	1	Spit	0.24m
MES44	ΑΠΚ108	17.7.92 ΔΤ3,square H (AB,14–15) 1.15m (Y27)	1	Curved knife blade	0.22
MES45	ΑΠΚ9	24.6.92 ΔΤ3,square E wall T1 0.60m (Y4/91)	2	Curved knife blades	a.0.21x0.02m b.0.12x0.02m
MES46	ΑΓΚ8662	30.6.92 ΔΤ3,square Δ (ΖΘ,4-6) 0.30m (Y17)	1	Arrowhead (pyramid in shape)	0.07m
MES47	ΑΠΚ11	22.6.92 ΔΤ3,square Δ between walls T6-T7 0.80m (Y23)	1	Spearhead	0.09m
MES48	ΑΠΚ190	24.7.91 ΔΤ3,square B (ΗΘ,12-14) 0.50m (Y2)	1	Arrowhead (pyramid in shape)	0.07m
MES49	ΑΓΚ8672	24.7.91 ΔΤ3,square B (ΗΘ,12–14) 0.50m (Y2)	1	Straight knife blade	0.07m
MES50	ΑΓΚ8661	28.7.92 ΔΤ3,square A (ΕΖ,7-9) 1.15m (Y24)	1	Straight knife blade	0.05m
MES51	ΑΠΚ2	13.6.95 Trench 1992 East of the area with amphorae	1	Straight knife blade	0.09m
MES53	ΑΠΚ82	29.6.95 ΔΤ4,square B 1st parallel road 0.50m (Y2)	1	Falx	0.08m
MES54	ΑΠΚ51	27.6.95 ΔΤ4,square Γ	1	Straight knife blade	0.14m
MES55		29.6.95 ΔΤ3,square B (AB,0–2) 0.20m (Y1)	1	Spearhead	0.14m
MES56		29.6.95 ΔΤ3,square B (ΒΓ,4–6)	1	Spit	0.53m
MES 62	ΑΠΚ299	30.7.96 ΔΤ4,square E, A vertical street		Arrowhead	0.098m
MES65	ΑΠΚ111	4.7.96 ΔΤ4,square B (0A,4-6) 1.20m (Y3/96)	1	Knife	0.15m

MES67	AΠK183	9.7.96 ΔT4,square B, area Γ (0A,6–8) 1.10m (Y4/95)	1	Falx	0.09x0.035m
MES68	AΠK344	7.8.96 from he excavation of 1992, surface find	1	Knife straight	0.15m
MES71	AΠK256	10.7.89 AT2, square Δ, depth 1.55m (Y8 □)	1	Nail	0.05m
MES72	AΠK255	10.7.89 AT2, square Δ, depth 1.55m	1	Nail	0.03m
MES73	AΠK264	18.7.89 AT2, square Δ, depth 1.80m (Y8β)	1	Nail	0.065m
MES74	AΠK207	3.8.89 AT2, square H, North of T2 depth 1.90m (Y10)	1	Nail	0.045m

Table 8. Avdera: list of industrial waste.

Sample no	Context	No of fragm.	Description	Weight
AV100	Trench E, area 4 AB 9 (Σχ 4)	1	Small fragment of amorphous, porous, magnetic slag	
AV101	Trench E, area 4 AB 9 (Σχ 4)	1	Small fragment of amorphous, porous, magnetic slag	
AV102	Trench M Section M1 ABΔ 76, 28	1	Small fragment of amorphous, porous, magnetic slag	
AV103	Trench M Section M1 AB 76–43	1	Small fragment of amorphous, porous, magnetic slag	
AV104	Trench A Section MA ABΔ 76–65	1	Small fragment of amorphous, porous, magnetic slag	
AV105	Trench Γ NA gate	1	Small fragment of amorphous, porous, magnetic slag	
AV106	West city wall Δτ β	1	Small amorphous, porous, magnetic slag	
AV107	Trench in the west city wall	1	Small fragment of amorphous, porous, magnetic slag	
AV108	1.6.98 Roman house Area Λ	1	Slag, large, shaped, heavy, magnetic.	1257gr

AV109	12.6.91 Trench E, E7 2.50 NE-1.90 SE layer 3rd	1	Slag, large, shaped, heavy, magnetic.	402gr
AV110 AV111	28.5.91 Classical enclosure, area of West Gate, Section Δ–Δ14 Layer 2nd	2	Small fragments of amorphous, porous, magnetic slag	155gr
AV112 AV113	8 and 15.5.91 West gate	2	Small fragment magnetic slag	227gr
AV114	4.6.91 Classical enclosure, West Gate road Area of the Dolphins house, layer 2nd	1	Medium size, shaped, porous, magnetic slag.	285gr
AV115 AV116 AV117 AV118 AV119	29.5.91 Classical enclosure, West Gate road Area of the Dolphins house, layer 1st	7	Small fragments of amorphous, porous, magnetic slag.	514gr
AV120 AV121	14.5. 91 Classical enclosure Trench Γ, section Γ17 Layer 1st	5	Small fragments of amorphous, porous, magnetic slag.	202gr
AV122	20.5.91 Classical enclosure Trench Δ, section Δ8 Layer 2nd	2	Small fragments of amorphous, porous, magnetic slag.	132gr
AV123	8.5.91 Area of the West Gate, trench Δ, section Δ5, layer 1st	1	Slag, large, shaped, heavy, magnetic.	929gr
AV124	22.5.91 Area of the West Gate, trench Δ, section Δ6, layer 3rd	6	Small fragments of amorphous, porous, magnetic slag.	111gr
AVD125	13.5.91 Area of the West Gate, trench Δ, section Δ7, Layer 2nd	7	Small fragments of amorphous, porous, magnetic slag.	435gr
AV126	14.5.91 Classical enclosure Area of the West Gate, trench Δ, section Δ9, Layer 2nd	2	Small fragments of amorphous, porous, magnetic slag.	76gr
AV127	5.11.91 Classical enclosure Γ, B_2 Layer 3rd	1	Small fragment of amorphous, porous, magnetic slag.	183gr

AV128	4.11.91 Classical enclosure Γ35 Layer 1st	1	Medium size, shaped, porous, magnetic slag.	303gr
AV129	5.11.91 Classical enclosure Γ35 Layer 1st	6	Small fragments of amorphous, porous, magnetic slag.	96gr
AV130	11 and 13.11.91 Classical enclosure Γ36	3	Small fragments of amorphous, porous, magnetic slag.	98gr
AV131	14.7.92 Classical enclosure Roman house Trench ΣT, section Θ	7	Small fragments of amorphous, porous, magnetic slag.	176gr
AV132	13.6.91 Classical enclosure Area of the West Gate, trench Δ, section Δ4, Layers 1st – 11th	4	Small fragments of amorphous, porous, magnetic slag.	110gr
AV133	7.7.92 Classical enclosure Roman house Trench ΣT	24	Small fragments of amorphous, porous, magnetic slag.	400gr
AVD134	9.7.92 Classical enclosure Trench ΣT, E Section H	1	Medium size, amorphous, porous, magnetic slag.	156gr
AV135	8.7.92 Classical enclosure Roman house Trench ΣT, section E	Many	Small and medium size fragments of amorphous, porous, magnetic slag.	154gr
AV136	23.5.91 Classical enclosure Area of the West Gate, trench Δ section Δ5	7	Small and medium size fragments of amorphous, porous	650gr
AV137	12.6.91 Classical enclosure West Gate road, trench Δ –Y1 Layer 5th	3	Small fragments of amorphous, porous, magnetic slag.	182gr
AV138 AV139 AV140	11.6.91 Classical enclosure Section Γ–Γ5 (NE) Layer 5th	many	Small and medium size fragments of amorphous, porous, magnetic slag.	1268gr
AV141	6.7.92 Classical enclosure Roman house Trench ΣT, section E		Small and medium size fragments of amorphous, porous, magnetic slag.	753gr

Table 9. Avdera: list of artefacts.

Sample no	Museum no	Context	Date	No of fragm	Description	Measurements
AV1	MA 974	2.7.92 Classical enclosure Trench ΣΤ, Δ1	Classical	1	Sheet	0.037x0.021m
AV2	MA 772	Classical enclosure Trench Γ, section Γ1, layer 3rd	Classical	1	Sheet	0.057x0.034x 0.005m
AV3	MA 1738	Lakiotis field Area enagismon Cemetery	4th B. C.	1	Sheet	
AV4	1986	'Museum' trench surface find	Roman	1	Spatula	0.083m
AV5	MA 358	Touzla Giol Burial mount Burial 7	4th B. C.	1	Striggil	
AV6	MA 314	Touzla Giol Burial mount Burial 2	4th B. C.	1	Striggil	
AV7	MA 1706	Lakiotis field Cremation-burial 5	4th B. C.	1	Striggil	
AV8	MA 1690	Lakiotis field Burial 7	4th B. C.	1	Striggil	
AV9	MA 1712	Lakiotis field Cremation-burial 16	4th B. C.	1	Striggil	
AV10		4.10.96 Kourou Tsesme Site 96.05 Burial II, pithos AB κτε (96.05) 10			knife with two pointed handle	
AV 11	MA4849	27.10.95 Paliochora N. Valsamas/F. Kissas field Site 95.18 Burial I, chest AB κτε (95.18) 32			knife with two pointed handle	
AV12	MA 6032	20.9.95 Kourou Tsesme El. Raxionis field Site 95.5 cremation-burial	Hellenistic	2	knife with two pointed handle	a.0.11x 0.01-0.015m b.0.03x0.01m
AV13		4.10.96 Kourou Tsesme-Veloni Site 96.05, burial II, pithos	Hellenistic	3	knife with two pointed handle	a.0.055x0.015m b.0.04x0.01m c.0.04x0.01m
AV14	MA 5873	22.9.79 House of mosaic North part ΑΒΔ 79		1	Fragment of chisel or scalpel with triangular profile.	0.07 x 0.35m
AV15	MA1301	3.11.81 West Gate Trench B, section A Depth 0.30-0.60m		1	Curved knife blade	0.11x0.016m

AV 16	MA 797	13.11.91 Classical enclosure Trench Γ, section Γ3, layer 8th AB91 (Γ3) 739		2	Straight knife blade.	a.0.04 x0.035m b.0.10x0.018m
AV17	MA 4454a	13.8.76 Trench A, square A20 Depth 0.06m		1	Nail	0.23m
AV 18	MA797	13.11.91 Classical enclosure Trench Γ, square Γ 3, layer 8th	Roman	2	Straight knife blade.	a. 0.04x0.013x 0.035m b. 0.10x0.018x 0.035m
AV19	MA3221	1.9.78 Trench B, West city-wall Area ΑΒΓΔ depth 0.50-0.70m	Roman	1	Chisel or scalpel	0.75x0.02x0.01 m
AV20	MA6574	14.7.82 Tsacal Tepe Burial mount 7 Cremation N'			Ring	0.025m
AV21	MA 6575	19.7.82 Tsacal T Burial mount 7 cremation P'		1	Striggil	0.26m
AV22	MA6573	14.7.82 Tsacal Tepe Burial mount 7 Cremation N'		1	Striggil	0.26m
AV23	MA6599	31.8.92 North enclosure Valta Zampaki Field Avrapoloulou Square ΛΔ, depth 10.93m	5th BC	1	Sickle	0.23x0.16m
AV24		21.9.88 Trench II Valta Zampaki, Kiosses field Section KA, depth 1.05m		1	Spit	0.36m
AV25	MA 6621	23.8.90 Trench II Valta Zampaki Avramoglou field	Early classical - archaic	1	clamp in double T shape (twheel)	0.25m
AV26		23. 5.91 Classical enclosure Area of West Gate Section Δ, layer 3rd		1	Saw	0.28 x0.11m
AV27		9.12.1977 Tsacal Tepe	4th B.C.	1	Straight knife blade	0.06x0.018m
AV28		1.11.75 Trench in the West city-wall, section ΔΤβ ΑΒΔ75 3ι		1	Scissors (small triangular blade)	0.08 x 0.02m
AV29		31.8.93 Molos, Lakiotis field Enagismos XIII	4th B.C.		knife with two pointed handle	0.09 x 0.01m

AV30		1975 Trench Loutra		1	Axe	0.04X0.06X0.00 8m
AV31		24.10.91 Classical enclosure Section Γ51 2.80Nx1.60E layer 5th	Roman	1	Straight knife blade	0.11x0.018m
AV32		16.5.91 Classical enclosure Section Γ30 2.00Nx3.70E layer 1st	Roman	1	Scissors (small triangular blade)	0.08x 0.015m
AVD33		5.11.91 Classical enclosure Area of the West Gate Trench Δ, section Δ15 Layer under the roof tiles	Roman	1	Straight knife blade	0.11x0.03m
AV 34		7.6.91 Classical enclosure Area of the West Gate Trench Γ, section Γ5		1	Straight knife blade	0.065x0.025m
AV35		17.5.91 Classical enclosure Area of the West Gate Trench Δ		1	Arrowhead	0.07 x 0.02m
AV36		7.11.75, Trench Loutra	Roman	1	Straight knife blade	0.13x0.02m
AV37		14.5.91 Trench E, section E5 1.40SWx2.60NW layer 2nd	Roman	2	Curved knife blade	a. 0.10 x 0.02m b. 0.06 x 0.02m
AV38		30.8.96 Classical enclosure Roman house Trench ΣΤ, area Λ	Roman	1	Straight knife blade	0.20x0.03m
AV39		24.5.91 Trench E, section E6 1.16NWx2.94SW layer 4th		1	Axe	0.09 X 0.03m
AV40		29.10.91 Classical enclosure Area of the West Gate Trench Δ–Δ19	Roman	1	Billhook	0.09 x0.05m
AV41		21.5.91 Classical enclosure Roman house Trench Γ,section Γ29 0.50N x 0.50E layer 4th		1	Spatula	0.06x0.03m
AV43		7.6.91 Trench E, section E8 layer 4th		1	Spit	0.19m
AV44		24.5.91 Trench E, section E10 3.40SEx0.70SW Layer 3rd		1	Spit	0.17m

AV45		4.6.91 Trench E, section E10 1.80NEx3.24SE layer 5th	Roman	1	Wedge	0.09x0.15m
AV46		7.6.91 Trench E, section E8 1.63Nex2.25NE layer 3rd	Roman	2	Knife (maxaira)	a. 0.10x0.06m b. 0.12 x 0.05m
AV47		3.9.96 Classical enclosure Roman house Trench ΣΤ, area Λ	Roman	1	Straight knife blade	0.08x0.03m
AV48		1991 Classical enclosure Area of the West Gate Trench Δ, section Δ7 Layer 1st	Roman	1	Two pointed pitchfork	0.28x0.15m
AV49		14.5.91 Classical enclosure Area of the West Gate Trench Δ Layer 2nd		1	Wedge	0.095x 0.017m
AV50		23.10.91 Classical enclosure Area of the West Gate Trench Γ–Γ51 (south) Layer 4th		2	Spit	a. 0.20m b. 0.15m
AV51	MA 2828	12.9.91 Valta Zampaki Avramoglou field Square IA, between walls 18-19, depth 10.77m.	Early classical- archaic	1	Curved knife blade	0.17x0.045m
AV52		24.8.89 Valta Zampaki Avramoglou field Trench II, Square E Layer 58th, depth 9.93m		1	Sheet	0.075 x0.04m
AV53	MA 5154	21.9.61 Trench Γ, square 5		1	Striggil	0.14m
AV54		28.5.91 Classical enclosure Trench Γ, section Γ2 Layer 4th		21	Nails	0.02-0.08m
AV55		4.6.91 Classical enclosure Trench Γ, section Γ2 Layer 6th		8	Nails	0.02-0.08m
AV56		21.5.91 Classical enclosure Trench Γ, section Γ1 Layer 3rd		24	Nails	0.02-0.08m
AV57		20.5.91 Classical enclosure Trench Γ, section Γ1 Layer 2nd		17	Nails	0.02-0.08m
AV58		30.5.91 Classical enclosure Trench Γ, section Γ2		14	Nails	0.02-0.08m

AV59		16.5.91 Classical enclosure Trench Γ, section Γ17 Layer 1st		16	Nails	0.02-0.08m
AV60		17.5.91 Classical enclosure Trench Γ, section Γ2 Layer 2nd		16	Nails	0.02-0.08m
AV61	MA 797	13.11.91 Classical enclosure Trench Γ, square □ 3, layer 8th	Roman	1	Straight knife blade	0.14x0.003m
AV62	MA798	13.11.91 Classical enclosure Trench Γ, square Γ 37, layer 1st	Roman	1	Curved knife blade	0.14x0.018m
AV63	MA773	22.5.91 Classical enclosure Trench Γ, square Γ 2, layer 3rd	Roman	1	Sheet	0.096x0.014x 0.004m
AV64	MA5653	8.9.77 North of the West Gate Trench B	Roman	1	Large triangular knife blade	0.17x0.02m
AV65	MA4349	1991 Classical enclosure Area of the West Gate Trench Δ	Roman	1	Straight knife blade	0.072x0.02m
AV66	MA 917	10.10.91 Trench E, area 3 Layer 7th		1	Bell	0.14x0.028m
AV67	MA4485	28.7.76 Trench M, section MA	Roman	1	Curved knife blade	0.063x0.012m
AV68	MA4318	28.7.76 Trench M, section MA	Roman	1	Scissors	0.074x0.11m
AV69	MA5873	22.9.79 House of the mosaic North part of the house	3rd to 2nd BC	1	Large triangular blade	0.07x0.035m
AV70	MA1723	29.7.92 Lakiotis field Cremation-burial 28 depth 2.23m	4th BC	1	Dagger	0.097x0.011m
AV71	MA4458	3.9.90 Valta Zampaki Avramoglou field Square I, circular pit	3rd to 2nd BC	1	Straight knife blade	0.11x0.016m
AV72	MA2268	6.7.88 Valta Zampaki Avramoglou field Trench II, section IA layer 0.20-0.40m	Early classical-archaic	1	Straight knife blade	0.06m
AV73a	MA6053	21.11.97 Sazli Tsair Field of P. Taxiarchis Site 97.25, burial 1	Hellenistic	1	Dagger	0.26m
AV73b	MA6054	21.11.97 Sazli Tsair Field of P. Taxiarchis Site 97.25, burial 1	Hellenistic	1	Curry-comp	

AV74	MA1218	20.11.97 Sazli Tsair Field of P.Averopoulou Sandstone sarcophagus	Hellenistic	1	Sheet	0.065x0.039x0.0 02m
AVD75	MA1142	20.11.97 Sazli Tsair Field of P.Averopoulou Sandstone sarcophagus From the right solder of the diceased.	Hellenistic	1	Dagger	0.125x0.019m
AV76	MA4849	27.10.95 Paliochora N. Valsamas/F. Kissas field Site 95.18 Burial I, chest		1	Curved knife blade	0.128x0.016m
AV77	MA6227	12.9.91 Chatzopoulos field Ceramoskepis taphos T37.		1	Dagger	0.04x0.016m
AV78	MA6236	27.9.91 Valta Zampaki Field Chatzopoulos Keramoskkepis taphos	3-2nd BC	1	Dagger	
AV79	MA1141	25.10.82 Sazli Tsair Field of P.Averopoulou Sandstone sarcophagus From the right solder of the dead			Scissors	0.15m
AV80	MA6262	10.6.92 Chatzopoulos field Chest burial No48 Δ42 Depth 1.35m		1	Scissors	0.15m
AV81	MA1721	24.6.92 Molos Lakiotis Cremation XXIII Depth 2m	4th BC	1	Spatula	0.10m
AV82	MA1739	29.6.92 Molos Lakiotis Area of enagismon IV Depth 1.81m	4th BC	1	Spatula	0.04m
AV83	MA 4539	4.9.91 Valta Zampakis Avramoglou field Square ΚΣΤ, depth 10.84m	4th BC	1	Spatula	0.08x0.07m
AVD84	MA4539	4.9.91 Valta Zampaki Avramoglou field Square ΚΣΤ, depth 10.84m, αρ10	4th BC	1	Spatula	0.08x0.005m
AV85	MA4454b	13.8.76 Trench A Square A20, layer 0.06m.		1	Spatula	0.08x0.005m
AV86	MA1496	13.8.91 Valta Zampaki	4th BC	1	Wedge	0.041x0.009m

		Avramoglou field Between squares K-KZ				
AV87	MA1496	13.8.91 Valta Zampaki Avramoglou field Between squares K-KZ	4th BC	1	Wedge	0.062x0.085m
AV88	MA1496	13.8.91 Valta Zampaki Avramoglou field Between squares K-KZ	4th BC	1	Wedge	0.049x0.006m
AV89	MA1496	13.8.91 Valta Zampaki Avramoglou field Between squares K-KZ	4th BC	1	Wedge	0.055x0.005m
AV90	MA2267	20.7.88 Valta Zampaki Avramoglou field Trench II, section IE Depth 1.30-1.40m	Early Classical-archaic	1	Wedge	0.072m
AV91	MA4364	19.7.76 Trench A, section A3 Layer 0.03m		1	Spatula	0.10m
AV92	MA830	13.11.91 Classical enclosure Area of the West Gate Trench Δ, square Δ15 (under the layer with the roof tiles)		1	Chisel or scalpel	0.12m
AV93	MA766	Avdera 1963 South enclosure Amphorae storage room Depth 1-1.30m		1	Double axe	0.13x0.11m
AVD94	MA4443	Classical enclosure West Gate Trench Δ, layer 2nd, depth 3.16m		1	Double axe	0.25x0.024m
AV95	MA2269	11.7.88 Valta Zampaki Avramoglou field Trench II, section IΔ Depth 0.80-0.90m	Early Classical-archaic	1	Double axe	0.14x0.05m
AV96	MA1495	9.8.91 Valta Zampaki Avramoglou field Between squares ΚΣΤ–ΙΘ, αρ.6		1	Spit	0.045x0.014m
AV97	MA4454	13.8.76 Trench A, square A20 Layer 0.06m		1	Spit	0.026m
AV98	MA4454	13.8.76 Trench A, square A20		1	Spit	0.08m
AV99	MA6576	21.9.88 Valta Zampaki Kiosses field no 46 Section IA, layer 1		1	Spit	0.36m
AV100	MA971	29.5.91 Classical enclosure Roman house Trench ΣΤ, area Γ	Roman	1	Arrowhead	0.072x0.013m

AV201	MA1026	8.10.92 Ag. Panteleimonas West part of the South enclosure Trench Θ	Classical	1	Arrowhead (similar to AV35)	0.038x0.014m
AV202	MA1076	3.11.92 Ag. Panteleimonas West part of the South enclosure Trench Θ, ΔΤ7	Classical	1	Arrowhead (from catapult)	0.076x0.016m
AV203	MA1494	9.8.91 Valta Zampaki Avramoglou field Between squares ΚΣΤ–ΙΘ, αρ.6	Classical	1	Arrowhead (Av35)	0.04m
AV204	MA2085	10.6.92 Chatzopoulos field North enclosure	Archaic	1	Arrowhead	0.022m
AV205	MA4039	24.9.94 Valta Zampaki East of the North enclosure	Archaic	1	Spearhead	0.17x0.057m
AV206	MA5515	1952 Hellenistic building	Hellenistic	1	Spearhead	0.23x0.018m
AV207	MA5868	12.8.76 Trench M, section MZ		1	Spearhead	0.054m
AV208	MA6079	Δ11, cremation 1		1	Spearhead	0.147x0.033m
AV209	MA6100	M50a, burial 14		1	Large curved knife (machaira)	0.045x0.015m
AV210	MA6101	M59b, burial 14		1	Large curved knife (maxaira)	0.05x0.015m
AV211	MA6117	Δ186, Αβ.Τ22		1	Spearhead	0.029m
AV212	MA6295	8.9.87 North enclosure Trench II, section II A-B/25-26 Depth 1.45m	Archaic	1	Javelins' point	0.155m
AV213	MA 2701	19.8.78 Trench B, west city wall, section Λ Layer 0.30-0.50m		1	Key (Laconic type)	0.053m
AV214	MA2834	29.4.91 Trial trench 1 Surface find		1	Key	0.089m
AV215	MA4348	5.11.91 Classical enclosure Square Γ36 , layer 1st	Roman	1	Key	0.047x0.008m
AV216	MA4486	5.6.91 Trench E, square E8 Depth 3.22m		1	Key	0.065x0.023m
AV217	MA5293	12.10.83 Roman house Trench Β–Γ/2–3 Squares 20-22 Layer 0.35-0.70m	Roman	1	Key	

AV218	MA6246	T45, Δ12		1	Key	
AV219	MA6678	6.6.91 Classical enclosure Road in the west gate House of the dolphins Layer 1st	Hellenistic	1	Key	0.092x0.023m
AV220	MA801	13.11.91 Classical enclosure Trench Γ, square □37 Layer 1st		1	Jointer	0.064m
AV221	MA802	5-11.11.91 Classical enclosure Trench Γ, square Γ37 Layer 1st		4	Chain rings	0.03-0.07m
AV222	MA807	5.11.91 Classical enclosure Trench Γ, square Γ37 Layer 1st		2	Jointers	a.0.076m b.0.034m
AV223	MA897	24.5.91 Trench E, square E11 Layer 2nd		1	Chain ring	0.03m
AV224	MA905	5.6.91 Trench E, square E9 Layer 3rd		1	Bell	0.072x0.011m
AV225	MA923	21.11.91 Trench E, square E13 Layer 4th		1	Chain ring	0.081x0.009m
AV226	MA1152	19.12.84 Sazli Tsair Field of Metaxa-Georgiadis Burial mount T228 Sandstone sarcophagus		1	Jointer	0.051x0.04m
AV227	MA1322	14.9.89 Trench II, Valta Zampaki Avramoglou field, mo 58 Square KH Surface layer		1	Chain ring	0.036m
AV228	MA1702	18.6.92 Molos Lakioti Burial XV		6	Jointers	0.01-0.027m
AV229	MA 4350	17.9.76 Trench M, section MA5 Layer 0.40-0.60m		1	Jointer	0.064m
AV230	MA5854	23.9.77 Trench B. North tower		1	Chain ring	0.052m
AV231	MA800	13.11.91 Classical enclosure Trench Γ, square Γ3 Layer 8th		1	Striggil	0.08x0.016m
AV232	MA1139	13.7.82 Touzla Gkiol A, Bambakidis field Burial mount, chest burial		1	Striggil	0.22x0.032m
AV233	MA1143	12.12.83 Gkiouvem Tsair		1	Striggil	0.22x0.04m

		K. Imamidis field Chest burial				
AV234	MA1706	1.11.90 Molos, Lakiotis field Cremation no V	4th BC	1	Striggil	0.12x0.01m
AV235	MA1706	1.11.90 Molos, Lakiotis field Cremation no V	4th BC	1	Striggil	0.43x0.012m
AV236	MA1718	Lakiotis field Cremation-burial 22	4th BC	1	Striggil	
AV237		25.10.82 Sazli Tsair P. Averopoulos field Sandstone sarcophagus (from the right shoulder of the second individual)		1	Striggil	0.25x0.026m
AV238	MA2833	1982 Trench A, burial	4th BC	1	Striggil	
AV239	MA3089	1973 Burial pithos		1	Striggil	
AV240	MA3274	26.11.78 Touzla Gkiol, cemetery Chest burial		1	Striggil	0.072m
AV241	MA4831	18.11.94 E. Raxionis field NW part of the burial		1	Striggil	0.039x0.025m
AV242	MA5154	21.9.61 Trench Γ, square Γ5		1	Striggil	0.139x0.014m
AV243	MA353	24.10.90 Touzla Gkiol Burial mount, burial no V Sandstone sarcophagus		1	Ring	0.022m
AV244	MA786	10.6.91 Classical enclosure Trench G, square G5 Layer 5th		1	Ring	0.04m
AV245	MA1155	8.4.77 Ampelotopoi, cemetery B.Karapanagiotis field Burial mount T195 Sandstone sarcophagus		1	Ring	0.025m
AV246	MA6180	Δ123, T23		1	Ring	0.02m
AV247	MA6217	10.9.91 Valta Zampaki Chatzopoulos field Free standing burial, T34 Δ37		1	Ring	
AV248	MA6242	8.6.92 Valta Zampaki Chatzopoulos field T45 Δ8		1	Ring	0.021m
AV249	MA6762	27.6.96 T53, M34 From the left hand of the diceased.		1	Ring	0.027m
AV250	MA354-355	24.10.90 Touzla Gkiol Burial mount, burial no		1	Pin	0.07m

		V Sandstone sarcophagus				
AV251	MA373	1990 Touzla Gkiol Burial mount burial no X		2	Pin	a. 0.034m b. 0.031m
AV252	MA761	14.5.91 Classical enclosure Trench G, square G2 Layer 1st		1	Pin	0.04m
AV253	MA1077	2.10.92 Ag.Panteleemonas Trench Θ		1	Pin	0.008m
AV254	MA1145	12.12.83 Gkiouvem Tsair K. Imamidis field Chest burial		1	Pin	0.055m
AV255	MA6208	9.9.91 Valta Zampaki Chatzopoulos field Δ30 T32 keramoskephs		1	Pin	0.14m
AV256	MA6647	19.9.95 Kourou Tsesme, site 95.9 El.Raxionis field, Pithos burial		1	Pin	0.03m
AV257	MA350	1990 Touzla Gkiol Burial Mount, burial IV Chest burial		1	Nail	0.02m
AV258	MA759	10.5.91 Classical enclosure Trench Γ, square Γ17 Surface layer		3	Nail	0.02-0.03m
AV259	MA780-781	4.6.91 Classical enclosure Trench Γ, square Γ52 Surface layer		3	Nail	0.02-0.03m
AV260	MA783	5.6.91 Classical enclosure Trench Γ, square Γ3 Layer 5th		6	Nails	0.024-0.075m
AV261	MA787	7.7.92 Classical enclosure Trench Γ, square Γ1–Γ2 Layer 7th		6	Nails	0.023-0.093m
AV262	MA808	7-9.7.92 Classical enclosure Trench Γ, squares Γ3–Γ7–Γ9		8	Nails	0.036-0.078m
AV263	MA829	4.11.91 Classical enclosure, West Gate, Trench Δ, square Δ18, layer 5th		1	Nail	0.079m

Table 10. Kalyva- Kastro: list of industrial waste.

Sample no	Excav. No.	Context	Measurements	Description
KL 70	ΚΛΒ Ο7γ	13. 7. 88 square E3, North, out of tower Z.	d= 0.07 m w=153.4 gr.	Medium size, round shaped, heavy and magnetic slag
KL 71		15. 9. 89 square Δ6γ area δ	d=0.15 m w=1664.1gr	Large, round shaped, porous, heavy, magnetic. Bloom
KL 72	ΚΛΒ 036α	28. 8. .90 square Δ3γ area A1 stratum 6th	d=0..09m w=201. 5gr	Medium amorphous, porous, magnetic slag fragment.
KL 73	ΚΛΒ 026α	22. 8. 90 squares Δ7γ–Δ7δ area B2	d=0.07m w=13. 3gr	Medium size, amorphous, porous, magnetic slag fragment.
KL 74	ΚΛΒ 016β	21. 8. .90	d=0.03m w=14.7gr.	Small amorphous, no magnetic fragment of vitrified ceramic.
KL 75	ΚΛΒ 04β	22. 8. 90 squares Δ7γ–Δ7δ area B2 stratum 8th (east)	d=0.045m w=43.7gr	Large, shaped, porous, heavy. Bloom
KL 76	ΚΛΒ 04β	8. 8. 90 square Δ6α		Small, flat and compact, magnetic. Ingot fragment
KL 77	ΚΛΒ 030α	24. 8.90 square Δ7γ area A1	d=0.08m w=144.3gr	Small amorphous, porous, magnetic, slag fragment.
KL 78a	ΚΛΒ 024β	22. 8. 90 squares Δ7γ–Δ7δ area B2 stratum 10th (west).	d=0.13m	Large, shaped, porous, heavy, magnetic. Bloom
KL 78b	ΚΛΒ 024β	22. 8. 90 squares Δ7γ–Δ7δ area B2 stratum 10th (west).	d=0.03 - 0.07m Total weight=1391.9gr	13 small and medium size amorphous, porous, magnetic fragments.
KL 79	ΚΛΒ 025δ	4. 9. 90 square Γ5Γ	d=0.09 w=287.7gr.	Medium size, amorphous, porous, magnetic slag fragment.
KL 80	ΚΛΒ 032α	27. 8. 90 squares Δ7γ–Δ7δ area B2	d=0.16m w=1259..5gr.	Large, shaped, porous, heavy, magnetic. Bloom
KL 81		1990 surface find west of tower Γ	d=017m. w=967gr.	Large amorphous, porous, heavy, magnetic. Bloom
KL 82	ΚΛΒ 017α	20. 8. 90 squares Δ7γ–Δ7δ area B2 stratum 8th	d=0.04m w=15. 8gr	Small amorphous, porous, magnetic, slag fragment.
KL 83	ΚΛΒ 031α	24. 8.90 squares	d=0.03-0.10m	8 small and medium size amorphous, porous, magnetic fragments.

		Δ7γ–Δ7δ area B2 stratum 12th		
KL 84	ΚΛΒ 026γ	23. 8. 90 squares Δ7γ–Δ7δ area B2	d=0.01-0.04m Total weight=105. 4gr	6 small amorphous, porous, magnetic fragments.
KL 85	ΚΛΒ 016α	20. 8. 90 squares Δ7γ–Δ7δ area B2 stratum 8th (east)	d=0.02m w=5gr	Small amorphous, not magnetic fragment of vitrified ceramic.
KL 86	ΚΛΒ 9β	3. 9. 91 section between squares Γ5α–Γ5γ	d=0.08m w=253.8gr.	Medium, shaped, heavy and magnetic slag fragment.
KL 87	ΚΛΒ 032γ	22. 7. 91 Tower A stratum 14th (1. 30m from P14B)	d=0.09m d=0.02-0.05m Total weight = 512gr	1 shaped and 17 small, heavy and magnetic slag fragments.
KL 88	ΚΛΒ 013γ	8. 7. 91 square Γ5β area α4	d=0.04m w=42.5gr.	Small amorphous, porous, magnetic.
KL 89		1991	d=0.02-0.07m	5 small and medium size, amorphous, porous, magnetic slag fragments.
KL 90	ΚΛΒ 023ε	15. 7. 91 square Δ8γ area A	d=0.02-0.03m w=31gr	2 small amorphous, porous, magnetic slag fragments.
KL 91	ΚΛΒ 036Β	22. 7. 91 square Δ8δ area A	d=0.06m w=65. 7gr	Small amorphous, porous, magnetic.
KL 92	ΚΛΒ 014	5..9. 91 section between squares Γ5α–Γ4γ- east	d=0.03-0.05m	2 small amorphous, porous, magnetic slag fragments.
KL 93	ΚΛΒ 08γ	3. 7. 91 square Γ5β area α4	d=0.01 - 0.07m Total weight = 281.5gr	3 small amorphous, porous, magnetic slag fragments.
KL 94	ΚΛΒ 041	30. 7. 91 square Δ8δ area A	d=0.02 - 0.06m Total weight = 116.8gr	4 small amorphous, porous, magnetic slag fragments.
KL 95	ΚΛΒ 017γ	10. 7. 91 square Γ5β area α4	d=0.04 and 0.06m w= 105. 7gr	2 small amorphous, porous, magnetic slag fragments.
KL96	ΚΛΒ 041α	30. 7. 91 square Δ8δ area A	d=0.02 - 0.06m w= 80gr	3 small amorphous, porous, magnetic slag fragments.
KL 97	ΚΛΒ 011β	25. 6. 91 square Γ5β– 1	d=0.08m w=194.2gr.	Medium size amorphous, porous, magnetic slag fragment.
KL 98	ΚΛΒ 037β	23. 7. 91 square Δ8δ area A	d=0.05-0.06m w=88.4gr	2 medium size, shaped, heavy and magnetic slag fragments.
KL 99a	ΚΛΒ 01	4. 9. 91 section between squares Γ5α–Γ4γ	d=0.09m w=331. 8gr	Large shaped, porous, magnetic slag.
KL 99b	ΚΛΒ 01	4. 9. 91 section between	d=0.08m w=127.3gr	Medium size, amorphous, porous, magnetic slag.

		Context		Description
		squares Γ5α–Γ4γ		
KL 100	ΚΛΒ 05στ	28. 6. 91 square Γ5β area β3	d=0.5m	Small amorphous, porous, magnetic slag fragment.
KL 101	ΚΛΒ 02β	31. 7. 91 square Γ7β area α	d=0.06m w=200. 7gr	Medium, amorphous, compact, heavy and magnetic slag.
KL 102a	ΚΛΒ 03β	2. 8. 91 section between squares Γ5β–Γ4δ	d=0.08 w=96.1gr	Medium, amorphous, compact, heavy and magnetic slag.
KL 102b	ΚΛΒ 03β	2. 8. 91 section between squares Γ5β–Γ4δ	d=0.13	Large, shaped, porous, heavy, magnetic.

Table 11. Kalyva- Kastro: list of artefacts.

Sample no	Excavation no	Context	No of frag.	Description	Measurements
KL 1	ΚΛΒ 76, 14λγ (ΑΓΚ 2964)	9.7.76 Cistern	1	Single edged curved knife blade	0.14 x 0.25m
KL 2	ΚΛΒ5γ (ΑΓΚ 4747)	7.11.84 Cistern	2	Hooks	0.30 x 0.19m
KL 3	ΚΛΒ 5ι (ΑΓΚ 4743)	8.11.84 Cistern	1	Sickle	0.13 x 0.14m
KL 4	ΚΛΒ 3ε (ΑΓΚ 4752)	26.10.84 Cistern	2	Single edged curved knife blade	a. 0.09 x 0.015m b. 0.08 x 0.01m
KL 5	ΚΛΒ 5ξ (ΑΓΚ 4755)	9.11.84 Cistern	1	Bar	0.29 x 0.10m
KL 6	ΚΛΒ 5ν	9.11.84 Cistern	1	Bar	0.28 x 0.10m
KL 7	ΚΛΒ 5δ (ΑΓΚ 4745)	7.11.84 Cistern	1	Hook	0.35 x 0.03m
KL 8	ΚΛΒ 10β	2.10.86 room I NA wall	1	Arrowhead	0.05 x 0.02m
KL 9	ΚΛΒ 16γ	3.9.87 square Δ8α	2	Nails	a. 0.13 x 0.025m b. 0.07 x 0.03m
KL 10	ΚΛΒ18β	7.9.87 square Δ8α 1	1	Single edged curved knife blade	0.20 x 0.07m
KL 11	ΚΛΒ 16ζ	4.9.87 square Δ8α	1	Horse bit	0.10 x0.025m
KL 12	ΚΛΒ 17δ	4.9.87 square Δ8α	1	Single edged curved knife blade	0.13 x 0.015m
KL 13	ΚΛΒ 17α	4.9.87 square Δ8α	1	Key	0.15m
KL 14	ΚΛΒ 34β	16.9.87 square E9 SW gate	1	Single edged curved knife blade	0.14 x 0.025m
KL 15	ΚΛΒ 18α	7.9.87 square Δ8α 1	1	Jointer	0.15 x 0.005m
KL 16	ΚΛΒ 16ιγ	4.9.87 square Δ8α	1	Single edged Straight knife blade	0.15 x 0.045m

Sample no	Excav. no	Context	No of frag.	Description	Measurements
KL 17	KΛB 07δ	13.7.88 square E3 N of tower Z	1	Arrowhead	0.06 x 0.023m
KL 18	KΛB 07β	13.7.88 square E3 N of tower Z	1	Jointer	0.015 x 0.01m
KL 19	KΛB 28β	21.7.88 square B4 N area	1	Nail	0.19 x 0.02m
KL 20	KΛB 010γ	14.9.86 square Δ6γ area γ	1	Single edged Straight knife blade	0.13 x 0.035m
KL 21	KΛB 013β	14.9.86 square Δ6γ area δ	1	Pin	0.052 X 0.01m
KL 22	KΛB 048β	28.9.89 square Δ7α area δ, ζ	1	Single edged Straight knife blade	0.13 x 0.07m
KL 23	KΛB 015γ	22.9.89 square Δ7 γ, δ area B	1	Single edged curved knife blade	0.0155 x 0.015m
KL 24	KΛB 011α	4.9.89 square Δ7α area γ north	1	Key	0.08 x 0.07m
KL 25	KΛB 035α	15.9.89 square Δ7β area B	1	Sheep bell	0.12 x 0.07m
KL 26	KΛB 023γ	21.9.89 square Δ6γ area γ	1	Hook of chain	0.068m
KL 27	KΛB 017γ	26.9.89 square Δ + δ	1	Arrowhead	0.053 x 0.012m
KL 28	KΛB034α	13.9.89 square Δ7α area γ 2	1	Hook	0.038m
KL 29	KΛB 035β	15.9.89 square Δ7β area β	1	Single edged small triangular blade (scissors)	0.095 x 0.023m
KL 30	KΛB 049B	28.9.89 square Δ7α wall Δ	1	Clamps	0.0125 x 0. 03m
KL 31	KΛB 017β	22.9.89 square Δ7δ area Γ	1	Jointer	0.08 x 0. 04m
KL 32	KΛB 046B	27.9.89 square Δ7α–β wall Γ	1	Single edged Straight knife blade	0.07 x 0.015m
KL 33	KΛB 052α	29.9.89 square Δ6γ area ε (pithos)	1	Sickle	0.16 x 0.35m
KL 34	KΛB 12η	29.9.89 square Γ4γ	1	Key	0.07 x 0.02m
KL 35	KΛB 036β	18.9.89 square Δ7β area B	1	Hook	0.045m

KL 36	ΚΛΒ 020α	16. 8. 90 square Δ6α area B	2	Sheets	a. 0.16 x 0.02m b. 0.058 x 0.02m
KL 37	ΚΛΒ 02γ	21. 8. 90 square Δ5γ 1	1	Single edged Straight knife blade	0.06 x 0. 035m
KL 38	ΚΛΒ 024α	3.9. 90 square Δ5γ 5	1	Key	0.08 x 0.02m
KL 39	ΚΛΒ 016β	14.8. 90 square Δ6α 5	1	Single edged Straight knife blade	0.09 x 0.025m
KL 40	ΚΛΒ 024δ	3.9. 90 square Δ5γ 5	1	Jointer	0.11 x 0.1m
KL 41	ΚΛΒ 016δ	14.8. 90 square Δ6α 5	1	Single edged Straight knife blade	0.075 x 0.015m
KL 42	ΚΛΒ 07α	24.8. 90 square Δ5δ 1	1	Hook	0.065m
KL 43	ΚΛΒ 07α	10. 8. 90 square Γ6β 3 area B	1	Sheet	0.45 x 0.02m
KL 44	ΚΛΒ 24α	5.9. 90 square E10β	1	Key	0.12 x 0.25m
KL 45	ΚΛΒ 17α	28.8. 90 square Z10α	1	Pin	0.067 x 0.08m
KL 46	ΚΛΒ 06β	22.8. 90 square Δ5γ 2	1	Single edged Straight knife blade	0.14 x 0.035m
KL47	ΚΛΒ 010β	10.8. 90 square Δ6α 4	1	Hook	0.04 x 0.01m
KL 48	ΚΛΒ 03ε	10.9. 91 square Δ6B layer 2	1	Single edged Straight knife blade	0.11 x 0.015m
KL 49	ΚΛΒ 03β	18.7.91 Tower A Layer 11th depth 1.19m	1	Single edged small triangular blade (scissors)	0.0105 x 0.03m
KL 50	ΚΛΒ 02α	17.7.91 Tower A (north) Layer 8th depth 0. 97m	1	Single edged Straight knife blade	0.13 x 0.03m
KL 51	ΚΛΒ 5α	24.7.91 square E9γ	4	Nails	0.13 - 0.08 x 0.007m
KL 52	ΚΛΒ 019α	11.7.91 square Γ5β 6	1	Sheet	0.08 x 0.06m
KL 53	ΚΛΒ 024γ	9.9.91 square Δ6β 1	1	Key	0.08 x 0.07m
KL 54	ΚΛΒ 04ε	11.9.91 square Δ6B layer 3	1	Nail	0.11 x 0.04m
KL 55	ΚΛΒ 03στ	18.7.91 Tower A Layer 12 depth 1.34m	1	Axe	0.08 x 0.06m
KL 56	ΚΛΒ 4δ	27.7.91 square E9γ 1	1	Hook	0.05m
KL 57	ΚΛΒ 04α	22.7.91 Tower A, Gate Layer 6th depth 1.35m	1	Clamp	0.12 x 0.015m

KL 58	ΚΛΒ 01α	25.6.91 square Δ8δ 1 area A	1	Nail	0.14 x 0. 01m
KL 59	ΚΛΒ 088γ	16.7.91 square Δ8α area A	1	Single edged small triangular blade (scissors)	0.07 x 0.02m
KL 60	ΚΛΒ 1ε	28.8.91 square Δ9β	1	Spearhead	0.18 x 0.02m
KL 61	ΚΛΒ 3α	18.7.91 Tower A Layer 11th depth 1.22m	1	Single edged Straight knife blade	0.0145 x 0.015m
KL 62	ΚΛΒ 2β	2.9.91 square Δ9β	1	Single edged Straight knife blade	0.08 x 0.015m
KL 63	ΚΛΒ 06α	4.8.92 square E8γ 3	1	Key	0.15m
KL 64	ΚΛΒ 05στ	3.8.92 square E8β 3	1	Billhook	0.12 x 0.05m
KL 65	ΚΛΒ 010α	6.8.92 square E8γ 4	2	Single edged curved knife blade	0.12 x 0.15m
KL 66	ΚΛΒ 05η	3.8.92 square E8γ 3	1	Single edged large triangular knife blade	0.17 x 0.04m
KL 67	ΚΛΒ 05α	3.8.92 square E8γ 3	1	Single edged large triangular knife blade	0.31 x 0.14m
KL 68	ΚΛΒ 092β	18.11.85 Cistern	1	Axe	0.25 x 0.06m

Table 12. SEM data from Messemvria – Zone (all analyses in weight %).

Analyses	**Na_2O**	**K_2O**	**Al_2O_3**	**SiO_2**	**MgO**	**CaO**	**TiO_2**	**MnO**	**FeO**	**P_2O_5**	**SO_3**	**Cr_2O_3**	**As_2O_3**	**NiO**	**ZnO**	**CuO**	**BaO**
MES 101																	
A	0.15	0.97	6.18	26.65	0.27	4.67	0.16	0.32	60.10	0.11	0.15	0.07	n.d.	0.07	0.02	0.07	0.46
A	0.04	0.80	5.04	26.41	0.34	3.90	0.19	0.26	62.61	0.06	0.13	0.09	n.d.	0.02	0.05	0.06	n.d.
mean A	0.09	0.89	5.61	26.53	0.31	4.29	0.18	0.29	61.36	0.08	0.14	0.08	n.d.	0.05	0.04	0.06	0.23
W	n.d.	n.d.	1.77	0.62	0.08	0.06	0.70	0.07	96.37	n.d.	0.07	0.11	n.d.	n.d.	n.d.	n.d.	n.d.
W	0.25	0.03	0.21	1.36	0.02	0.11	0.19	0.02	97.30	0.03	0.01	0.12	0.13	n.d.	0.01	0.06	0.15
mean W	0.13	0.02	0.99	0.99	0.05	0.09	0.45	0.05	96.84	0.01	0.04	0.12	0.06	n.d.	n.d.	0.03	0.07
F	0.01	0.01	0.14	26.16	0.46	1.06	0.02	0.56	71.35	0.05	0.03	0.07	n.d.	n.d.	n.d.	0.03	0.06
F	0.07	n.d.	0.23	25.95	0.39	1.22	n.d.	0.49	70.98	0.27	0.07	0.06	n.d.	n.d.	0.07	n.d.	0.21
mean F	0.04	n.d.	0.18	26.05	0.43	1.14	0.01	0.52	71.16	0.16	0.05	0.06	n.d.	n.d.	0.03	0.01	0.13
Ti, Al rich F	0.03	0.16	13.59	24.29	0.04	13.67	0.90	0.18	46.09	0.36	0.09	0.01	0.20	n.d.	n.d.	n.d.	0.41
Ti, Al rich F	n.d.	0.46	14.60	27.36	n.d.	14.16	1.28	0.05	40.92	0.44	0.08	0.05	0.02	n.d.	0.14	n.d.	0.44
mean TI, Al rich F	0.01	0.31	14.09	25.82	0.02	13.92	1.09	0.11	43.50	0.40	0.09	0.03	0.11	n.d.	0.07	n.d.	0.42
I	0.32	4.83	13.07	39.20	n.d.	16.54	0.17	0.11	24.28	0.87	0.37	0.01	n.d.	n.d.	0.06	0.06	0.11
I	0.17	5.83	14.24	39.47	n.d.	16.27	0.17	0.12	21.64	1.64	0.60	0.08	0.03	n.d.	n.d.	0.07	0.25
mean I	0.25	5.33	13.66	39.33	n.d.	16.40	0.17	0.11	22.96	1.26	0.49	0.05	0.02	n.d.	n.d.	0.06	0.18
matte	n.d.	0.36	2.64	6.33	n.d.	2.72	0.11	0.03	16.04	0.03	29.79	0.02	0.04	0.12	n.d.	41.74	0.04
matte	n.d.	0.09	0.15	0.67	n.d.	0.43	0.02	0.02	16.31	0.06	37.29	n.d	n.d.	0.10	n.d.	44.80	0.05
mean matte	n.d.	0.22	1.40	3.50	n.d.	1.58	0.07	0.03	16.17	0.04	33.54	0.01	0.02	0.11	n.d.	43.27	0.05
MES 111																	
A	n.d.	1.43	4.10	21.52	0.76	3.03	n.d.	2.88	64.79	0.85	0.10	0.06	n.d.	n.d.	0.05	0.07	0.38
A	0.08	0.49	3.19	18.45	0.71	3.86	n.d.	2.23	69.30	1.04	0.13	0.03	0.05	n.d.	0.03	0.15	0.26
mean A	0.04	0.96	3.64	19.98	0.73	3.45	n.d.	2.55	67.04	0.95	0.11	0.04	0.02	n.d.	0.04	0.11	0.32
F	n.d.	0.41	0.59	27.67	2.76	2.12	n.d.	6.72	58.60	0.85	0.06	0.04	0.02	n.d.	n.d.	n.d.	0.19
F	0.12	0.55	0.86	27.98	1.54	3.86	n.d.	6.08	58.12	0.50	0.02	n.d	0.02	n.d.	0.07	n.d.	0.15
mean F	0.06	0.48	0.72	27.82	2.15	2.99	n.d.	6.40	58.36	0.67	0.04	0.02	0.02	n.d.	0.04	n.d.	0.17
I	0.83	7.97	10.15	34.14	n.d.	1.16	0.33	1.62	27.29	3.10	0.49	0.03	n.d.	n.d.	n.d.	0.14	0.76
I	0.96	8.30	12.57	35.23	0.03	12.52	0.31	1.24	24.23	3.30	0.53	0.05	0.01	0.01	n.d.	0.01	0.71
mean I	0.89	8.14	11.36	34.69	0.01	6.84	0.32	1.43	25.76	3.20	0.51	0.04	n.d.	n.d.	n.d.	0.07	0.74
Ore	0.03	0.21	1.26	9.07	0.41	1.41	n.d.	1.12	85.18	0.57	0.15	0.12	n.d.	n.d.	0.18	0.15	0.15
Prill	0.04	0.03	n.d.	0.08	0.15	0.02	0.01	0.06	99.39	n.d.	0.07	n.d	0.06	n.d.	n.d.	0.10	n.d.

Analyses	Na_2O	K_2O	Al_2O_3	SiO_2	MgO	CaO	TiO_2	MnO	FeO	P_2O_5	SO_3	Cr_2O_3	As_2O_3	NiO	ZnO	CuO	BaO
MES 124																	
A	0.47	0.48	2.70	10.83	1.72	12.67	n.d.	2.32	67.31	0.93	0.13	0.01	0.21	0.01	n.d.	0.04	0.16
A	0.29	0.56	3.41	12.33	1.85	14.68	0.08	2.11	63.21	1.08	0.25	n.d	n.d.	0.02	0.05	0.02	0.06
mean A	0.38	0.52	3.06	11.58	1.79	13.67	0.04	2.21	65.26	1.01	0.19	n.d	0.11	0.02	0.03	0.03	0.11
W	0.13	n.d.	0.48	0.07	2.25	0.30	0.06	3.03	93.32	n.d.	0.04	n.d	0.22	0.09	n.d.	n.d.	0.03
W	0.30	0.01	0.63	0.19	2.07	0.27	n.d.	2.72	93.32	0.01	0.12	n.d	0.33	n.d.	n.d.	n.d.	0.14
mean W	0.22	n.d.	0.55	0.13	2.16	0.28	0.03	2.88	93.32	n.d	0.08	n.d	0.27	0.04	n.d	n.d	0.08
F	0.36	0.87	10.20	26.58	2.15	37.87	0.14	1.55	18.42	1.07	0.54	0.05	n.d.	0.05	n.d.	0.09	0.07
F	0.19	0.05	0.31	28.56	2.43	37.82	0.04	2.41	26.87	1.04	n.d.	0.06	n.d.	n.d.	0.06	0.05	0.11
mean F	0.28	0.46	5.25	27.57	2.29	37.85	0.09	1.98	22.64	1.06	0.27	0.05	n.d.	0.03	0.03	0.07	0.09
I	1.22	12.77	17.62	23.24	0.71	11.67	0.46	0.79	27.49	1.05	1.82	0.03	0.15	n.d.	0.01	0.09	0.51
I	0.52	0.97	21.41	26.36	0.99	41.11	0.05	0.43	7.08	1.05	0.20	0.05	0.11	0.05	n.d.	0.08	0.18
mean I	0.87	6.87	19.51	24.80	0.85	26.39	0.26	0.61	17.28	1.05	1.01	0.04	0.13	0.03	n.d.	0.09	0.35
Prill	0.14	n.d.	n.d.	0.10	0.14	0.15	n.d.	0.02	97.64	1.05	n.d.	0.03	0.50	0.91	n.d.	0.35	n.d.
MES 125																	
A	0.19	2.39	6.28	37.63	0.40	4.16	0.29	1.01	46.27	0.85	0.12	n.d	n.d.	0.09	n.d.	0.04	0.28
A	0.21	2.54	6.97	39.78	0.23	3.45	0.35	0.94	44.21	0.77	0.12	0.01	n.d.	n.d.	0.04	0.12	0.27
mean A	0.20	2.47	6.63	38.70	0.32	3.81	0.32	0.98	45.24	0.81	0.12	n.d.	n.d.	0.04	0.02	0.08	0.28
W																	
F	n.d.	0.21	0.36	27.73	0.87	0.69	0.02	1.87	67.40	0.34	0.02	0.07	0.36	n.d.	0.04	0.09	n.d.
F	0.02	n.d.	n.d.	27.14	1.61	0.39	n.d.	1.95	68.47	0.16	0.08	0.08	n.d.	n.d.	0.05	n.d.	0.04
mean F	0.01	0.11	0.18	27.43	1.24	0.54	0.01	1.91	67.93	0.25	0.05	0.08	0.18	n.d.	0.05	0.05	0.02
I	0.54	5.66	13.19	51.44	n.d.	8.91	0.47	0.28	17.10	1.35	0.39	0.04	n.d.	0.05	0.02	0.06	0.51
I	0.40	4.16	12.68	51.99	n.d.	8.18	0.27	0.53	19.37	1.67	0.24	n.d	n.d.	n.d.	n.d.	n.d.	0.51
mean I	0.47	4.91	12.93	51.71	n.d.	8.54	0.37	0.41	18.23	1.51	0.32	0.02	n.d.	0.03	0.01	0.03	0.51
MES 129																	
A	0.30	0.27	1.10	2.15	0.30	0.40	n.d.	69.15	2.01	0.04	0.07	0.61	0.30	0.15	0.11	n.d.	23.08
A	n.d	0.64	2.76	33.16	n.d.	4.13	n.d.	41.81	1.89	0.30	0.01	0.27	n.d.	n.d.	n.d	0.13	14.90
A	0.19	3.82	9.18	72.40	0.44	8.47	0.53	0.29	2.59	1.18	0.16	n.d.	n.d.	n.d.	0.20	0.06	0.49
mean A	0.16	1.58	4.35	35.90	0.25	4.33	0.18	37.08	2.16	0.51	0.08	0.29	0.10	0.05	0.11	0.06	12.82
MES 130																	
A	0.16	2.28	4.18	17.86	0.93	26.58	0.07	0.26	46.14	0.75	0.08	0.09	0.05	n.d.	n.d	0.29	0.29
A	0.24	2.27	4.34	18.02	0.96	26.76	0.09	0.19	45.67	0.82	0.13	0.05	0.19	n.d.	n.d	0.08	0.18
mean A	0.20	2.28	4.26	17.94	0.95	26.67	0.08	0.22	45.90	0.79	0.11	0.07	0.12	n.d.	n.d	0.19	0.23
W	0.08	0.09	0.94	0.25	1.15	0.48	0.24	0.49	95.99	n.d.	0.03	0.09	0.05	n.d.	0.04	0.05	0.06

Analyses	Na_2O	K_2O	Al_2O_3	SiO_2	MgO	CaO	TiO_2	MnO	FeO	P_2O_5	SO_3	Cr_2O_3	As_2O_3	NiO	ZnO	CuO	BaO
W	0.19	0.06	0.83	0.26	1.28	0.47	0.09	0.53	95.81	0.05	0.05	0.07	0.22	n.d.	n.d	n.d.	0.10
mean W	0.13	0.08	0.88	0.25	1.21	0.47	0.17	0.51	95.90	0.03	0.04	0.08	0.13	n.d.	0.02	0.02	0.08
F	0.68	20.59	18.09	24.98	0.38	3.84	0.41	0.11	30.11	0.24	0.04	0.02	n.d.	0.07	n.d	0.12	0.34
I	n.d.	0.57	0.09	30.45	1.60	59.25	0.03	0.05	7.02	0.84	0.07	0.02	n.d.	n.d.	n.d	n.d.	n.d
I	0.08	0.64	0.08	30.57	1.74	58.44	n.d.	0.13	7.03	0.99	n.d.	0.04	n.d.	0.02	n.d	0.09	0.15
mean I	0.04	0.60	0.08	30.51	1.67	58.85	0.02	0.09	7.03	0.92	0.04	0.03	n.d.	0.01	n.d	0.05	0.07
MES 146b																	
A	0.94	0.88	15.29	60.26	1.68	4.64	1.04	0.19	14.91	n.d.	0.04	n.d.	n.d.	0.04	n.d	0.08	n.d
A	0.91	1.22	14.32	65.40	1.61	5.27	1.05	0.47	9.57	0.14	n.d.	n.d.	n.d.	0.05	n.d	n.d.	n.d
mean A	0.93	1.05	14.80	62.83	1.64	4.95	1.04	0.33	12.24	0.07	0.02	n.d.	n.d.	0.05	n.d.	0.04	n.d.
Prill	n.d	n.d	n.d	0.22	0.06	n.d.	0.06	n.d.	96.91	1.35	0.02	n.d.	0.49	0.36	0.11	0.43	n.d
Prill	0.13	0.01	n.d	0.24	0.14	0.09	n.d.	n.d.	96.72	1.35	0.01	0.03	0.49	0.24	n.d	0.44	0.11
mean prill	0.07	n.d	n.d	0.23	0.10	0.04	0.03	n.d.	96.82	1.35	0.02	0.02	0.49	0.30	0.06	0.43	0.06
MES 146c																	
A	0.32	2.36	6.96	38.77	0.47	3.48	0.37	1.03	44.95	0.63	0.16	0.05	0.14	n.d.	n.d	0.04	0.26
A	0.18	2.27	6.28	35.81	0.27	4.10	0.21	1.06	48.42	0.77	0.16	0.06	n.d.	n.d.	0.12	0.06	0.24
mean A	0.25	2.31	6.62	37.29	0.37	3.79	0.29	1.05	46.68	0.70	0.16	0.05	0.07	n.d.	0.06	0.05	0.25
F	0.02	0.56	0.61	28.71	0.65	1.46	0.05	1.92	65.47	0.34	n.d.	0.02	0.02	n.d.	n.d	0.10	0.08
F	0.15	0.03	0.06	26.63	1.13	0.48	n.d.	1.73	69.22	0.16	0.06	0.05	0.07	n.d.	0.14	0.04	0.07
mean F	0.08	0.29	0.33	27.67	0.89	0.97	0.03	1.82	67.34	0.25	0.03	0.03	0.05	n.d.	0.07	0.07	0.07
I	0.46	5.27	13.50	53.03	n.d.	9.43	0.09	0.34	15.50	1.44	0.21	0.02	0.04	0.03	n.d	0.23	0.42
I	0.19	4.36	12.93	54.24	0.06	8.68	0.13	0.36	16.25	1.56	0.47	0.12	n.d.	n.d.	0.07	0.05	0.55
mean I	0.33	4.81	13.21	53.64	0.03	9.05	0.11	0.35	15.88	1.50	0.34	0.07	0.02	0.01	0.04	0.14	0.48
MES 146d																	
A	0.12	4.35	11.56	52.79	0.89	20.17	0.55	0.51	7.93	0.55	0.04	0.11	n.d.	n.d.	0.18	n.d.	0.27
A	0.25	4.98	10.67	52.25	0.97	7.10	0.46	1.30	20.65	0.95	n.d.	n.d.	n.d.	0.04	0.16	n.d.	0.23
mean A	0.19	4.66	11.12	52.52	0.93	13.63	0.51	0.90	14.29	0.75	0.02	0.05	n.d.	0.02	0.17	n.d.	0.25
W	0.15	1.26	4.75	12.76	2.02	1.37	0.38	2.62	74.13	0.23	n.d.	0.06	n.d.	n.d.	0.02	n.d.	0.26
W	0.15	1.57	6.43	19.68	2.06	2.00	0.36	2.53	64.46	0.31	n.d.	0.05	0.05	n.d.	n.d.	0.05	0.29
mean W	0.15	1.42	5.59	16.22	2.04	1.69	0.37	2.57	69.29	0.27	n.d.	0.06	0.02	n.d.	0.01	0.03	0.27
I	0.17	5.24	12.68	57.39	1.09	10.55	0.39	1.09	9.60	1.26	0.06	0.03	n.d.	0.13	n.d.	n.d.	0.32
I	n.d.	5.35	12.15	56.07	0.93	9.63	0.42	1.32	12.63	1.24	n.d.	0.02	0.04	n.d.	n.d.	n.d.	0.20
mean I	0.08	5.30	12.41	56.73	1.01	10.09	0.41	1.20	11.11	1.25	0.03	0.03	0.02	0.07	n.d.	n.d.	0.26

Analyses	Na_2O	K_2O	Al_2O_3	SiO_2	MgO	CaO	TiO_2	MnO	FeO	P_2O_5	SO_3	Cr_2O_3	As_2O_3	NiO	ZnO	CuO	BaO
MES 147b																	
A	0.59	1.77	5.37	30.39	0.81	3.79	0.18	0.59	54.91	0.90	0.24	0.07	n.d.	n.d.	0.13	0.15	0.11
A	0.35	1.57	4.55	29.99	0.94	3.43	0.10	0.60	57.16	0.87	0.16	0.08	0.06	n.d.	n.d.	0.09	0.06
mean A	0.47	1.67	4.96	30.19	0.88	3.61	0.14	0.60	56.04	0.89	0.20	0.07	0.03	n.d.	0.06	0.12	0.08
F	n.d.	0.03	0.09	27.01	1.14	0.84	n.d.	0.78	69.62	0.20	n.d.	0.06	0.10	n.d.	n.d.	n.d.	0.16
F	n.d.	0.03	0.10	27.18	1.76	0.73	n.d.	0.75	68.97	0.13	n.d.	0.04	0.30	n.d.	n.d.	n.d.	0.02
mean F	n.d.	0.03	0.09	27.10	1.45	0.78	n.d.	0.77	69.29	0.16	n.d.	0.05	0.20	n.d.	n.d.	n.d.	0.09
I	1.09	5.06	12.04	37.07	0.05	11.35	0.60	0.17	28.77	2.27	0.74	0.11	n.d.	n.d.	0.06	n.d.	0.63
I	1.13	3.93	10.22	35.45	n.d.	7.54	0.36	0.35	38.39	1.71	0.37	0.07	n.d.	n.d.	0.04	0.04	0.42
mean I	1.11	4.50	11.13	36.26	0.03	9.45	0.48	0.26	33.58	1.99	0.55	0.09	n.d.	n.d.	0.05	0.02	0.53
MES 21																	
W	0.02	0.05	0.45	0.40	0.13	0.09	0.53	0.51	97.46	n.d.	n.d.	0.11	0.05	0.15	0.04	0.01	n.d.
F	0.26	0.05	0.59	24.97	0.97	0.76	n.d.	2.09	69.66	0.14	n.d.	n.d.	n.d.	0.15	0.09	0.09	0.17
F	0.09	0.03	0.52	25.51	0.71	0.69	0.08	2.09	69.96	0.05	n.d.	0.04	0.16	0.07	n.d.	0.02	n.d.
mean F	0.17	0.04	0.55	25.24	0.84	0.72	0.04	2.09	69.81	0.09	n.d.	0.02	0.08	0.11	0.05	0.06	0.09
I	n.d.	0.11	2.54	11.39	0.10	1.24	0.07	1.02	82.05	0.39	0.74	n.d.	n.d.	0.18	0.10	0.07	n.d.
I	n.d.	0.10	2.09	19.37	0.09	0.33	n.d.	0.47	76.32	0.06	0.62	n.d.	0.32	0.07	n.d.	n.d.	0.09
mean I	n.d.	0.10	2.31	15.38	0.09	0.79	0.04	0.75	79.18	0.22	0.68	n.d.	0.16	0.13	0.05	0.04	0.04
M	0.09	0.02	n.d	0.06	0.13	n.d.	n.d.	n.d.	99.08	n.d.	n.d.	0.20	0.17	0.15	0.06	0.17	0.07
MES 23																	
W	0.24	0.27	1.91	3.06	0.12	0.32	0.53	0.12	93.19	0.10	0.07	0.02	n.d.	n.d.	0.06	n.d.	n.d.
W	0.06	0.17	0.96	1.59	n.d.	0.15	0.57	n.d.	95.98	n.d.	0.81	0.04	0.21	n.d.	0.07	0.12	n.d.
mean W	0.15	0.22	1.44	2.32	0.06	0.24	0.55	0.06	94.58	0.05	0.44	0.03	0.11	n.d.	0.07	0.06	n.d.
F	n.d.	0.13	1.39	24.19	0.67	0.80	n.d.	0.31	72.44	0.02	0.01	0.01	n.d.	0.02	0.01	n.d.	n.d.
F	0.09	0.09	0.90	24.67	0.57	0.95	0.11	0.26	71.97	0.16	0.03	n.d.	0.02	0.05	n.d.	0.12	n.d.
mean F	0.05	0.11	1.15	24.43	0.62	0.87	0.06	0.29	72.20	0.09	0.02	n.d.	0.01	0.04	n.d.	0.06	n.d.
I	0.20	2.46	5.21	25.53	0.45	3.11	0.10	0.17	61.11	1.20	0.15	0.11	0.03	0.13	0.02	n.d.	0.03
M	0.06	n.d	0.02	0.09	n.d.	n.d.	0.03	n.d.	99.51	0.05	0.06	n.d.	0.02	0.11	n.d.	0.05	n.d.
MES 25																	
W	n.d.	0.10	0.73	0.99	0.28	0.48	0.30	0.10	96.51	0.20	0.11	n.d.	0.02	0.13	n.d.	n.d.	0.04
F	n.d.	0.13	0.53	27.49	3.55	2.70	n.d.	0.79	64.52	0.20	0.05	n.d.	n.d.	n.d.	0.05	n.d.	n.d.
I	0.37	3.33	6.13	26.65	0.61	10.52	0.07	0.46	47.26	3.08	0.96	0.13	0.07	0.08	n.d.	0.01	0.29
M	n.d.	n.d.	n.d.	0.12	0.02	0.05	n.d.	0.04	99.40	n.d.	0.02	n.d.	0.04	0.06	n.d.	0.20	0.05

Analyses	Na_2O	K_2O	Al_2O_3	SiO_2	MgO	CaO	TiO_2	MnO	FeO	P_2O_5	SO_3	Cr_2O_3	As_2O_3	NiO	ZnO	CuO	BaO
MES 33																	
W	n.d.	0.32	0.40	1.88	0.28	0.55	0.02	0.24	95.90	0.01	0.10	n.d.	0.10	n.d.	n.d.	n.d.	0.20
F	0.24	3.57	4.39	33.10	0.68	8.31	0.01	0.60	46.01	1.14	1.25	0.02	n.d.	0.13	0.05	0.02	0.50
I	0.25	4.81	5.47	43.70	1.26	9.87	0.12	1.01	32.50	n.d.	0.14	0.07	n.d.	n.d.	n.d.	0.01	0.80
M	0.10	0.06	n.d.	0.17	0.24	0.02	n.d.	n.d.	98.47	0.05	0.09	n.d.	0.72	n.d.	0.04	n.d.	0.05
MES 44																	
I	0.33	2.06	4.37	30.75	0.36	17.43	0.15	0.22	43.66	0.11	0.17	0.05	n.d.	0.07	0.02	0.05	0.17
I	n.d.	1.54	3.27	21.28	0.23	7.98	0.08	1.61	57.55	5.88	0.17	0.05	0.07	n.d.	n.d.	0.07	0.23
mean I	0.16	1.80	3.82	26.02	0.29	12.70	0.12	0.92	50.60	3.00	0.17	0.05	0.03	0.03	0.01	0.06	0.20
M	0.05	n.d.	n.d.	0.13	0.04	n.d.	n.d.	n.d.	97.48	0.05	0.04	0.04	0.33	0.03	0.04	1.78	n.d.
M	n.d.	n.d.	n.d.	0.04	n.d.	0.41	n.d.	n.d.	98.43	n.d.	0.06	0.08	n.d.	n.d.	0.05	1.31	n.d.
mean M	0.03	n.d.	n.d.	0.09	0.02	0.21	n.d.	n.d.	97.96	0.02	0.05	0.06	0.16	0.02	0.04	1.54	n.d.
MES 53																	
I	1.61	7.78	8.60	43.69	3.06	16.66	0.27	4.56	13.31	0.15	0.17	0.02	n.d.	n.d.	n.d.	n.d.	0.10
I	1.92	6.59	7.92	39.64	2.44	14.75	0.29	4.59	21.15	0.33	0.14	n.d.	0.15	n.d.	n.d.	0.03	0.08
mean I	1.76	7.18	8.26	41.66	2.75	15.70	0.28	4.58	17.23	0.24	0.15	0.01	0.07	n.d.	n.d.	0.02	0.09
M	0.20	n.d.	0.05	0.09	0.31	n.d.	n.d.	n.d.	99.04	0.04	0.05	0.06	0.01	0.02	n.d.	0.10	n.d.
M	n.d.	n.d.	n.d.	0.02	n.d.	0.04	n.d.	n.d.	99.82	0.01	0.01	n.d.	n.d.	0.05	0.05	n.d.	n.d.
mean M	0.10	n.d.	0.03	0.05	0.16	0.02	n.d.	n.d.	99.43	0.03	0.03	0.03	n.d.	0.03	0.02	0.05	n.d.
MES 62																	
W	0.14	0.08	0.40	0.33	0.19	0.08	0.22	1.29	96.96	n.d.	0.11	0.02	n.d.	0.04	0.05	0.09	n.d.
I	0.58	3.64	5.63	31.05	0.39	7.70	0.03	3.22	44.01	2.83	0.28	n.d.	0.26	0.07	0.02	n.d.	0.30

Area (A) = mean over 2 areas x100
Wustite (W)= mean over 2 spots x2K
Fayalite (F) = mean over 2 spots x2K
Interstitial (I)= mean over 2 spots x2K
Metal (M) = mean over 2 spots x2K
Other (O)= mean over 2 spots x2K
n.d. = not determoned

Table 12b. SEM data fromMessemvria-Zone (summary of maen analyses from table 12a).

Analyses	Na_2O	K_2O	Al_2O_3	SiO_2	MgO	CaO	TiO_2	MnO	FeO	P_2O_5	SO_3	As_2O_3	NiO	ZnO	CuO	BaO
MES 101																
A	0.09	0.89	5.61	26.53	0.31	4.29	0.18	0.29	61.36	0.08	0.14	n.d.	0.05	0.04	0.06	0.23
W	0.13	0.02	0.99	0.99	0.05	0.09	0.45	0.05	96.84	0.01	0.04	0.06	n.d.	n.d.	0.03	0.07
F	0.04	n.d.	0.18	26.05	0.43	1.14	0.01	0.52	71.16	0.16	0.05	n.d.	n.d.	0.03	0.01	0.13
TI, Al rich F	0.01	0.31	14.09	25.82	0.02	13.92	1.09	0.11	43.50	0.40	0.09	0.11	n.d.	0.07	n.d.	0.42
I	0.25	5.33	13.66	39.33	n.d.	16.40	0.17	0.11	22.96	1.26	0.49	0.02	n.d.	n.d.	0.06	0.18
matte	n.d.	0.22	1.40	3.50	n.d.	1.58	0.07	0.03	16.17	0.04	33.54	0.02	0.11	n.d.	43.27	0.05b
MES 111																
A	0.04	0.96	3.64	19.98	0.73	3.45	n.d.	2.55	67.04	0.95	0.11	0.02	n.d.	0.04	0.11	0.32
F	0.06	0.48	0.72	27.82	2.15	2.99	n.d.	6.40	58.36	0.67	0.04	0.02	n.d.	0.04	n.d.	0.17
I	0.89	8.14	11.36	34.69	0.01	6.84	0.32	1.43	25.76	3.20	0.51	n.d.	n.d.	n.d.	0.07	0.74
Ore	0.03	0.21	1.26	9.07	0.41	1.41	n.d.	1.12	85.18	0.57	0.15	n.d.	n.d.	0.18	0.15	0.15
Prill	0.04	0.03	n.d.	0.08	0.15	0.02	0.01	0.06	99.39	n.d.	0.07	0.06	n.d.	n.d.	0.10	n.d.
MES 124																
A	0.38	0.52	3.06	11.58	1.79	13.67	0.04	2.21	65.26	1.01	0.19	0.11	0.02	0.03	0.03	0.11
W	0.22	n.d.	0.55	0.13	2.16	0.28	0.03	2.88	93.32	n.d	0.08	0.27	0.04	n.d	n.d	0.08
F	0.28	0.46	5.25	27.57	2.29	37.85	0.09	1.98	22.64	1.06	0.27	n.d.	0.03	0.03	0.07	0.09
I	0.87	6.87	19.51	24.80	0.85	26.39	0.26	0.61	17.28	1.05	1.01	0.13	0.03	n.d.	0.09	0.35
Prill	0.14	n.d.	n.d.	0.10	0.14	0.15	n.d.	0.02	97.64	1.05	n.d.	0.50	0.91	n.d.	0.35	n.d.
MES 125																
A	0.20	2.47	6.63	38.70	0.32	3.81	0.32	0.98	45.24	0.81	0.12	n.d.	0.04	0.02	0.08	0.28
F	0.01	0.11	0.18	27.43	1.24	0.54	0.01	1.91	67.93	0.25	0.05	0.18	n.d.	0.05	0.05	0.02
I	0.47	4.91	12.93	51.71	n.d.	8.54	0.37	0.41	18.23	1.51	0.32	n.d.	0.03	0.01	0.03	0.51
MES 129																
A	0.16	1.58	4.35	35.90	0.25	4.33	0.18	37.08	2.16	0.51	0.08	0.10	0.05	0.11	0.06	12.82
MES 130																
A	0.20	2.28	4.26	17.94	0.95	26.67	0.08	0.22	45.90	0.79	0.11	0.12	n.d.	n.d	0.19	0.23
W	0.13	0.08	0.88	0.25	1.21	0.47	0.17	0.51	95.90	0.03	0.04	0.13	n.d.	0.02	0.02	0.08
F	0.68	20.59	18.09	24.98	0.38	3.84	0.41	0.11	30.11	0.24	0.04	n.d.	0.07	n.d	0.12	0.34
I	0.04	0.60	0.08	30.51	1.67	58.85	0.02	0.09	7.03	0.92	0.04	n.d.	0.01	n.d	0.05	0.07
MES 146c																
A	0.25	2.31	6.62	37.29	0.37	3.79	0.29	1.05	46.68	0.70	0.07	n.d.	0.06	0.05	0.25	
F	0.08	0.29	0.33	27.67	0.89	0.97	0.03	1.82	67.34	0.25	0.05	n.d.	0.07	0.07	0.07	
I	0.33	4.81	13.21	53.64	0.03	9.05	0.11	0.35	15.88	1.50	0.02	0.01	0.04	0.14	0.48	
MES 146d																
A	0.19	4.66	11.12	52.52	0.93	13.63	0.51	0.90	14.29	0.75	0.02	n.d.	0.02	0.17	n.d.	0.25
W	0.15	1.42	5.59	16.22	2.04	1.69	0.37	2.57	69.29	0.27	n.d.	0.02	n.d.	0.01	0.03	0.27
I	0.08	5.30	12.41	56.73	1.01	10.09	0.41	1.20	11.11	1.25	0.03	0.02	0.07	n.d.	n.d.	0.26
MES 147b																
A	0.47	1.67	4.96	30.19	0.88	3.61	0.14	0.60	56.04	0.89	0.20	0.03	n.d.	0.06	0.12	0.08
F	n.d.	0.03	0.09	27.10	1.45	0.78	n.d.	0.77	69.29	0.16	n.d.	0.20	n.d.	n.d.	n.d.	0.09
I	1.11	4.50	11.13	36.26	0.03	9.45	0.48	0.26	33.58	1.99	0.55	n.d.	n.d.	0.05	0.02	0.53
MES 21																
W	0.02	0.05	0.45	0.40	0.13	0.09	0.53	0.51	97.46	n.d.	n.d.	0.05	0.15	0.04	0.01	n.d.
F	0.17	0.04	0.55	25.24	0.84	0.72	0.04	2.09	69.81	0.09	n.d.	0.08	0.11	0.05	0.06	0.09
I	n.d.	0.10	2.31	15.38	0.09	0.79	0.04	0.75	79.18	0.22	0.68	0.16	0.13	0.05	0.04	0.04
M	0.09	0.02	n.d	0.06	0.13	n.d.	n.d.	n.d.	99.08	n.d.	n.d.	0.17	0.15	0.06	0.17	0.07

Analyses	Na_2O	K_2O	Al_2O_3	SiO_2	MgO	CaO	TiO_2	MnO	FeO	P_2O_5	SO_3	As_2O_3	NiO	ZnO	CuO	BaO
MES 23																
W	0.15	0.22	1.44	2.32	0.06	0.24	0.55	0.06	94.58	0.05	0.44	0.11	n.d.	0.07	0.06	n.d.
F	0.05	0.11	1.15	24.43	0.62	0.87	0.06	0.29	72.20	0.09	0.02	0.01	0.04	n.d.	0.06	n.d.
I	0.20	2.46	5.21	25.53	0.45	3.11	0.10	0.17	61.11	1.20	0.15	0.03	0.13	0.02	n.d.	0.03
M	0.06	n.d	0.02	0.09	n.d.	n.d.	0.03	n.d.	99.51	0.05	0.06	0.02	0.11	n.d.	0.05	n.d.
MES 25																
W	n.d.	0.10	0.73	0.99	0.28	0.48	0.30	0.10	96.51	0.20	0.11	0.02	0.13	n.d.	n.d.	0.04
F	n.d.	0.13	0.53	27.49	3.55	2.70	n.d.	0.79	64.52	0.20	0.05	n.d.	n.d.	0.05	n.d.	n.d.
I	0.37	3.33	6.13	26.65	0.61	10.52	0.07	0.46	47.26	3.08	0.96	0.07	0.08	n.d.	0.01	0.29
M	n.d.	n.d.	n.d.	0.12	0.02	0.05	n.d.	0.04	99.40	n.d.	0.02	0.04	0.06	n.d.	0.20	0.05
MES 33																
W	n.d.	0.32	0.40	1.88	0.28	0.55	0.02	0.24	95.90	0.01	0.10	0.10	n.d.	n.d.	n.d.	0.20
F	0.24	3.57	4.39	33.10	0.68	8.31	0.01	0.60	46.01	1.14	1.25	n.d.	0.13	0.05	0.02	0.50
I	0.25	4.81	5.47	43.70	1.26	9.87	0.12	1.01	32.50	n.d.	0.14	n.d.	n.d.	n.d.	0.01	0.80
M	0.10	0.06	n.d.	0.17	0.24	0.02	n.d.	n.d.	98.47	0.05	0.09	0.72	n.d.	0.04	n.d.	0.05
MES 44																
I	0.16	1.80	3.82	26.02	0.29	12.70	0.12	0.92	50.60	3.00	0.17	0.03	0.03	0.01	0.06	0.20
M	0.03	n.d.	n.d.	0.09	0.02	0.21	n.d.	n.d.	97.96	0.02	0.05	0.16	0.02	0.04	1.54	n.d.
MES 53																
I	1.76	7.18	8.26	41.66	2.75	15.70	0.28	4.58	17.23	0.24	0.15	0.07	n.d.	n.d.	0.02	0.09
M	0.10	n.d.	0.03	0.05	0.16	0.02	n.d.	n.d.	99.43	0.03	0.03	n.d.	0.03	0.02	0.05	n.d.
MES 62																
W	0.14	0.08	0.40	0.33	0.19	0.08	0.22	1.29	96.96	n.d.	0.11	n.d.	0.04	0.05	0.09	n.d.
I	0.58	3.64	5.63	31.05	0.39	7.70	0.03	3.22	44.01	2.83	0.28	0.26	0.07	0.02	n.d.	0.30

Area (A) = mean over 2 areas x100
Wustite (W) = mean over 2 spots x2K
Fayalite (F) = mean over 2 spots x2K
Interstitial (I)= mean over 2 spots x2K
Metal (M) = mean over 2 spots x2K
Other (O) = mean over 2 spots x2K
n.d. = not determoned

Table 13a. SEM data from Avdera (all analyses in weight %).

Analyses	Na_2O	K_2O	Al_2O_3	SiO_2	MgO	CaO	TiO_2	MnO	FeO	P_2O_5	SO_3	Cr_2O_3	As_2O_3	NiO	ZnO	CuO	BaO
AVD 103																	
A	1.81	1.05	2.85	22.04	1.27	5.26	n.d.	0.23	46.27	0.32	7.20	n.d.	1.02	0.03	1.01	0.03	9.60
A	1.86	1.18	2.72	22.65	1.27	5.38	n.d.	0.19	44.81	0.24	7.70	n.d.	1.34	n.d.	1.15	0.16	9.37
mean A	1.83	1.12	2.78	22.34	1.27	5.32	n.d.	0.21	45.54	0.28	7.45	n.d.	1.18	0.02	1.08	0.10	9.49
W	0.63	n.d.	0.30	0.37	1.22	0.11	n.d.	0.13	95.98	0.14	0.12	0.05	0.02	n.d.	0.74	n.d.	0.19
W	0.96	0.06	2.66	0.66	0.60	0.21	0.35	0.11	92.48	n.d.	0.50	0.01	0.18	0.18	0.72	0.04	0.29
mean W	0.79	0.03	1.48	0.51	0.91	0.16	0.17	0.12	94.23	0.07	0.31	0.03	0.10	0.09	0.73	0.02	0.24
F	0.50	0.80	2.39	15.16	0.85	2.78	n.d.	0.17	64.06	0.24	3.58	0.02	1.22	n.d.	1.07	0.02	7.16
I	1.68	1.45	3.80	33.72	1.02	7.87	n.d.	0.50	28.85	0.33	4.19	n.d.	1.16	0.13	1.27	0.16	13.88
Prill	n.d.	n.d.	1.53	0.25	0.22	0.03	n.d.	0.22	2.43	n.d.	48.60	n.d.	45.82	0.11	0.33	0.30	0.17
Prill	n.d.	n.d.	1.14	n.d.	n.d.	n.d.	n.d.	n.d.	2.49	n.d.	49.47	n.d.	46.00	0.15	0.42	0.18	0.16
mean prill	n.d.	n.d.	1.33	0.12	0.11	0.02	n.d.	0.11	2.46	n.d.	49.03	n.d.	45.91	0.13	0.37	0.24	0.17
AVD 108																	
A	0.03	0.05	1.61	5.47	0.33	2.54	0.14	n.d.	88.89	0.40	0.22	0.01	0.19	0.84	0.03	n.d.	n.d.
A	0.10	n.d.	1.84	5.82	0.40	1.91	0.44	n.d.	88.50	0.33	0.20	0.19	0.16	n.d.	n.d.	0.08	0.04
mean A	0.06	0.02	1.73	5.64	0.36	2.23	0.29	n.d.	88.69	0.37	0.21	0.10	0.18	0.42	0.01	0.04	0.02
W	0.31	n.d.	0.36	0.18	0.69	3.99	0.05	n.d.	93.85	0.02	0.01	0.17	0.11	0.07	0.02	n.d.	0.19
W	0.03	n.d.	0.97	0.39	n.d.	0.08	n.d.	n.d.	98.52	n.d.	0.67	n.d.	0.17	n.d.	n.d.	n.d.	0.09
mean W	0.17	n.d.	0.67	0.28	0.34	2.04	0.02	n.d.	96.18	0.01	0.34	0.08	0.14	0.03	0.01	n.d.	0.14
F	0.22	0.04	1.95	12.57	n.d.	0.24	n.d.	n.d.	84.19	0.41	0.05	n.d.	0.27	0.07	n.d.	n.d.	n.d.
F	0.67	0.06	1.26	3.82	0.78	1.03	0.13	n.d.	90.20	0.22	0.77	0.09	n.d.	n.d.	n.d.	0.65	0.33
mean F	0.44	0.05	1.60	8.19	0.39	0.64	0.06	n.d.	87.19	0.31	0.41	0.05	0.13	0.03	n.d.	0.33	0.17
I	n.d.	0.02	5.80	8.21	0.02	0.84	5.44	n.d.	77.38	2.08	0.07	0.07	n.d.	0.02	0.01	0.06	n.d.
AVD 112																	
A	0.93	1.83	6.16	23.27	0.56	5.71	0.27	0.04	60.15	0.82	0.15	0.04	n.d.	0.07	n.d.	n.d.	n.d.
A	1.01	2.09	7.28	28.03	0.64	7.42	0.23	0.08	51.67	1.17	0.18	n.d.	0.03	n.d.	0.09	n.d.	0.09
mean A	0.97	1.96	6.72	25.65	0.60	6.57	0.25	0.06	55.91	0.99	0.17	0.02	0.02	0.03	0.04	n.d.	0.04
W	0.20	0.09	0.57	0.44	0.13	0.33	0.86	0.09	97.05	n.d.	n.d.	0.06	0.05	0.06	0.05	n.d.	0.03
W	0.09	0.07	0.63	0.50	0.03	0.21	0.90	n.d.	97.34	n.d.	0.01	0.03	n.d.	0.04	n.d.	n.d.	0.14
mean W	0.14	0.08	0.60	0.47	0.08	0.27	0.88	0.04	97.20	n.d.	n.d.	0.04	0.03	0.05	0.03	n.d.	0.08
F	0.01	0.02	0.22	26.68	2.00	1.59	0.07	0.14	69.00	0.06	n.d.	0.07	n.d.	0.11	0.03	n.d.	n.d.
F	0.21	0.04	0.17	26.48	2.26	1.64	0.02	0.17	68.62	n.d.	0.02	0.03	0.08	0.09	0.08	0.04	0.05
mean F	0.11	0.03	0.19	26.58	2.13	1.62	0.05	0.15	68.81	0.03	0.01	0.05	0.04	0.10	0.06	0.02	0.03
I	1.64	5.32	17.16	34.60	0.09	15.91	0.58	0.01	21.13	3.14	0.10	0.10	n.d.	n.d.	0.14	n.d.	0.08
I	1.76	6.89	15.47	37.50	n.d.	13.89	0.54	0.01	20.55	2.83	0.19	0.06	n.d.	n.d.	n.d.	0.12	0.19
mean I	1.70	6.11	16.31	36.05	0.04	14.90	0.56	0.01	20.84	2.99	0.14	0.08	n.d.	n.d.	0.07	0.06	0.14
Metal	0.33	n.d.	0.02	0.15	0.13	0.02	n.d.	n.d.	99.08	n.d.	n.d.	n.d.	n.d.	0.01	0.07	0.08	0.12

Analyses	Na_2O	K_2O	Al_2O_3	SiO_2	MgO	CaO	TiO_2	MnO	FeO	P_2O_5	SO_3	Cr_2O_3	As_2O_3	NiO	ZnO	CuO	BaO
AVD 114																	
A	0.83	2.35	7.83	29.44	0.65	4.09	0.15	0.95	52.48	0.58	0.21	0.02	n.d.	0.08	0.17	0.04	0.14
A	0.89	2.64	8.34	29.91	0.49	4.58	0.18	1.04	50.66	0.71	0.14	0.04	0.03	0.03	0.12	0.15	0.06
mean A	0.86	2.50	8.08	29.67	0.57	4.33	0.16	1.00	51.57	0.65	0.17	0.03	0.02	0.05	0.15	0.09	0.10
W	n.d.	0.12	0.87	0.62	0.15	0.19	0.56	0.33	96.84	n.d.	n.d.	0.13	n.d.	0.08	0.04	0.04	0.04
W	0.08	0.46	1.46	2.66	0.12	0.49	0.72	0.36	92.97	0.08	0.04	0.10	0.39	n.d.	n.d.	0.08	n.d.
mean W	0.04	0.29	1.16	1.64	0.13	0.34	0.64	0.34	94.91	0.04	0.02	0.11	0.20	0.04	0.02	0.06	0.02
F	0.19	0.03	0.23	26.35	1.35	0.87	n.d.	1.54	69.09	0.28	0.01	n.d.	n.d.	n.d.	0.03	0.03	0.01
F	0.02	0.47	1.89	27.15	0.22	1.88	0.11	1.39	66.52	0.24	0.03	n.d.	n.d.	0.01	0.04	n.d.	0.04
mean F	0.10	0.25	1.06	26.75	0.78	1.37	0.05	1.46	67.80	0.26	0.02	n.d.	n.d.	n.d.	0.03	0.01	0.02
I	1.74	5.88	16.23	35.91	0.02	10.10	0.28	0.52	26.59	1.71	0.32	n.d.	0.17	n.d.	0.18	0.11	0.23
I	1.90	5.79	15.27	36.88	0.15	9.10	0.23	0.65	27.23	1.59	0.40	0.02	0.25	n.d.	n.d.	0.15	0.38
mean I	1.82	5.84	15.75	36.40	0.08	9.60	0.26	0.58	26.91	1.65	0.36	n.d.	0.21	n.d.	0.09	0.13	0.31
Prill	0.29	0.22	2.03	40.91	2.83	0.96	n.d.	1.04	49.88	0.40	0.35	n.d.	n.d.	n.d.	0.38	0.51	0.21
Prill	1.59	4.01	13.32	33.52	0.25	7.18	0.36	0.49	37.20	1.25	0.29	n.d.	0.13	0.04	0.09	0.12	0.16
mean prill	0.94	2.12	7.67	37.22	1.54	4.07	0.18	0.77	43.54	0.82	0.32	n.d.	0.06	0.02	0.23	0.32	0.19
AVD 116																	
A	0.68	1.66	4.82	19.93	0.75	14.93	0.05	n.d.	56.39	0.54	0.02	0.06	0.03	0.06	n.d.	n.d.	0.09
A	0.73	1.24	4.11	15.76	0.89	11.12	0.10	0.09	64.89	0.31	0.11	0.05	0.18	0.02	0.18	0.06	0.17
mean A	0.71	1.45	4.47	17.84	0.82	13.02	0.08	0.05	60.64	0.43	0.06	0.05	0.11	0.04	0.09	0.03	0.13
W	0.14	n.d.	0.62	0.26	0.60	0.17	0.13	0.21	97.78	n.d.	n.d.	n.d.	n.d.	n.d.	0.08	n.d.	0.02
W	n.d.	n.d.	0.60	0.10	0.67	0.19	0.21	n.d.	97.99	n.d.	n.d.	0.04	0.07	0.08	n.d.	0.06	n.d.
mean W	0.07	n.d.	0.61	0.18	0.63	0.18	0.17	0.10	97.88	n.d.	n.d.	0.02	0.04	0.04	0.04	0.03	0.01
F	0.02	n.d.	0.22	27.96	1.70	32.31	n.d.	0.09	36.47	0.63	0.05	n.d.	0.26	0.04	n.d.	0.15	0.12
F	0.15	0.05	0.14	26.58	1.53	28.41	0.17	0.12	42.16	0.63	n.d.	n.d.	n.d.	0.02	n.d.	0.06	n.d.
mean F	0.09	0.02	0.18	27.27	1.62	30.36	0.08	0.11	39.31	0.63	0.02	n.d.	0.13	0.03	n.d.	0.10	0.06
I	2.40	8.49	18.82	33.56	0.29	7.59	0.09	0.03	27.19	0.82	0.16	n.d.	n.d.	0.06	n.d.	0.02	0.50
I	1.66	8.22	16.44	36.91	0.14	14.59	0.21	0.01	18.96	1.32	0.30	n.d.	0.47	0.06	n.d.	0.15	0.57
mean I	2.03	8.36	17.63	35.24	0.21	11.09	0.15	0.02	23.07	1.07	0.23	n.d.	0.23	0.06	n.d.	0.09	0.53
AVD 123																	
A	1.34	2.29	4.39	18.60	1.59	4.49	0.06	1.74	64.67	0.82	n.d.	n.d.	n.d.	n.d.	n.d.	n.d.	0.02
A	1.15	2.19	4.43	19.07	1.73	4.56	0.08	1.90	64.09	0.62	0.16	n.d.	n.d.	n.d.	n.d.	n.d.	0.01
mean A	1.25	2.24	4.41	18.84	1.66	4.52	0.07	1.82	64.38	0.72	0.08	n.d.	n.d.	n.d.	n.d.	n.d.	0.01
W	n.d.	0.10	0.47	0.49	0.64	0.12	0.11	1.37	96.63	0.06	n.d.	n.d.	n.d.	n.d.	n.d.	n.d.	0.03
W	0.63	0.02	0.47	0.40	0.58	0.11	0.06	1.29	96.37	n.d.	0.03	n.d.	n.d.	n.d.	n.d.	n.d.	0.05
mean W	0.31	0.06	0.47	0.44	0.61	0.11	0.08	1.33	96.50	0.03	0.01	n.d.	n.d.	n.d.	n.d.	n.d.	0.04
F	n.d.	0.27	0.58	28.99	3.45	4.47	n.d.	3.23	58.52	0.38	0.04	n.d.	n.d.	n.d.	n.d.	n.d.	0.08
F	n.d.	0.19	0.56	30.07	3.99	3.54	0.09	3.39	57.75	0.40	0.02	n.d.	n.d.	n.d.	n.d.	n.d.	n.d.

Analyses	Na_2O	K_2O	Al_2O_3	SiO_2	MgO	CaO	TiO_2	MnO	FeO	P_2O_5	SO_3	Cr_2O_3	As_2O_3	NiO	ZnO	CuO	BaO
mean F	n.d.	0.23	0.57	29.53	3.72	4.01	0.05	3.31	58.13	0.39	0.03	n.d.	n.d.	n.d.	n.d.	n.d.	0.04
I	3.29	10.28	14.14	33.55	0.41	10.90	0.18	0.99	22.71	2.90	0.54	n.d.	n.d.	n.d.	n.d.	n.d.	0.11
I	3.14	6.37	11.80	35.23	0.43	11.18	0.12	1.45	27.69	2.26	0.27	n.d.	n.d.	n.d.	n.d.	n.d.	0.06
mean I	3.21	8.33	12.97	34.39	0.42	11.04	0.15	1.22	25.20	2.58	0.41	n.d.	n.d.	n.d.	n.d.	n.d.	0.09
A ore	1.80	1.97	3.90	16.60	1.46	3.95	0.10	1.68	67.64	0.81	0.10	n.d.	n.d.	n.d.	n.d.	n.d.	n.d.
Prill	n.d.	0.03	0.07	0.21	0.20	0.07	n.d.	n.d.	99.38	0.01	n.d.	n.d.	n.d.	n.d.	n.d.	n.d.	0.03
AVD 129																	
A	0.55	0.63	3.10	13.27	0.52	1.60	0.11	0.36	78.44	0.59	0.44	0.01	n.d.	0.13	0.10	0.13	0.01
A	0.45	0.81	3.35	14.02	0.32	1.79	0.02	0.47	77.75	0.44	0.21	n.d.	0.29	0.03	n.d.	0.01	0.03
mean A	0.50	0.72	3.23	13.65	0.42	1.70	0.07	0.41	78.09	0.51	0.32	n.d.	0.15	0.08	0.05	0.07	0.02
W	0.03	0.04	0.53	0.54	0.44	0.05	0.11	0.41	97.48	0.07	0.07	n.d.	0.02	0.10	0.06	0.06	n.d.
F	0.12	0.04	0.40	26.29	1.08	1.19	0.01	0.98	67.54	0.13	0.08	n.d.	0.06	0.07	n.d.	0.01	n.d.
I	1.64	4.92	16.00	31.49	0.25	10.77	0.19	0.24	30.52	2.70	0.73	0.02	0.11	0.07	0.03	0.08	0.25
AVD 130																	
A	1.02	3.24	16.46	65.20	1.92	3.32	1.01	0.15	6.81	n.d.	0.05	n.d.	0.51	n.d.	0.05	0.25	n.d.
A	0.81	3.67	16.85	52.27	2.63	5.66	2.47	0.30	14.62	0.09	0.20	0.02	n.d.	0.02	0.07	0.34	n.d.
mean A	0.91	3.45	16.65	58.73	2.27	4.49	1.74	0.22	10.71	0.04	0.13	0.01	0.25	0.01	0.06	0.29	n.d.
I	0.68	2.50	11.56	74.89	1.18	2.16	0.93	0.22	5.38	n.d.	0.17	n.d.	0.17	0.02	n.d.	0.13	n.d.
I	1.33	5.40	13.88	67.57	0.94	2.66	0.80	0.12	7.02	n.d.	n.d.	0.61	0.01	n.d.	0.02	0.03	0.17
mean I	1.01	3.95	12.72	71.23	1.06	2.41	0.87	0.17	6.20	n.d.	0.09	0.31	0.09	0.01	0.01	0.08	0.09
AVD 136																	
A	0.56	0.01	0.25	0.17	0.24	0.06	n.d.	0.04	97.38	n.d.	0.08	0.25	0.41	0.28	n.d.	n.d.	0.26
AVD 18																	
I	0.45	5.25	6.73	36.73	1.57	11.76	0.12	0.06	36.28	0.53	0.05	n.d.	0.24	0.03	0.08	0.04	0.07
I	0.05	6.26	5.65	30.92	0.84	10.26	n.d.	0.03	44.48	1.14	n.d.	0.07	0.10	0.02	n.d.	n.d.	0.16
I	n.d.	5.69	6.74	42.29	1.74	12.96	0.18	0.24	29.17	0.42	n.d.	0.39	n.d.	n.d.	0.08	0.01	0.10
mean I	0.17	5.73	6.37	36.65	1.38	11.66	0.10	0.11	36.64	0.69	0.02	0.15	0.11	0.02	0.05	0.02	0.11
AVD 30																	
I	n.d.	3.39	8.99	29.38	1.03	16.50	3.00	2.13	30.36	4.55	0.19	0.18	n.d.	0.10	n.d.	0.07	0.13
I	0.56	4.39	8.98	33.78	1.17	23.76	5.44	0.79	17.84	2.70	0.16	0.01	0.19	n.d.	0.11	0.03	0.09
mean I	0.28	3.89	8.98	31.58	1.10	20.13	4.22	1.46	24.10	3.62	0.17	0.09	0.10	0.05	0.05	0.05	0.11
AVD 53																	
I	0.45	7.02	12.78	52.38	3.08	20.95	0.17	0.16	2.38	0.25	0.01	0.10	n.d.	0.03	0.11	n.d.	0.13
I	0.36	7.19	10.59	41.52	3.88	32.46	0.43	0.18	2.65	0.38	n.d.	n.d.	0.12	n.d.	0.03	0.14	0.07
meanI	0.40	7.10	11.68	46.95	3.48	26.70	0.30	0.17	2.52	0.31	0.01	0.05	0.06	0.01	0.07	0.07	0.20

Table 13b. SEM data from Avdera (summary of mean analyses from table 13a).

Analyses	Na_2O	K_2O	Al_2O_3	SiO_2	MgO	CaO	TiO_2	MnO	FeO	P_2O_5	SO_3	As_2O_3	NiO	ZnO	CuO	BaO
AVD 103																
A	1.83	1.12	2.78	22.34	1.27	5.32	n.d.	0.21	45.54	0.28	7.45	1.18	0.02	1.08	0.10	9.49
W	0.79	0.03	1.48	0.51	0.91	0.16	0.17	0.12	94.23	0.07	0.31	0.10	0.09	0.73	0.02	0.24
F	0.50	0.80	2.39	15.16	0.85	2.78	n.d.	0.17	64.06	0.24	3.58	1.22	n.d.	1.07	0.02	7.16
I	1.68	1.45	3.80	33.72	1.02	7.87	n.d.	0.50	28.85	0.33	4.19	1.16	0.13	1.27	0.16	13.88
P	n.d.	n.d.	1.33	0.12	0.11	0.02	n.d.	0.11	2.46	n.d.	49.03	45.91	0.13	0.37	0.24	0.17
AVD 108																
A	0.06	0.02	1.73	5.64	0.36	2.23	0.29	n.d.	88.69	0.37	0.21	0.18	0.42	0.01	0.04	0.02
W	0.17	n.d.	0.67	0.28	0.34	2.04	0.02	n.d.	96.18	0.01	0.34	0.14	0.03	0.01	n.d.	0.14
F	0.44	0.05	1.60	8.19	0.39	0.64	0.06	n.d.	87.19	0.31	0.41	0.13	0.03	n.d.	0.33	0.17
I	n.d.	0.02	5.80	8.21	0.02	0.84	5.44	n.d.	77.38	2.08	0.07	n.d.	0.02	0.01	0.06	n.d.
AVD 112																
A	0.97	1.96	6.72	25.65	0.60	6.57	0.25	0.06	55.91	0.99	0.17	0.02	0.03	0.04	n.d.	0.04
W	0.14	0.08	0.60	0.47	0.08	0.27	0.88	0.04	97.20	n.d.	n.d.	0.03	0.05	0.03	n.d.	0.08
F	0.11	0.03	0.19	26.58	2.13	1.62	0.05	0.15	68.81	0.03	0.01	0.04	0.10	0.06	0.02	0.03
I	1.70	6.11	16.31	36.05	0.04	14.90	0.56	0.01	20.84	2.99	0.14	n.d.	n.d.	0.07	0.06	0.14
M	0.33	n.d.	0.02	0.15	0.13	0.02	n.d.	n.d.	99.08	n.d.	n.d.	n.d.	0.01	0.07	0.08	0.12
AVD 114																
A	0.86	2.50	8.08	29.67	0.57	4.33	0.16	1.00	51.57	0.65	0.17	0.02	0.05	0.15	0.09	0.10
W	0.04	0.29	1.16	1.64	0.13	0.34	0.64	0.34	94.91	0.04	0.02	0.20	0.04	0.02	0.06	0.02
F	0.10	0.25	1.06	26.75	0.78	1.37	0.05	1.46	67.80	0.26	0.02	n.d.	n.d.	0.03	0.01	0.02
I	0.07	0.27	1.11	14.20	0.46	0.86	0.35	0.90	81.36	0.15	0.02	0.10	n.d.	0.09	0.04	0.31
P	0.94	2.12	7.67	37.22	1.54	4.07	0.18	0.77	43.54	0.82	0.32	0.06	0.02	0.23	0.32	0.19
AVD 116																
A	0.71	1.45	4.47	17.84	0.82	13.02	0.08	0.05	60.64	0.43	0.06	0.11	0.04	0.09	0.03	0.13
W	0.35	n.d.	0.61	8.92	0.41	6.51	0.04	0.02	30.32	n.d.	n.d.	0.05	0.04	0.04	0.03	0.07
F	0.09	0.02	0.18	27.27	1.62	30.36	0.08	0.11	39.31	0.63	0.02	0.13	0.03	n.d.	0.10	0.06
I	2.03	8.36	17.63	35.24	0.21	11.09	0.15	0.02	23.07	1.07	0.23	0.23	0.06	n.d.	0.09	0.53
AVD 123																
A	1.25	2.24	4.41	18.84	1.66	4.52	0.07	1.82	64.38	0.72	0.08	n.d.	n.d.	n.d.	n.d.	0.01
W	0.31	0.06	0.47	0.44	0.61	0.11	0.08	1.33	96.50	0.03	0.01	n.d.	n.d.	n.d.	n.d.	0.04
F	n.d.	0.23	0.57	29.53	3.72	4.01	0.05	3.31	58.13	0.39	0.03	n.d.	n.d.	n.d.	n.d.	0.04
I	3.21	8.33	12.97	34.39	0.42	11.04	0.15	1.22	25.20	2.58	0.41	n.d.	n.d.	n.d.	n.d.	0.09
A ore	1.80	1.97	3.90	16.60	1.46	3.95	0.10	1.68	67.64	0.81	0.10	n.d.	n.d.	n.d.	n.d.	n.d.
P	n.d.	0.03	0.07	0.21	0.20	0.07	n.d.	n.d.	99.38	0.01	n.d.	n.d.	n.d.	n.d.	n.d.	0.03
AVD 129																
A	0.50	0.72	3.23	13.65	0.42	1.70	0.07	0.41	78.09	0.51	0.32	0.15	0.08	0.05	0.07	0.02
W	0.03	0.04	0.53	0.54	0.44	0.05	0.11	0.41	97.48	0.07	0.07	0.02	0.10	0.06	0.06	n.d.
F	0.12	0.04	0.40	26.29	1.08	1.19	0.01	0.98	67.54	0.13	0.08	0.06	0.07	n.d.	0.01	n.d.
I	1.64	4.92	16.00	31.49	0.25	10.77	0.19	0.24	30.52	2.70	0.73	0.11	0.07	0.03	0.08	0.25

Analyses	Na_2O	K_2O	Al_2O_3	SiO_2	MgO	CaO	TiO_2	MnO	FeO	P_2O_5	SO_3	As_2O_3	NiO	ZnO	CuO	BaO
AVD 130																
A	0.91	3.45	16.65	58.73	2.27	4.49	1.74	0.22	10.71	0.04	0.13	0.25	0.01	0.06	0.29	n.d.
I	1.01	3.95	12.72	71.23	1.06	2.41	0.87	0.17	6.20	n.d.	0.09	0.09	0.01	0.01	0.08	0.09
AVD 136																
A	0.56	0.01	0.25	0.17	0.24	0.06	n.d.	0.04	97.38	n.d.	0.08	0.41	0.28	n.d.	n.d.	0.26
AVD 18																
I	0.17	5.73	6.37	36.65	1.38	11.66	0.10	0.11	36.64	0.69	0.02	0.11	0.02	0.05	0.02	0.11
AVD 30																
I	0.28	3.89	8.98	31.58	1.10	20.13	4.22	1.46	24.10	3.62	0.17	0.10	0.05	0.05	0.05	0.11
AVD 53																
I	0.40	7.10	11.68	46.95	3.48	26.70	0.30	0.17	2.52	0.31	0.01	0.06	0.01	0.07	0.07	0.20

Area (A) = mean over 2 areas x100
Wustite (W) = mean over 2 spots x2K
Fayalite (F) = mean over 2 spots x2K
Interstitial (I) = mean over 2 spots x2K
Metal (M) = mean over 2 spots x2K
Other (O) = mean over 2 spots x2K
nd = not determoned

Table 14a. SEM data from Kalyva-Kastro (all analyses in weight %).

Analyses	Na_2O	K_2O	Al_2O_3	SiO_2	MgO	CaO	TiO_2	MnO	FeO	P_2O_5	SO_3	BaO	Cr_2O_3	As_2O_3	NiO	ZnO	CuO
KL 70																	
A	0.44	1.21	2.96	12.35	0.38	3.25	nd	0.10	78.87	0.32	0.00	nd	nd	nd	nd	nd	nd
A	0.60	0.74	2.42	10.09	0.57	2.37	nd	0.14	82.48	0.30	0.22	nd	nd	nd	nd	nd	nd
A	0.42	1.34	3.73	14.80	0.60	3.16	nd	nd	75.30	0.44	0.14	nd	nd	nd	nd	nd	nd
mean A	0.49	1.09	3.03	12.41	0.51	2.93	nd	0.08	78.88	0.35	nd	nd	nd	nd	nd	nd	nd
W	0.27	nd	0.28	0.33	0.90	0.07	nd	0.10	97.96	nd	nd	0.08	nd	nd	nd	nd	nd
W	0.40	nd	0.76	0.14	0.62	nd	nd	nd	98.74	nd	nd	nd	nd	nd	nd	nd	nd
mean W	0.34	nd	0.52	0.24	0.76	0.04	nd	0.05	98.35	nd	nd	0.04	nd	nd	nd	nd	nd
Mg, P rich F	0.14	0.37	0.12	29.98	1.25	29.97	nd	0.09	36.86	1.05	nd	0.14	nd	nd	nd	nd	nd
Al, P rich I	1.32	18.57	18.68	38.22	0.06	4.16	0.01		16.19	2.05	0.24	0.50	nd	nd	nd	nd	nd
KL 73																	
A	0.10	1.90	3.50	20.20	0.20	4.90	nd	0.30	67.70	0.30	0.20	0.20	nd	nd	nd	nd	nd
A	0.50	1.70	3.20	19.00	0.10	4.50	nd	0.50	70.00	0.10	nd	0.10	nd	nd	nd	nd	nd
A	0.30	1.70	4.90	25.00	0.20	5.20	nd	0.50	60.20	1.40	0.20	nd	nd	nd	nd	nd	nd
mean A	0.30	1.80	3.80	21.40	0.16	4.80	nd	0.40	66.00	0.60	0.16	0.10	nd	nd	nd	nd	nd
W	0.30	nd	nd	0.20	0.30	nd	nd	0.40	98.00	nd	0.20	nd	nd	nd	nd	nd	nd
W	nd	nd	nd	0.30	0.50	nd	nd	0.70	97.60	nd	nd	nd	nd	nd	nd	nd	nd
mean W	0.15	nd	nd	0.25	0.40	nd	nd	0.55	97.80	nd	0.10	nd	nd	nd	nd	nd	nd
I	0.60	7.00	6.00	36.70	nd	19.70	nd	0.50	27.70	0.90	nd	0.10	nd	nd	nd	nd	nd
I	0.50	6.70	6.60	37.00	nd	20.20	nd	0.30	27.20	0.80	nd	0.10	nd	nd	nd	nd	nd
I	0.70	6.50	5.60	36.40	nd	21.40	nd	0.50	26.60	1.00	0.40	0.20	nd	nd	nd	nd	nd
mean I	0.60	6.70	6.00	36.60	nd	20.40	nd	0.40	27.10	0.90	0.10	0.10	nd	nd	nd	nd	nd
Fe prill	nd	nd	nd	nd	nd	nd	nd	nd	99.45	0.10	nd	0.10	nd	nd	nd	nd	nd
KL 74																	
A	0.80	4.20	17.00	58.20	0.40	9.80	1.00	0.20	7.60	0.20	0.10	nd	nd	nd	nd	nd	nd
A	nd	3.60	23.00	45.70	0.50	12.70	1.90	nd	11.70	0.30	nd	nd	nd	nd	nd	nd	nd
A	0.10	2.20	21.00	42.40	0.30	21.40	0.90	0.60	10.40	0.30	0.20	nd	nd	nd	nd	nd	nd
mean A	0.50	2.90	22.00	44.00	0.40	17.00	1.40	0.30	11.00	0.30	0.10	nd	nd	nd	nd	nd	nd
W	nd	0.50	9.70	11.50	1.60	2.70	1.60	0.30	71.60	0.10	nd	nd	nd	nd	nd	nd	nd
Al rich I	0.30	0.40	30.70	45.80	nd	20.70	nd	nd	1.70	nd	0.10	nd	nd	nd	nd	nd	nd
Al rich I	0.30	0.40	31.00	45.30	nd	21.10	0.10	nd	1.30	0.10	nd	nd	nd	nd	nd	nd	nd
mean	0.30	0.40	30.80	45.50	nd	20.90	0.50	nd	1.50	0.50	nd	nd	nd	nd	nd	nd	nd
Al, Mg, Ti rich F	0.60	3.20	16.20	45.20	0.90	18.30	1.20	0.60	12.40	0.30	0.10	nd	nd	nd	nd	nd	nd
Al, Mg, Ti rich F	0.40	4.30	10.90	43.30	1.60	18.50	1.80	0.80	17.30	0.60	0.30	nd	nd	nd	nd	nd	nd
Al, Mg, Ti rich F	0.20	5.70	20.30	45.30	1.30	5.20	1.80	0.30	19.20	0.40	nd	nd	nd	nd	nd	nd	nd
mean Al, Mg, Ti rich F	0.40	4.40	15.80	44.60	1.30	14.00	1.60	0.60	16.30	0.40	0.10	nd	nd	nd	nd	nd	nd
KL 79																	
A	nd	nd	nd	1.06	0.32	1.43	0.04	0.08	95.92	0.69	0.46	nd	nd	nd	nd	nd	nd

Analyses	Na_2O	K_2O	Al_2O_3	SiO_2	MgO	CaO	TiO_2	MnO	FeO	P_2O_5	SO_3	BaO	Cr_2O_3	As_2O_3	NiO	ZnO	CuO
A	0.13	0.13	0.18	1.27	0.33	3.02	0.00	0.06	91.16	3.50	0.03	0.17	nd	nd	nd	nd	nd
mean A	0.65	0.65	0.09	1.16	0.32	2.23	0.02	0.07	93.54	2.95	0.23	0.08	nd	nd	nd	nd	nd
Fe prill	0.28	1.62	7.54	9.37	0.56	3.57	1.85	0.08	72.96	1.83	0.08	0.27	nd	nd	nd	nd	nd
KL 82																	
A	0.30	1.10	1.90	10.80	0.40	8.10	nd	0.20	76.00	0.40	0.30	0.10	nd	nd	nd	nd	nd
A	0.90	1.50	2.70	15.00	0.40	9.60	nd	0.20	68.50	0.40	nd	0.10	nd	nd	0.10	nd	nd
A	0.40	1.10	2.20	11.10	0.50	7.60	nd	0.20	75.80	0.20	0.40	0.10	nd	nd	nd	nd	nd
mean A	0.53	1.23	2.27	12.30	0.43	8.43	nd	0.20	73.43	0.33	0.23	0.10	nd	nd	nd	nd	nd
W	1.00	0.70	2.30	10.40	0.70	8.00	nd	0.30	75.00	0.10	0.30	0.10	nd	nd	nd	0.10	nd
W	nd	1.20	3.10	8.30	0.40	3.50	nd	0.10	82.70	nd	nd	0.20	nd	nd	0.10	nd	nd
mean W	0.50	0.95	2.70	9.35	0.55	5.75	nd	0.20	78.85	0.05	0.15	0.15	nd	nd	0.05	0.05	nd
F	0.80	5.20	5.10	31.50	0.50	24.40	nd	0.10	30.90	0.70	0.10	nd	nd	0.20	nd	nd	nd
F	0.30	5.00	5.70	32.40	0.60	22.90	nd	0.20	32.10	0.40		0.20	nd	nd	nd	nd	nd
F	0.40	3.90	4.00	30.20	nd	33.80	nd	0.10	25.50	1.10	0.20	0.10	nd	nd	nd	nd	nd
mean F	0.50	4.70	4.93	31.37	0.37	27.03	nd	0.13	29.50	0.73	0.10	0.10	nd	nd	nd	nd	nd
I	1.10	9.00	12.80	37.80	nd	11.00	0.20	0.10	26.00	0.30	0.40	0.50	nd	0.20	nd	0.10	0.10
I	0.90	2.40	4.70	37.20	0.60	30.80	0.20	0.20	22.30	0.10	nd	nd	nd	nd	nd	nd	nd
I	1.30	9.40	14.60	42.10	nd	10.00	0.50	0.10	20.20	0.50	0.30	nd	nd	nd	nd	nd	0.20
mean I	1.10	10.40	10.70	39.03	0.20	17.27	0.30	0.13	22.83	0.30	0.23	0.17	nd	0.07	nd	0.03	0.10
Mg rich F	nd	0.20	nd	22.00	1.20	26.50	nd	0.30	41.90	0.50	nd	nd	nd	nd	nd	nd	nd
Mg rich F	0.90	1.00	0.50	30.30	1.60	27.30	nd	0.30	37.30	0.40	nd	nd	nd	nd	nd	nd	nd
mean Mg rich F	0.45	0.15	0.25	26.15	1.40	26.90	nd	0.30	39.60	0.45	nd	nd	nd	nd	nd	nd	nd
KL 85																	
A	1.50	6.20	13.00	74.10	0.30	1.90	0.30	nd	2.00	0.10	nd	0.20	nd	nd	0.10	nd	nd
A	1.10	6.10	11.60	48.80	0.40	6.90	0.30	0.10	24.00	0.40	nd	nd	nd	nd	nd	nd	nd
A	1.70	3.50	5.70	32.20	0.80	11.40	0.10	0.10	43.70	0.50	nd	nd	nd	nd	nd	nd	nd
A	0.40	9.90	17.40	67.00	0.20	1.50	0.10	0.10	2.60	0.10	nd	nd	nd	nd	nd	nd	nd
mean A	1.18	6.43	11.93	55.53	0.43	5.43	0.20	0.08	18.08	0.28	nd	nd	nd	nd	nd	nd	nd
W	1.10	2.20	2.80	14.70	0.60	4.50	0.20	0.20	72.80	0.30	nd	nd	nd	nd	nd	nd	nd
W	nd	1.20	1.50	7.60	0.20	2.40	0.20	0.20	86.10	nd	nd	nd	nd	nd	nd	nd	nd
W	nd	2.50	6.50	16.60	nd	0.30	nd	nd	70.30	2.90	0.10	nd	nd	0.20	0.20	nd	nd
mean W	0.37	1.97	3.60	12.97	0.27	2.40	0.13	0.13	76.40	1.07	nd	nd	nd	nd	nd	nd	nd
F	0.60	3.70	6.00	38.10	0.50	15.50	nd	0.10	34.30	0.70	nd	0.10	nd	nd	nd	nd	nd
F	0.80	4.50	6.30	38.90	0.80	15.00	0.10	0.30	32.50	0.50	nd	0.10	nd	nd	nd	nd	nd
mean F	0.70	4.10	6.15	38.50	0.65	15.25	0.05	0.20	33.40	0.60	nd	0.10	nd	nd	nd	nd	nd
I	0.30	11.60	19.30	63.60	0.10	1.20	0.90	0.10	2.30	nd	nd	0.10	nd	nd	0.10	nd	nd
I	0.40	11.70	20.70	61.90	nd	1.70	0.50	nd	2.50	nd	nd	0.20	nd	nd	nd	nd	nd
mean I	0.35	11.65	20.00	62.75	0.05	1.45	0.70	0.05	2.40	nd	nd	0.15	nd	nd	0.05	nd	nd
Fe prill (Ti,Ni,P rich)	0.70	10.80	9.50	51.20	nd	1.80	1.10	nd	7.50	14.90	nd	nd	nd	nd	1.10	nd	nd
KL 86																	
A	0.17	0.24	1.36	6.71	0.33	3.49	nd	0.10	86.99	0.40	0.15	0.06	nd	nd	nd	nd	nd
A	0.29	0.83	1.78	7.26	0.37	3.80	0.03	0.10	85.04	0.44	nd	0.05	nd	nd	nd	nd	nd
mean A	0.15	0.61	1.70	8.78	0.32	5.20	0.01	0.08	82.40	0.56	0.09	0.09	nd	nd	nd	nd	nd
W	0.15	0.02	0.16	0.11	0.43	0.31	0.04	0.10	98.51	nd	nd	0.15	nd	nd	nd	nd	nd

Analyses	Na_2O	K_2O	Al_2O_3	SiO_2	MgO	CaO	TiO_2	MnO	FeO	P_2O_5	SO_3	BaO	Cr_2O_3	As_2O_3	NiO	ZnO	CuO
F	nd	0.12	0.40	49.01	0.46	5.97	nd	nd	40.95	2.80	0.13	0.17	nd	nd	nd	nd	nd
I	0.02	0.09	nd	57.97	0.12	4.83	nd	0.10	34.91	1.73	0.16	0.06	nd	nd	nd	nd	nd
KL 87																	
A	0.30	0.50	1.20	6.10	0.40	2.90	nd	nd	87.20	nd	nd	nd	nd	nd	nd	0.10	0.60
A	0.20	2.20	3.50	19.30	0.40	9.80	nd	0.10	62.90	0.80	nd	0.30	nd	nd	nd	nd	nd
mean A	0.25	1.35	2.35	12.70	0.40	6.35	nd	0.05	75.05	0.40	nd	0.15	nd	nd	nd	nd	nd
W	nd	nd	0.50	0.40	nd	nd	nd	0.20	97.80	nd	0.10	nd	nd	nd	nd	0.10	nd
W	nd	nd	0.40	0.40	0.40	nd	nd	0.20	98.20	nd	nd	nd	nd	nd	nd	nd	nd
W	nd	nd	0.50	nd	nd	nd	nd	nd	97.90	nd	nd	nd	nd	nd	nd	nd	nd
mean W	nd	nd	0.47	0.27	0.13	nd	nd	0.13	97.97	nd	nd	nd	nd	nd	nd	nd	nd
F	0.70	0.10	nd	23.80	1.80	23.20	nd	nd	49.30	0.50	nd	nd	nd	nd	nd	nd	nd
F	0.70	nd	nd	29.30	1.90	28.60	nd	nd	38.30	nd	nd	nd	nd	nd	nd	nd	nd
mean F	0.70	0.05	nd	26.55	1.85	25.90	nd	nd	43.80	0.25	nd	nd	nd	nd	nd	nd	nd
I	1.40	15.00	20.30	33.20	nd	9.00	nd	nd	17.10	3.20	nd	0.20	nd	nd	nd	nd	0.20
I	1.20	27.30	29.00	39.40	nd	nd	nd	nd	2.30	nd	nd	0.30	nd	nd	nd	nd	nd
mean I	1.30	21.15	24.65	36.30	nd	4.50	nd	nd	9.70	1.60	nd	0.25	nd	nd	nd	nd	nd
Mg rich F	0.20	0.60	nd	27.80	1.60	10.90	nd	0.20	56.90	1.20	nd	nd	nd	nd	nd	nd	nd
Mg rich F	0.70	0.20	nd	29.20	1.40	27.40	nd	0.40	39.90	0.80	nd	nd	nd	nd	nd	nd	nd
Mg rich F	0.45	0.40	nd	28.50	1.50	19.15	nd	0.30	48.40	1.00	nd	nd	nd	nd	nd	nd	nd
Al, Mg, P rich F	0.20	nd	13.30	30.40	1.50	6.00	nd	0.10	40.30	5.70	0.20	0.30	nd	nd	nd	nd	1.00
KL 89																	
A	0.58	3.13	4.69	18.62	0.58	10.07	0.03	0.51	60.53	0.40	0.19	0.03	nd	nd	nd	nd	nd
A	0.57	3.44	4.79	22.59	1.04	14.25	0.11	0.73	51.91	0.46	nd	0.10	nd	nd	nd	nd	nd
mean A	0.58	3.29	4.74	20.61	0.81	12.16	0.07	0.62	56.22	0.43	0.10	0.07	nd	nd	nd	nd	nd
W	0.35	0.04	0.24	0.22	0.48	0.30	0.14	0.57	97.50	nd	0.10	0.06	nd	nd	nd	nd	nd
F	nd	0.13	nd	28.53	1.54	28.17	nd	1.06	39.91	0.66	nd	nd	nd	nd	nd	nd	nd
F	0.13	0.04	nd	28.87	1.69	29.49	0.02	0.96	38.90	0.78	0.01	nd	nd	nd	nd	nd	nd
F	0.07	0.09	nd	28.70	1.62	28.83	0.01	1.01	39.41	0.72	0.01	nd	nd	nd	nd	nd	nd
I	0.72	31.97	28.11	37.25	nd	0.04	0.02	nd	1.75	nd	nd	0.13	nd	nd	nd	nd	nd
I	0.79	30.26	28.17	38.33	nd	0.23	nd	nd	2.03	nd	0.02	0.17	nd	nd	nd	nd	nd
mean I	0.76	31.12	28.14	37.79	nd	0.14	0.01	nd	1.89	nd	0.01	0.15	nd	nd	nd	nd	nd
KL 90																	
A	0.15	0.28	3.54	14.72	0.37	1.49	0.02	0.06	78.84	0.39	0.16	0.02	nd	nd	nd	nd	nd
A	0.09	0.57	5.58	22.87	0.37	3.22	nd	n.d.	66.27	0.69	0.14	0.20	nd	nd	nd	nd	nd
A	0.18	0.97	3.77	15.62	0.29	4.53	nd	0.19	73.63	0.47	0.05	0.29	nd	nd	nd	nd	nd
mean A	0.14	0.60	4.30	17.74	0.43	3.08	nd	0.08	72.91	0.52	0.12	0.17	nd	nd	nd	nd	nd
W	0.17	0.04	0.70	0.17	0.86	0.22	0.02	0.28	97.51	n.d.	n.d.	0.01	nd	nd	nd	nd	nd
F	0.14	3.34	2.70	30.64	0.10	30.94	0.09	0.03	29.78	0.93	0.05	0.08	nd	nd	nd	nd	nd
F	0.21	0.76	0.94	29.55	1.12	34.56	nd	0.28	31.29	1.03	0.06	0.19	nd	nd	nd	nd	nd
mean F	0.18	2.05	1.82	30.10	0.61	32.75	0.05	0.16	30.54	0.98	0.06	0.14	nd	nd	nd	nd	nd
I	1.15	21.20	22.72	34.66	nd	6.87	0.16	0.09	12.57	0.44	0.04	0.09	nd	nd	nd	nd	nd
I	2.29	14.62	23.03	34.50	nd	13.25	0.12	0.14	10.43	0.67	0.25	0.69	nd	nd	nd	nd	nd
mean I	1.72	17.91	22.88	34.58	nd	10.06	0.14	0.12	11.50	0.56	0.15	0.39	nd	nd	nd	nd	nd
KL 91																	
A	0.76	1.15	3.53	15.95	0.46	9.52	0.04	0.23	66.56	1.22	0.08	nd	nd	0.23	0.06	0.07	0.09
A	0.63	1.00	3.98	18.87	0.78	6.49	0.03	0.39	66.34	0.85	0.13	nd	nd	0.26	0.30	0.19	nd
mean A	0.70	1.08	3.76	17.41	0.62	8.01	0.04	0.31	66.45	1.04	0.11	nd	nd	0.25	0.18	0.13	0.05
W	nd	0.18	0.51	0.53	0.93	0.66	0.01	0.45	96.46	nd	nd	nd	nd	0.12	nd	nd	0.10
W	nd	0.30	0.26	1.17	0.06	0.39	nd	0.13	97.48	0.18	0.13	nd	nd	nd	nd	nd	0.02
W	nd	nd	nd	0.26	nd	0.04	0.02	0.65	99.28	0.07	0.05	nd	nd	nd	nd	0.09	nd
mean W	nd	0.16	0.26	0.65	0.33	0.36	0.01	0.41	97.74	0.08	0.06	nd	nd	0.04	nd	0.03	0.04
F	0.25	5.95	7.54	29.00	0.95	22.80	0.12	0.40	32.12	0.42	0.12	0.45	nd	0.12	nd	nd	nd
F	nd	1.11	5.39	41.54	0.34	10.13	0.02	0.20	36.08	4.61	0.10	0.33	nd	0.09	nd	nd	nd

Analyses	Na_2O	K_2O	Al_2O_3	SiO_2	MgO	CaO	TiO_2	MnO	FeO	P_2O_5	SO_3	BaO	Cr_2O_3	As_2O_3	NiO	ZnO	CuO
mean F	0.13	3.53	6.47	35.27	0.65	16.47	0.07	0.30	34.10	2.52	0.11	0.39	nd	0.11	nd	nd	nd
I	nd	0.34	1.38	33.68	0.34	9.66	nd	0.35	51.76	2.13	0.02	0.27	nd	0.03	nd	nd	nd
I	nd	0.71	9.60	51.16	0.62	6.40	0.23	0.33	29.50	0.80	0.17	nd	nd	0.16	nd	nd	0.17
mean I	nd	0.53	5.49	42.42	0.48	8.03	0.12	0.34	40.63	1.47	0.10	0.14	nd	0.10	nd	nd	0.09
KL 95																	
A	0.47	0.28	0.73	4.22	0.14	1.25	nd	0.04	91.26	0.20	0.09	nd	nd	0.14	0.14	0.06	0.92
A	0.53	0.35	0.84	5.46	0.31	2.51	nd	0.08	88.32	0.18	0.12	nd	nd	0.34	nd	0.40	0.85
mean A	0.50	0.32	0.79	4.84	0.23	1.88	nd	0.06	89.79	0.19	0.11	nd	nd	0.24	0.07	0.23	0.89
W	0.42	0.10	0.30	0.16	0.23	0.06	0.08	0.09	98.35	0.02	0.01	nd	nd	nd	0.04	nd	0.05
F	nd	2.55	2.37	46.84	0.11	9.25	nd	0.02	30.47	3.38	0.27	0.29	nd	0.29	0.06	0.13	3.92
F	0.25	0.70	0.56	34.98	0.29	15.84	0.06	0.08	41.34	2.35	0.21	0.14	nd	0.11	0.02		2.90
mean Al, P,Cu rich F	0.13	1.63	1.47	40.91	0.20	12.55	0.03	0.05	35.91	2.87	0.24	0.22	nd	0.20	0.04	0.07	3.41
Al, P rich I	0.74	16.19	18.40	39.62	0.05	6.27	nd	nd	16.09	1.03	0.21	0.20	nd	0.19	0.09	nd	0.09
Al, P rich I	1.07	15.16	18.42	39.99	nd	6.47	0.01	nd	16.55	1.09	0.08	0.08	nd	nd	nd	nd	1.01
mean Al, P rich I	0.91	15.68	18.41	39.81	0.03	6.37	0.01	nd	16.32	1.06	0.15	0.14	nd	0.10	0.05	nd	0.55
Cu rich I	0.02	1.37	1.50	66.34	0.38	7.04	0.10	0.12	18.47	2.60	0.23	nd	nd	0.26	nd	nd	1.52
Cu rich I	nd	1.10	0.91	44.68	0.17	24.09	nd	0.11	22.75	2.64	0.55	0.43	nd	1.13	0.08	0.20	1.28
mean Cu rich I	0.01	1.24	1.21	55.51	0.28	15.57	0.05	0.12	20.61	2.62	0.39	0.22	nd	0.70	0.04	0.10	1.40
KL 98																	
A	0.26	1.55	4.55	20.19	0.39	3.51	0.04	0.13	68.80	0.38	0.19	0.00	nd	nd	nd	nd	nd
A	0.59	1.96	4.75	20.94	0.60	4.78	0.00	0.10	65.54	0.50	0.18	0.05	nd	nd	nd	nd	nd
mean A	0.42	1.76	4.65	20.56	0.49	4.14	0.02	0.11	67.17	0.18	0.02	nd	nd	nd	nd	nd	nd
W	nd	0.05	0.39	0.51	0.37	0.27	0.25	0.09	98.05	0.00	0.00	0.00	nd	nd	nd	nd	nd
Mg rich F	0.34	0.42	0.80	30.24	1.76	21.16	0.10	0.40	44.31	0.43	0.03	0.00	nd	nd	nd	nd	nd
K, Al rich I	1.45	10.03	14.40	40.63	0.29	7.50	0.26	0.17	24.13	0.84	0.06	0.22	nd	nd	nd	nd	nd
KL 99a																	
A	0.39	1.27	4.01	17.25	0.52	4.17	0.18	0.13	71.86	0.15	0.08	nd	nd	nd	nd	nd	nd
A	0.41	1.04	3.23	13.61	0.51	3.45	0.09	0.04	77.21	0.18	0.13	nd	nd	nd	nd	nd	nd
mean A	0.40	1.15	3.62	15.43	0.51	3.81	0.13	0.08	74.54	0.16	0.10	nd	nd	nd	nd	nd	nd
W	0.13	0.05	0.59	0.38	0.37	0.05	0.14	0.08	98.17	nd	0.04	nd	nd	nd	nd	nd	nd
W	0.12	2.95	5.94	10.64	0.19	0.26	0.47	0.06	79.24	0.04	nd	nd	nd	nd	nd	nd	nd
mean W	0.13	1.50	3.27	5.51	0.28	0.16	0.31	0.07	88.71	0.02	0.02	nd	nd	nd	nd	nd	nd
F	0.65	2.92	4.85	32.04	0.63	14.24	nd	0.22	43.93	0.30	nd	0.23	nd	nd	nd	nd	nd
F	0.75	1.72	3.85	30.62	1.16	15.92	0.02	0.27	45.46	0.14	0.05	0.03	nd	nd	nd	nd	nd
F	0.03	0.15	ndn	28.23	1.49	15.09	0.02	0.27	54.45	0.26	nd	nd	nd	nd	nd	nd	nd
mean F	0.48	1.60	2.90	30.30	1.09	15.08	0.01	0.25	47.95	0.23	0.02	0.09	nd	nd	nd	nd	nd
I	2.48	7.21	15.75	37.25	0.03	10.96	0.06	0.06	25.40	0.48	0.09	0.20	nd	nd	nd	nd	nd
I	2.49	9.40	19.49	42.37	nd	5.42	0.03	n.d.	19.70	0.58	0.15	0.37	nd	nd	nd	nd	nd
mean I	2.49	8.31	17.62	39.81	0.02	8.19	0.05	0.03	22.55	0.53	0.12	0.29	nd	nd	nd	nd	nd
KL 99b																	
A	0.22	0.05	1.91	11.51	0.22	1.57	0.00	0.04	83.80	0.39	0.17	0.10	nd	nd	nd	nd	nd
A	0.26	0.32	1.69	11.36	0.55	5.49	0.05	0.13	79.09	0.77	0.25	0.04	nd	nd	nd	nd	nd
mean A	0.24	0.18	1.80	11.43	0.38	3.53	0.02	0.08	81.44	0.58	0.21	0.07	nd	nd	nd	nd	nd
W	0.25	0.04	0.24	0.25	0.13	0.34	0.09	0.00	98.52	0.06	0.08	0.00	nd	nd	nd	nd	nd
F	0.08	0.02	8.49	15.43	0.43	1.58	0.28	0.05	72.77	0.73	0.01	0.04	nd	nd	nd	nd	nd
I	0.06	0.15	0.00	29.58	0.47	39.08	0.00	0.03	30.34	0.28	0.00	0.00	nd	nd	nd	nd	nd
KL 101b																	
A	nd	1.60	4.20	16.60	0.30	7.90	0.10	9.30	52.60	1.10	4.00	0.20	0.60	0.80	0.10	nd	nd
A	nd	1.30	3.80	15.70	0.20	7.10	nd	9.10	54.70	1.40	3.60	0.30	1.10	0.10	nd	nd	nd

Analyses	Na_2O	K_2O	Al_2O_3	SiO_2	MgO	CaO	TiO_2	MnO	FeO	P_2O_5	SO_3	BaO	Cr_2O_3	As_2O_3	NiO	ZnO	CuO
A	nd	1.70	4.30	17.00	0.20	8.20	0.10	9.10	52.70	1.20	3.80	0.10	0.50	0.70	0.10	nd	nd
mean A	nd	1.53	4.10	16.43	0.23	7.73	0.07	9.17	53.33	1.23	3.80	0.20	0.73	0.53	0.07	nd	nd
W	nd	nd	0.40	0.40	nd	nd	0.30	6.40	90.90	nd	nd	nd	0.40	0.60	nd	nd	nd
W	0.40	nd	0.20	0.50	nd	0.20	0.10	6.30	90.90	nd	0.10	nd	0.20	0.40	nd	nd	nd
mean W	0.20	nd	0.30	0.45	nd	0.10	0.20	6.35	90.90	nd	0.05	nd	0.30	0.50	nd	nd	nd
Mn, S, P, Zn rich F	nd	nd	7.00	21.70	0.90	10.30	0.30	12.40	43.50	0.70	1.60	nd	nd	nd	nd	1.20	nd
Mn, S, P, Zn rich F	nd	nd	6.10	23.40	0.70	11.00	0.20	13.00	42.70	0.50	0.60	nd	nd	nd	nd	1.10	nd
mean Mn, S, P, Zn rich F	nd	nd	3.40	22.55	0.80	10.65	0.25	12.70	43.10	0.60	1.10	nd	nd	nd	nd	1.15	nd
KL 102a																	
A	0.65	3.51	5.99	30.55	0.51	10.58	0.07	0.11	46.88	0.31	0.40	0.22	nd	0.39	0.03	nd	0.08
A	0.08	2.71	5.34	24.69	0.56	7.65	0.04	nd	57.98	0.40	0.02	0.15	nd	nd	nd	0.11	0.23
A	1.24	3.31	5.76	29.23	0.61	8.82	0.08	0.12	49.93	0.34	0.06	0.06	nd	0.11	0.06	0.16	0.04
A	1.54	3.26	6.22	27.40	0.57	8.15	0.09	0.14	51.98	0.33	0.14	nd	nd	nd	0.07	0.04	0.01
mean A	0.88	3.20	5.83	27.97	0.56	8.80	0.07	0.09	51.69	0.35	0.16	0.11	nd	0.13	0.04	0.05	0.09
W	nd	0.16	0.68	1.86	0.19	0.47	0.07	0.08	96.27	nd	nd	nd	nd	0.17	0.01	nd	nd
W	0.69	0.58	1.65	13.85	0.82	3.67	0.08	0.14	77.15	0.30	0.04	0.05	nd	nd	0.12	nd	0.14
mean W	0.15	0.37	1.17	7.86	0.51	2.07	0.08	0.11	86.71	0.15	0.02	0.03	nd	0.09	0.07	nd	0.07
F	0.31	0.19	nd	29.18	0.87	18.30	0.02	0.08	50.12	0.45	0.08	0.04	nd	0.09	0.09	nd	0.12
F	0.07	0.15	nd	29.07	2.02	11.55	0.09	0.11	56.67	0.18	0.03	nd	nd	nd	nd	nd	nd
mean F	0.19	0.17	n.d.	29.13	1.45	14.93	0.06	0.10	53.40	0.32	0.06	0.04	nd	0.05	0.05	nd	0.06
I	0.86	4.60	6.70	37.83	0.43	13.37	nd	0.11	35.26	0.48	0.09	0.20	nd	nd	nd	nd	0.20
I	1.08	4.91	7.25	38.41	0.47	12.78	nd	0.01	34.21	0.38	0.09	0.19	nd	0.10	nd	nd	0.06
I	2.58	9.52	14.99	44.85	nd	7.05	0.15	0.03	19.92	0.63	0.05	0.04	nd	nd	nd	nd	0.15
I	2.06	11.23	16.17	48.48	nd	6.48	0.13	0.02	14.10	0.76	0.18	0.26	nd	nd	0.04	nd	0.40
mean I	1.65	7.57	11.28	42.39	0.23	9.92	0.07	0.04	25.87	0.56	0.10	0.17	nd	0.03	0.01	nd	0.20
Mn rich I	1.99	9.42	13.21	44.08	0.08	9.19	0.21	6.04	20.07	0.87	0.22	0.16	nd	nd	0.30	0.05	0.33
KL 4																	
I	0.07	0.57	2.26	7.35	0.51	3.00	n.d.	n.d.	84.75	0.41	0.20	nd	n.d.	0.14	0.37	0.35	0.02
I	0.30	0.04	0.25	0.22	n.d.	6.48	n.d.	0.18	91.94	0.10	0.17	nd	0.10	n.d.	0.15	n.d.	0.07
mean I	0.19	0.31	1.26	3.79	0.26	4.74	n.d.	0.09	88.35	0.26	0.19	nd	0.05	0.07	0.26	0.18	0.05
KL 33a																	
I	0.70	3.98	10.45	43.35	2.45	24.00	7.73	0.28	5.46	1.60	nd	nd	nd	nd	nd	nd	nd
I	0.64	4.00	9.75	43.30	2.70	23.92	7.69	0.07	5.75	1.56	nd	0.37	nd	0.25	nd	nd	nd
mean I	0.67	3.99	10.10	43.33	2.58	23.96	7.71	0.18	5.61	1.58	nd	0.19	nd	0.13	nd	nd	nd
KL 33b																	
I	0.15	4.30	9.27	40.09	2.29	19.04	7.08	0.62	15.27	1.89	nd	nd	nd	nd	nd	nd	nd
I	0.47	2.52	7.42	28.82	2.00	13.82	5.53	0.43	36.32	1.90	nd	0.37	nd	0.24	0.16	nd	nd
mean I	0.31	3.41	8.35	34.46	2.15	16.43	6.31	0.53	25.80	1.90	nd	0.19	nd	0.12	0.08	nd	nd

Table 14b. SEM data from Kalyva-Kastro (summary of mean analyses fromtable 14a).

Analyses	Na_2O	K_2O	Al_2O_3	SiO_2	MgO	CaO	TiO_2	MnO	FeO	P_2O_5	SO_3	BaO	Cr_2O_3	As_2O_3	NiO	ZnO	CuO
KL 70																	
A	0.49	1.09	3.03	12.41	0.51	2.93	nd	0.08	78.88	0.35	nd	nd	nd	nd	nd	nd	nd
W	0.34	n.d.	0.52	0.24	0.76	0.04	nd	0.05	98.35	nd	nd	0.04	nd	nd	nd	nd	nd
Mg, P rich F	0.14	0.37	0.12	29.98	1.25	29.97	nd	0.09	36.86	1.05	nd	0.14	nd	nd	nd	nd	nd
Al, P rich I	1.32	18.57	18.68	38.22	0.06	4.16	0.01	nd	16.19	2.05	0.24	0.50	nd	nd	nd	nd	nd
KL 73																	
A	0.30	1.80	3.80	21.40	0.16	4.80	nd	0.40	66.00	0.60	0.16	0.10	nd	nd	nd	nd	nd
W	0.15	nd	nd	0.25	0.40	nd	nd	0.55	97.80	nd	0.10	nd	nd	nd	nd	nd	nd
F																	
I	0.60	6.70	6.00	36.60	nd	20.40	nd	0.40	27.10	0.90	0.10	0.10	nd	nd	nd	nd	nd
Fe prill	nd	nd	nd	nd	nd	nd	nd	nd	99.45	0.10	nd	0.10	nd	nd	nd	nd	nd
KL 74																	
A	0.50	2.90	22.00	44.00	0.40	17.00	1.40	0.30	11.00	0.30	0.10	nd	nd	nd	nd	nd	nd
W		0.50	9.70	11.50	1.60	2.70	1.60	0.30	71.60	0.10	nd	nd	nd	nd	nd	nd	nd
Al, Mg, Ti rich F	0.40	4.40	15.80	44.60	1.30	14.00	1.60	0.60	16.30	0.40	0.10	nd	nd	nd	nd	nd	nd
Al rich I	0.30	0.40	30.80	45.50	nd	nd	nd	nd	nd	nd	nd	nd	nd	nd	nd	nd	nd
KL 79																	
A	0.65	0.65	0.09	1.16	0.32	2.23	0.02	0.07	93.54	2.95	0.23	0.08	nd	nd	nd	nd	nd
Fe prill (Ti rich)	0.28	1.62	7.54	9.37	0.56	3.57	1.85	0.08	72.96	1.83	0.08	0.27	nd	nd	nd	nd	nd
KL 82																	
A	0.53	1.23	2.27	12.30	0.43	8.43	nd	0.20	73.43	0.33	0.23	0.10	nd	nd	nd	nd	nd
W	0.50	0.95	2.70	9.35	0.55	5.75	nd	0.20	78.85	0.05	0.15	0.15	nd	nd	0.05	0.05	Nd
F	0.50	4.70	4.93	31.37	0.37	27.03	nd	0.13	29.50	0.73	0.10	0.10	nd	nd	nd	nd	nd
I	1.10	10.40	10.70	39.03	0.20	17.27	0.30	0.13	22.83	0.30	0.23	0.17	nd	0.07		0.03	0.10
Mg rich F	0.45	0.15	0.25	26.15	1.40	26.90	nd	0.30	39.60	0.45	nd	nd	nd	nd	nd	nd	nd
KL85																	
A	1.18	6.43	11.93	55.53	0.43	5.43	0.20	0.08	18.08	0.28	nd	nd	nd	nd	nd	nd	nd
W	0.37	1.97	3.60	12.97	0.27	2.40	0.13	0.13	76.40	1.07	nd	nd	nd	nd	nd	nd	nd
F	0.70	4.10	6.15	38.50	0.65	15.25	0.05	0.20	33.40	0.60	nd	0.10	nd	nd	nd	nd	nd
I	0.35	11.65	20.00	62.75	0.05	1.45	0.70	0.05	2.40	nd	nd	0.15	nd	nd	0.05	nd	nd
Fe prill (Ti,Ni,P rich)	0.70	10.80	9.50	51.20	nd	1.80	1.10	nd	7.50	14.90	nd	nd	nd	nd	1.10	nd	nd
KL86																	
A	0.15	0.61	1.70	8.78	0.32	5.20	0.01	0.08	82.40	0.56	0.09	0.09	nd	nd	nd	nd	nd
W	0.15	0.02	0.16	0.11	0.43	0.31	0.04	0.10	98.51	nd	nd	0.15	nd	nd	nd	nd	nd
KL 70																	
F	nd	0.12	0.40	49.01	0.46	5.97	nd	nd	40.95	2.80	0.13	0.17	nd	nd	nd	nd	nd
	0.02	0.09		57.97	0.12	4.83	nd	0.10	34.91	1.73	0.16	0.06	nd	nd	nd	nd	nd
KL87																	
A	0.25	1.35	2.35	12.70	0.40	6.35	nd	0.05	75.05	0.40	nd	0.15	nd	nd	nd	nd	nd
W	nd	nd	0.47	0.27	0.13	nd	nd	0.13	97.97	nd	nd	nd	nd	nd	nd	nd	nd
F	0.70	0.05	nd	26.55	1.85	25.90	nd	nd	43.80	0.25	nd	nd	nd	nd	nd	nd	nd
I	1.30	21.15	24.65	36.30	nd	4.50	nd	nd	9.70	1.60	nd	0.25	nd	nd	nd	nd	nd
Mg rich F	0.45	0.40	nd	28.50	1.50	19.15	nd	0.30	48.40	1.00	nd	nd	nd	nd	nd	nd	nd
Al,Mg,P rich F	0.20	nd	13.30	30.40	1.50	6.00		0.10	40.30	5.70	0.20	0.30	nd	nd	nd	nd	1.00

Analyses	Na_2O	K_2O	Al_2O_3	SiO_2	MgO	CaO	TiO_2	MnO	FeO	P_2O_5	SO_3	BaO	Cr_2O_3	As_2O_3	NiO	ZnO	CuO
KL 89																	
A	0.58	3.29	4.74	20.61	0.81	12.16	0.07	0.62	56.22	0.43	0.10	0.07	nd	nd	nd	nd	nd
W	0.35	0.04	0.24	0.22	0.48	0.30	0.14	0.57	97.50	nd	0.10	0.06	nd	nd	nd	nd	nd
F	0.07	0.09	nd	28.70	1.62	28.83	0.01	1.01	39.41	0.72	0.01	nd	nd	nd	nd	nd	nd
I	0.76	31.12	28.14	37.79	nd	0.14	0.01	nd	1.89	nd	0.01	0.15	nd	nd	nd	nd	nd
KL 90																	
A	0.14	0.60	4.30	17.74	0.43	3.08		0.08	72.91	0.52	0.12	0.17	nd	nd	nd	nd	nd
W	0.17	0.04	0.70	0.17	0.86	0.22	0.02	0.28	97.51	nd	nd	0.01	nd	nd	nd	nd	nd
F	0.18	2.05	1.82	30.10	0.61	32.75	0.05	0.16	30.54	0.98	0.06	0.14	nd	nd	nd	nd	nd
I	1.72	17.91	22.88	34.58	n.d.	10.06	0.14	0.12	11.50	0.56	0.15	0.39	nd	nd	nd	nd	nd
KL 91																	
A	0.70	1.08	3.76	17.41	0.62	8.01	0.04	0.31	66.45	1.04	0.11	nd	nd	0.25	0.18	0.13	0.05
W	nd	0.16	0.26	0.65	0.33	0.36	0.01	0.41	97.74	0.08	0.06	nd	nd	0.04	nd	0.03	0.04
F	0.13	3.53	6.47	35.27	0.65	16.47	0.07	0.30	34.10	2.52	0.11	0.39	nd	0.11	nd	nd	
I	nd	0.53	5.49	42.42	0.48	8.03	0.12	0.34	40.63	1.47	0.10	0.14	nd	0.10	nd	nd	0.09
KL 95																	
A	0.50	0.32	0.79	4.84	0.23	1.88	nd	0.06	89.79	0.19	0.11	nd	nd	0.24	0.07	0.23	0.89
W	0.42	0.10	0.30	0.16	0.23	0.06	0.08	0.09	98.35	0.02	0.01	nd	nd	nd	0.04	nd	0.05
Al,P,Cu rich F	0.13	1.63	1.47	40.91	0.20	12.55	0.03	0.05	35.91	2.87	0.24	0.22	nd	0.20	0.04	0.07	3.41
Al,P rich I	0.27	0.86	0.88	20.54	0.03	6.30	0.01		67.13	1.44	0.13	0.11	nd	0.10	0.05	nd	1.73
Cu rich I	0.01	1.24	1.21	55.51	0.28	15.57	0.05	0.12	20.61	2.62	0.39	0.22	nd	0.70	0.04	0.10	1.40
KL 98																	
A	0.42	1.76	4.65	20.56	0.49	4.14	0.02	0.11	67.17	0.18	0.02	nd	nd	nd	nd	nd	nd
W		0.05	0.39	0.51	0.37	0.27	0.25	0.09	98.05	nd	nd	nd	nd	nd	nd	nd	nd
Mg rich F	0.34	0.42	0.80	30.24	1.76	21.16	0.10	0.40	44.31	0.43	0.03	nd	nd	nd	nd	nd	nd
K, Al rich I	1.45	10.03	14.40	40.63	0.29	7.50	0.26	0.17	24.13	0.84	0.06	0.22	nd	nd	nd	nd	nd
KL 99a																	
A	0.40	1.15	3.62	15.43	0.51	3.81	0.13	0.08	74.54	0.16	0.10	nd	nd	nd	nd	nd	nd
W	0.13	1.50	3.27	5.51	0.28	0.16	0.31	0.07	88.71	0.02	0.02	nd	nd	nd	nd	nd	nd
F	0.48	1.60	2.90	30.30	1.09	15.08	0.01	0.25	47.95	0.23	0.02	0.09	nd	nd	nd	nd	nd
I	0.30	1.55	3.08	17.90	0.69	7.62	0.16	0.16	68.33	0.13	0.02	0.04	nd	nd	nd	nd	nd
KL 99b																	
A	0.24	0.18	1.80	11.43	0.38	3.53	0.02	0.08	81.44	0.58	0.21	0.07	nd	nd	nd	nd	nd
W	0.25	0.04	0.24	0.25	0.13	0.34	0.09	0.00	98.52	0.06	0.08	nd	nd	nd	nd	nd	nd
F	0.08	0.02	8.49	15.43	0.43	1.58	0.28	0.05	72.77	0.73	0.01	0.04	nd	nd	nd	nd	nd
I	0.06	0.15	0.00	29.58	0.47	39.08	0.00	0.03	30.34	0.28	0.00	0.00	nd	nd	nd	nd	nd
KL 101b																	
A	nd	1.53	4.10	16.43	0.23	7.73	0.07	9.17	53.33	1.23	3.80	0.20	0.73	0.53	0.07	nd	nd
Mn rich W	0.20	nd	0.30	0.45	nd	0.10	0.20	6.35	90.90	nd	0.05	nd	0.30	0.50	nd	nd	nd
Mn, S, P, Zn rich F	nd	nd	3.40	22.55	0.80	10.65	0.25	12.70	43.10	0.60	1.10	nd	nd	nd	nd	1.15	nd
KL 102a																	
A	0.88	3.20	5.83	27.97	0.56	8.80	0.07	0.09	51.69	0.35	0.16	0.11	nd	0.13	0.04	0.05	0.09
W	0.15	0.37	1.17	7.86	0.51	2.07	0.08	0.11	86.71	0.15	0.02	0.03	nd	0.26	0.07	nd	0.07
F	0.19	0.17		29.13	1.45	14.93	0.06	0.10	53.40	0.32	0.06	0.04	nd	0.05	0.05	nd	0.06
I	1.65	7.57	11.28	42.39	0.23	9.92	0.07	0.04	25.87	0.56	0.10	0.17	nd	0.03	0.01	nd	0.20
Mn rich I	1.99	9.42	13.21	44.08	0.08	9.19	0.21	6.04	20.07	0.87	0.22	0.16	nd	nd	0.30	0.05	0.33
KL 4																	
I	0.19	0.31	1.26	3.79	0.26	4.74	nd	0.09	88.35	0.26	0.19	nd	0.05	0.07	0.26	0.18	0.05

Analyses	Na_2O	K_2O	Al_2O_3	SiO_2	MgO	CaO	TiO_2	MnO	FeO	P_2O_5	SO_3	BaO	Cr_2O_3	As_2O_3	NiO	ZnO	CuO
KL 33a																	
I	0.67	3.99	10.10	43.33	2.58	23.96	7.71	0.18	5.61	1.58	nd	0.19	nd	0.13	nd	nd	Nd
KL 33b																	
I	0.31	3.41	8.35	34.46	2.15	16.43	6.31	0.53	25.80	1.90	nd	0.19	nd	0.12	0.08	nd	nd

Table15 . Metallography of the iron artefacts from Messemvria-Zone.

Typological Classification	Sample No	Date	Structure
Weapons Arrowheads	MES 29	Late 6th c BC	F with slag inclusions.
	MES 62	5th c BC	Edge: F and P, eutectic composition. Core: F, no welding lines.
	MES 1	4th c BC	F
	MES 2	4th c BC	F
Spearheads	MES 22	Late 6th c BC	Edge: P and F, eutectic composition. Core: F and P, medium carbon.steel, Welding lines.
	MES 23	Late 6th c BC	Edge: F and P, low carbon steel. Core: F
	MES 25	Late 6th BC	Edge: P and F, eutectic composition. Core: F and P, medium carbon steel, No welding lines.
	MES 10	4th c BC	Edge: TM Core: F and P, low carbon steel.
	MES 7	4th c BC	Edge: P and C, hyper-eutectic composition. Core: F and P, medium carbon steel.
Dagger	MES 16	Late 6th c BC	Edge: Lamellar P and F in Pro-eutectic composition. Handle: F and , medium carbon steel.
Spits	MES 31	Late 6th c BC	F and P, medium carbon steel P and C, hyper-eutectic composition
	MES 32	Late 6th c BC	F and P, medium carbon steel P and C, hyper-eutectic composition
	MES 33	Late 6th c BC	F F and P, low carbon steel
	MES 35	5th c BC	Edge: F and P, eutectic composition Core: F
Iron bar	Mes 21	Late 6th BC	F with slag inclusions.
Knives, straight	MES 17	Late 6th c BC	Edge: P and F, eutectic steel. Core: F and P, low carbon steel, welding lines.
	MES 9	4th c BC	P and spearodised C, uniform high carbon steel.
	MES 50	4th c BC	Edges: F Core and point of the edge: F and P, Low carbon steel.
Knives, curved	MES 65	5th c BC	F and P, welding of high carbon to medium carbon steel.
Sickles	MES 44	5th c BC	Welded sheets of F and P, low carbon steel that is Cu rich (different color under the microscope)
Falx	MES 53	5th c BC	Edge: medium carbon steel F and P. Core: low carbon F and P, welding

Table 16: Metallography of iron artefacts from Avdera.

Typological Classification	Sample No	Date	Structure
Straight knives	AVD 51	6th –5th c BC	Edge and core: F and P, hypo-eutectic Composition, low carbon steel.
	AVD 11	5th c BC	Edge: deformed F Core: F
	AVD 15		Edge: P and F, eutectic composition Core: F and P, medium carbon steel, carburization
	AVD 18	1st –2nd c AD	Edge: F and P, medium carbon steel Core: F, welding lines
	AVD 33	1st –2nd c AD	Edge and core: F and P, hypo-eutectic Composition, low carbon steel.
	AVD 36	1st –2nd c AD	Edge: P and some bainite Core: F and P, low carbon steel, welding lines
	AVD 53	1st –2nd c AD	Edge: P and F, eutectic composition And some bainite Core: F and welding lines
Curved knives	AVD 37	1st –2nd c AD	Edge: deformed F Core: F and slag inclusions
Sickles	AVD 23	5th c BC	Edge and core: F and P. hypo-eutectic Composition, low carbon steel
Falx	AVD 40		F and slag inclusions
Wedges	AVD 49	5th c BC	Edge and core: P and B
	AVD 61	1st –2nd cAD	Edge and core: P and F, eutectic Composition, some B
Axes	AVD 30		Edge: P and B Core: P and F, eutectic composition, Welding lines
	AND 39	2nd-1st c BC	Edge: F and P, hypo-eutectic Core: F
Two pointed mattock	AVD 48	1st –2nd c AD	Edge: eutectic P and some B Core: F and P, low carbon, a lot of Slag inclusions ad corrosion along the welding lines
Shaw	AVD 26	1st –2nd c AD	Edge: F and P, medium carbon Core: F and welding lines
Spits	AVD 43	1st –2nd cAD	F and P, uniform low carbon steel
	AVD 44	1st –2nd c AD	P, sphaerodised cementite and B Core: F and slag inclusions
	AVD 50	1st –2nd cAD	F with slag inclusions
Curry-comps	AVD 1	4th c BC	Edge: deformed F Core: F
	AVD 5	4th c BC	Edges: F Core: F and P, medium carbon steel, Slag inclusions

Table 17 : Metallography of the iron artefacts from Kalyva-Kastro.

Ty[pological Classification	Sample No	Date	Structure	HV (100p)
Sockets of spearheads (type 7)	KL 60	2nd BC	Pearlite.	233
	KL 17	3rd AD	Edge: Accicular Ferrite. Core: Pearlite.	181 202
	KL 27	2nd AD.	Edge: small grains of Ferrite and Pearlite. Core: large grains of Pearlite with Ferrite between the boundaries.	237 206
Arrowheads (type 7)	KL 8	2nd AD	Ferrite.	
knives (type 1) straight blades	KL 16	3rd AD	Core: Accicular ferrite and unresolvable Pearlite.	142
	KL 20	3rd AD	Edge: Tempered Martensite. Core: Pearlite and some Ferrite.	383 220
	KL 22	3rd AD	Edge: Ferrite.	119
	KL 37	3rd AD	Edge: Tempered Martensite.	299
	KL 39	3rd AD	Edge: Ferrite and Pearlite. Core: Ferrite.	235 114
	KL 41	3rd AD	Edge: Pearlite and Accicular Ferrite.	
	KL 46	3rd AD	Edge: Tempered Martensite.	357
	KL 48	3rd AD	Ferrite and Pearlite.	181
	KL 50	3rd AD		
	KL 62		Ferrite and some Pearlite.	130
knives (type2)				
curved blades	KL 4	3rd AD	Edge: Ferrite.	151
	KL 12		Edge: Ferrite and Pearlite. Core: Ferrite	142 123
	KL 14		Edge: Ferrite and Pearlite. Core: Ferrite and Pearlite.	205 180
	KL 23		Edge: Tempered Martensite. Core: Ferrite.	309 152
	KL 65		Edge: Pearlite and Cementite Core: Ferrite.	 131
knives (type 3) large triangular blades	KL 66		Edge: Ferrite and Pearlite.	128
	KL 67		Edge: Tempered Martensite. Core: Ferrite and Pearlite.	294 177
knives (typ4) small triangular blades				
	KL 59	3rd AD	Edge: Deformed Ferrite.	336
Sickles, billhooks, falces (type 5).	KL 3	3rd AD	Edge: Ferrite.	186
	KL 33	3rd AD	Edge: Tempered Martensite. Core: Pearlite.	317 222
	KL 64	3rd AD	Edge: Pearlite.	227
Axes (type 6)	KL 55	3rd AD	Edge: Tempered Martensite.	322
Jointers (type 9)	KL 21	2nd AD	Ferrite and Pearlite	
	KL 30	2nd BC	Ferrite	125
	KL 31	2nd BC	Pearlite and some Ferrite in the center	221
Nails (type 9)	KL 9b	3rd AD	Ferrite and Pearlite	

PLATE 1. MESSEMVRIA-ZONE, SMELTING SLAG-GROUP 1.

SEM photograph of MES130.	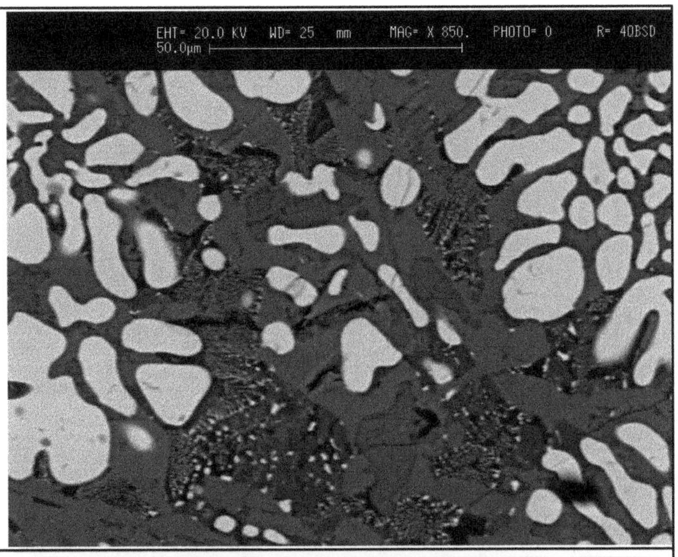
MES 137 x528, similar to MES130.	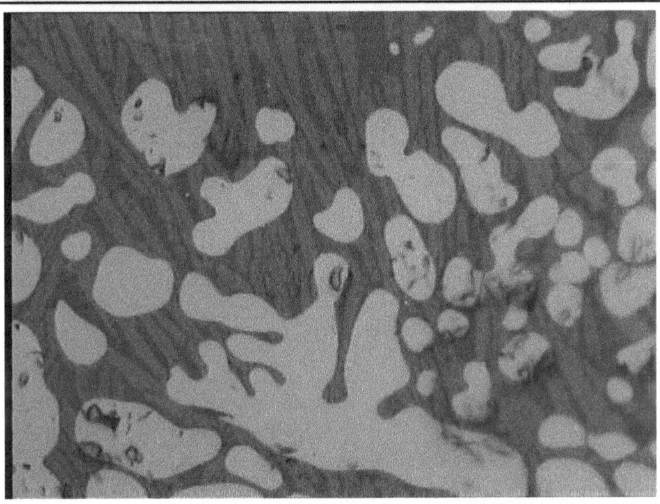

PLATE 2. MESSEMVRIA-ZONE, SMELTING SLAG-GROUP 2.

SEM photograph MES 147b, showing uniform distribution of long fayalite needles, and pores (black holes).	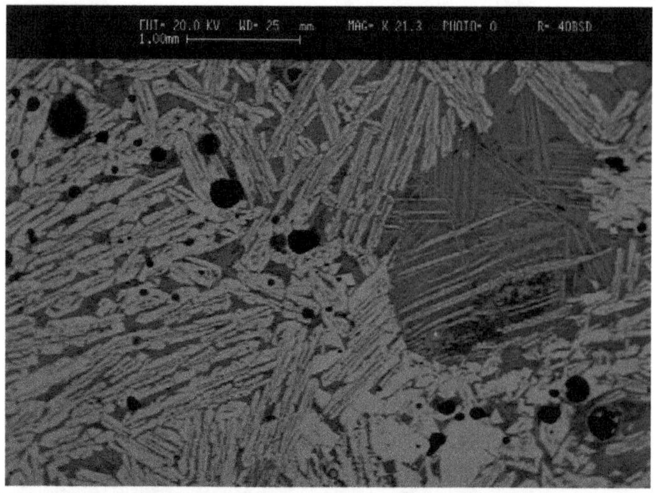
SEM photograph of MES 111, showing long fayalite needles rich in Mg (2.15%), Ca (2.99%) and Mn (6.40%). The prill is 99.39% iron. The very light gray areas on the corner of the picture are limonite from the weathering of the iron ore.	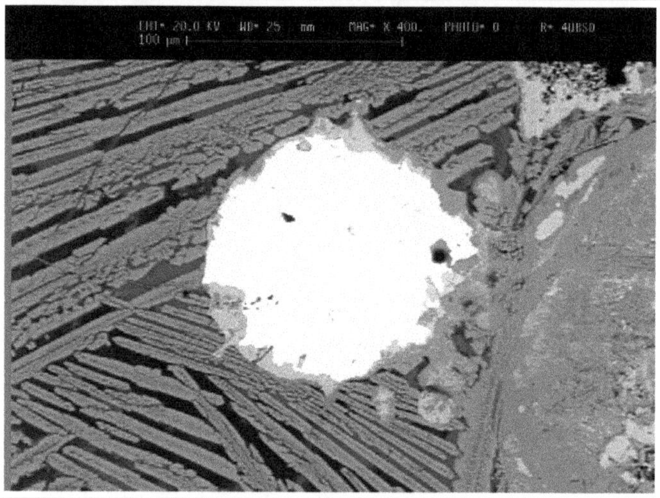
Photomicrograph of MES 125 x 264, the iron oxide formed little magnetite skeletons (white) that point to the smelt of a high quality ore. Fayalite is rich in Mg (1.24%) and Mn (1.91%).	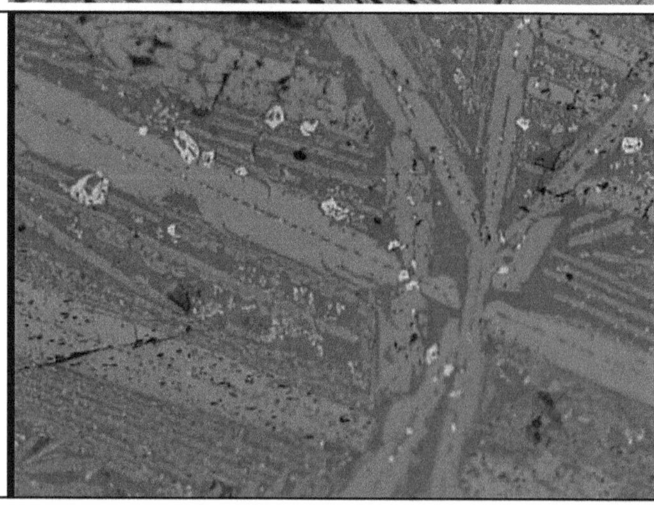

PLATE 3. MESSEMVRIA-ZONE. SMITHING SLAG.

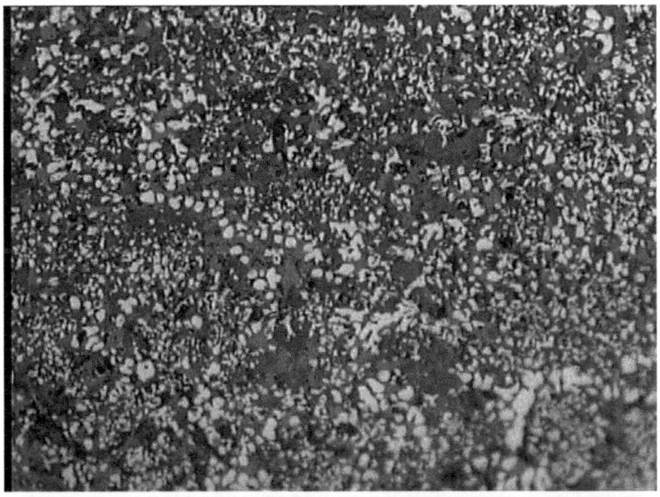

MES 143 x200, showing globules of wustite (white), and islands of fayalite (light gray) in a glassy matrix (dark gray). Black areas are holes.

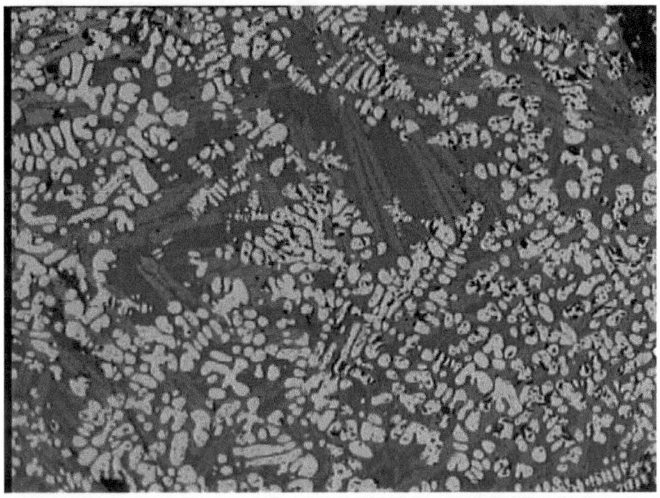

MES 106 x100, showing dentrites of wustite (white), and needles of fayalite (light gray) in a glassy matrix (dark gray). Black areas are holes.

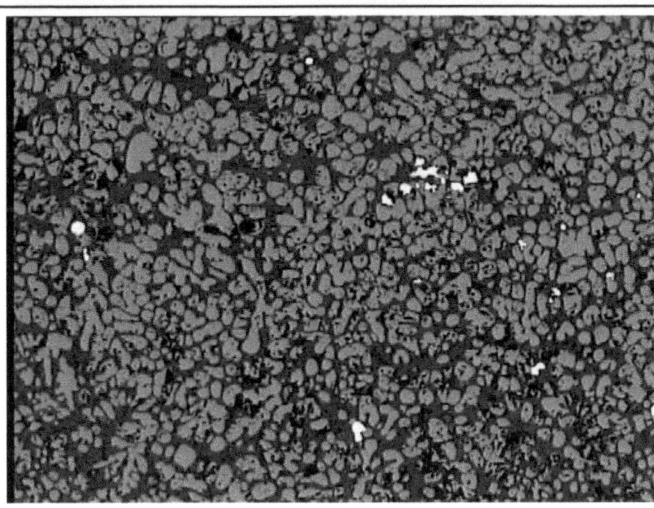

MES 124 x200, showing drops of metallic iron (white), and globules of wustite (light gray), in a glassy matrix (dark gray). Black areas are holes.

PLATE 4. MESSEMVRIA-ZONE, FURNACE LINING.

SEM photograph of MES 146c showing the distribution of two fayalitic phases in the interstitial glass..	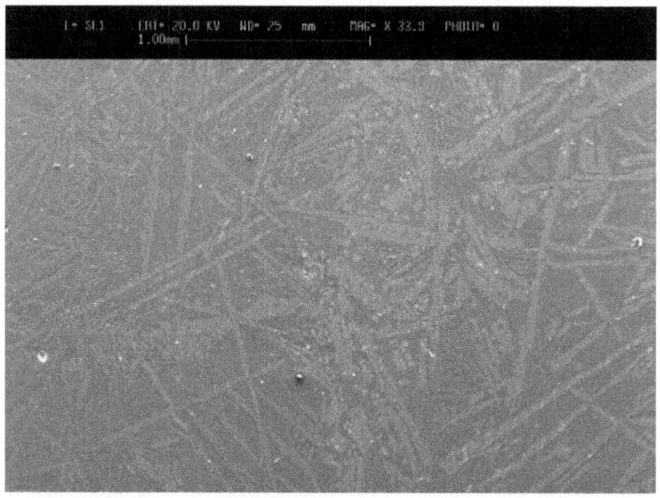
SEM photograph of MES 146d is a highly vitrified and porous material with also some cracks probably because of the thermal shock.	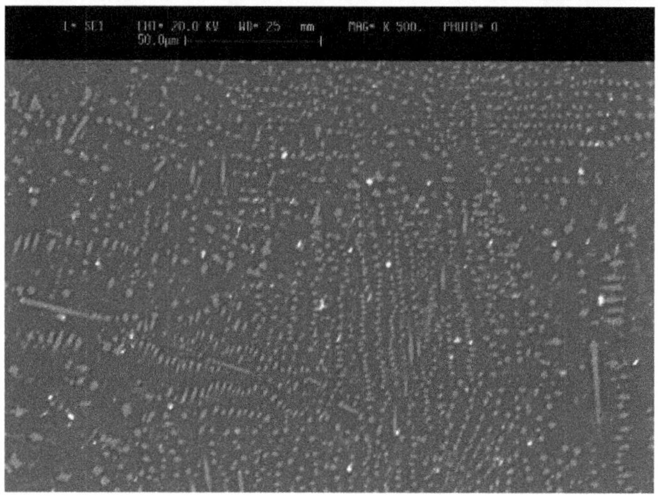

PLATE 5a. MESSEMVRIA-ZONE, ARTEFACTS OF FERRITIC STRUCTURE.

Spearheads

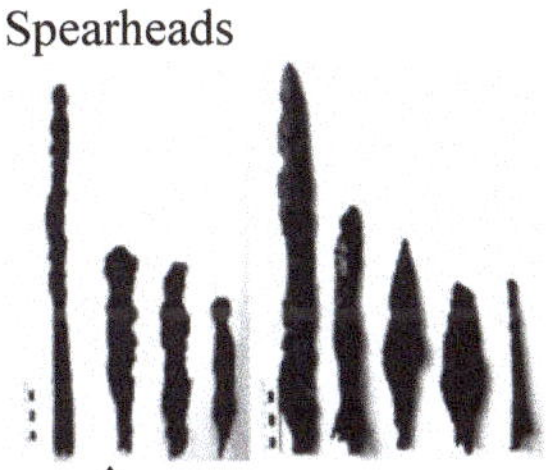

↑

MES 23 x66, edge. Very soft and slightly deformed ferrite.

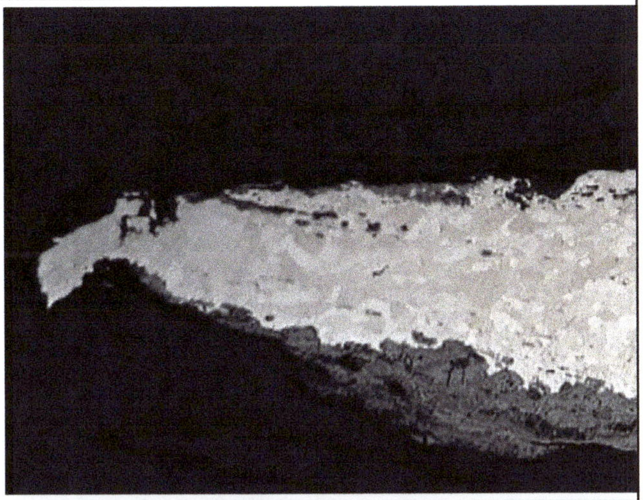

SEM photographs of slag inclusions of MES 23 showing Wustite, Fayalite and Interstitial glass. The absence of Mn points out to its not local origin.

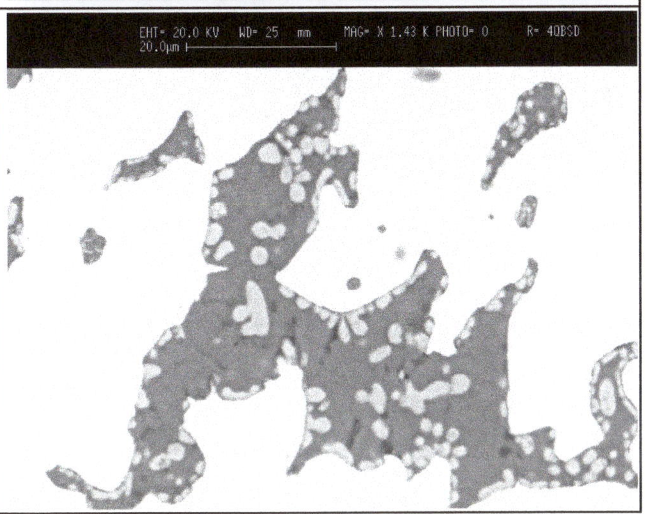

Analyses	Na_2O	K_2O	Al_2O_3	SiO_2	MgO	CaO	TiO_2	MnO	FeO	P_2O_5	SO_3
MES 23											
W	0.15	0.22	1.44	2.32	0.06	0.24	0.55	0.06	94.58	0.05	0.44
F	0.05	0.11	1.15	24.43	0.62	0.87	0.06	0.29	72.20	0.09	0.02
I	0.20	2.40	5.21	25.53	0.45	3.11	0.10	0.17	61.11	1.20	0.15
M	0.06	n.d	0.02	0.09	n.d.	n.d.	0.03	n.d.	99.51	0.05	0.06

PLATE 5b. MESSEMVRIA-ZONE, ARTEFACTS OF FERRITIC STRUCTURE.

Arrowheads

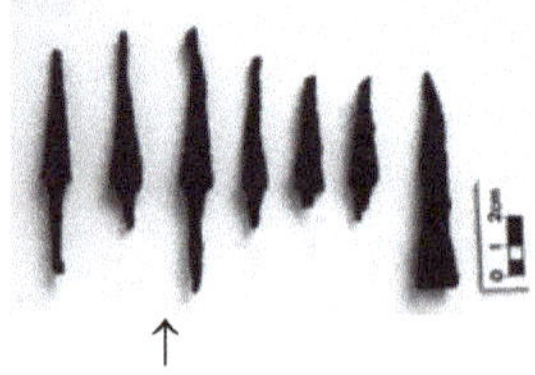

Photomicrograph of MES1 x66. Large ferrite grains with a lot of slag inclusions that are responsible for the corrosion.

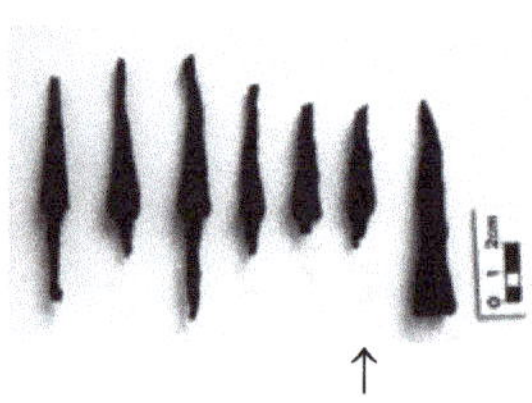

Photomicrograph of MES 2 x66. Ferrite with slag inclusions.

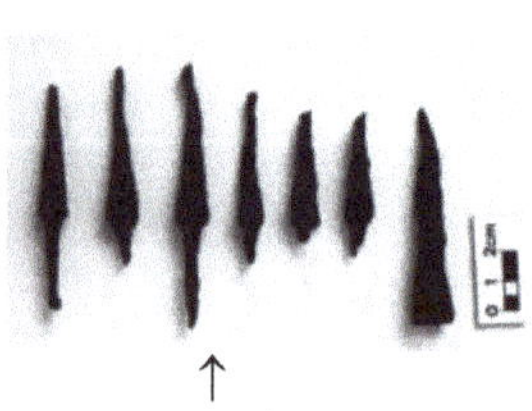

Photomicrograph of MES 29 x66. Ferrite with slag inclusions.

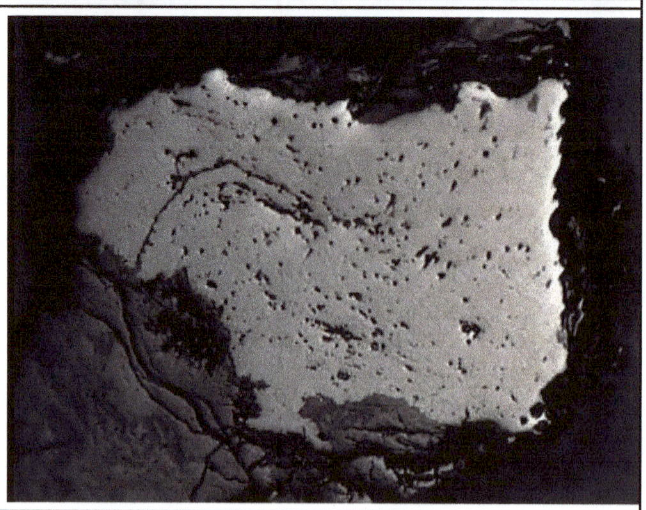

PLATE 6A. MESSEMVRIA-ZONE. ARTEFACTS OF SURFACE CARBURISATION.

Spearheads

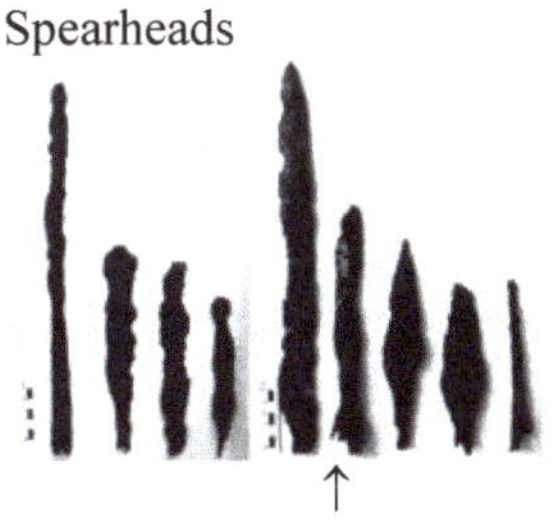

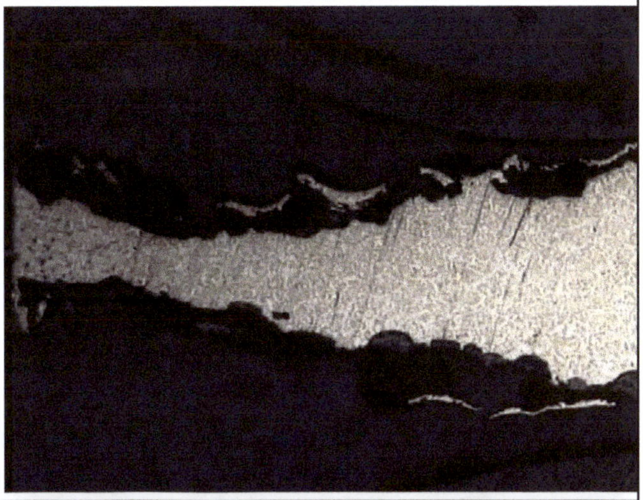

MES 7 x50. Medium carbon steel with carburization on the edge.

MES 7 x50. Slag inclusions are more in the core causing corrosion.

MES 62 x50. Low C steel with carburization along the edges. Slag inclusions contain two phases, wustite rich in Mn (1.29%) and interstitial glass rich in K (3.64%), Al (5.60%), Mn (3.22%) and P (2.80%).

PLATE 6b. MESSEMVRIA-ZONE. ARTEFACTS OF SURFACE CARBURISATION.

Spearheads

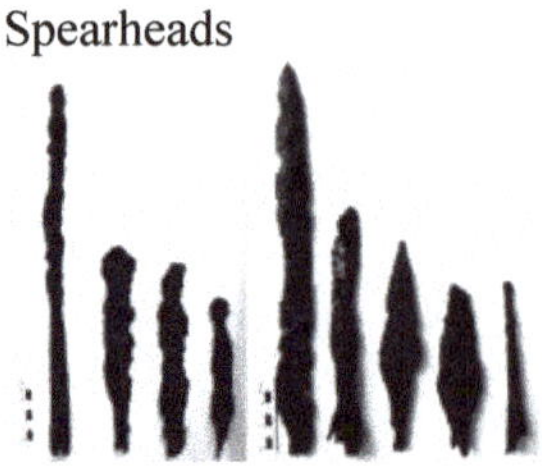

Photomicrograph of MES 25 x132. Edge made by successful carburisation of medium carbon steel as it is indicated by the nice diffusion o carbon
And the only few slag inclusions.

SEM photograph of MES 25 showing three phase slag inclusion containing wustite (96.51% FeO), fayalite and interstitial glass with no characteristic elements of local origin (note the absence of manganese oxide in the table below)

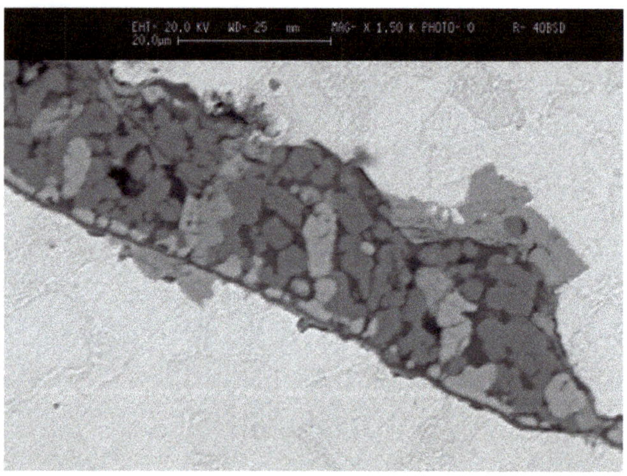

Analyses	Na_2O	K_2O	Al_2O_3	SiO_2	MgO	CaO	TiO_2	MnO	FeO	P_2O_5	SO_3
MES 25											
W	n.d.	0.10	0.73	0.99	0.28	0.48	0.30	0.10	96.51	0.20	0.11
F	n.d.	0.13	0.53	27.49	3.55	2.70	n.d.	0.79	64.52	0.20	0.05
I	0.37	3.33	6.13	26.65	0.61	10.52	0.07	0.46	47.26	3.08	0.96
M	n.d.	n.d.	n.d.	0.12	0.02	0.05	n.d.	0.04	99.40	n.d.	0.02

PLATE 7a. MESSEMVRIA-ZONE. ARTEFACTS OF WELDED STRUCTURE.

Spearheads  Photomicrograph of MES 10a x66, edge. Edge made by welding steel with eutectic composition (P and F at 0.8% carbon) to low carbon steel.	
Photomicrograph of MES 10 x66, back.. Corrosion has been developed along the welding lines.	
 Photomicrograph of MES 22 x66 showing welding of high carbon steel to low carbon steel. Corosion along the welding lines.	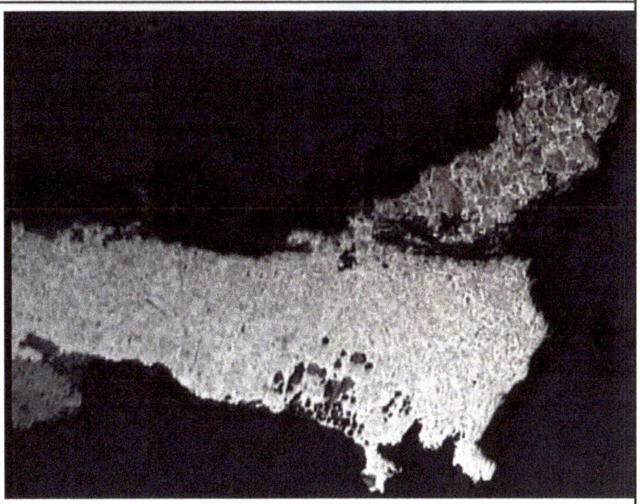

PLATE 7b. MESSEMVRIA-ZONE. ARTEFACTS OF WELDED STRUCTURE.

Spits

↑ (3rd from the last) Photomicrograph of MES 31 x66. Spit made by forging together and then twisting one high carbon (in the photograph on the right) sheet with one medium carbon seat.

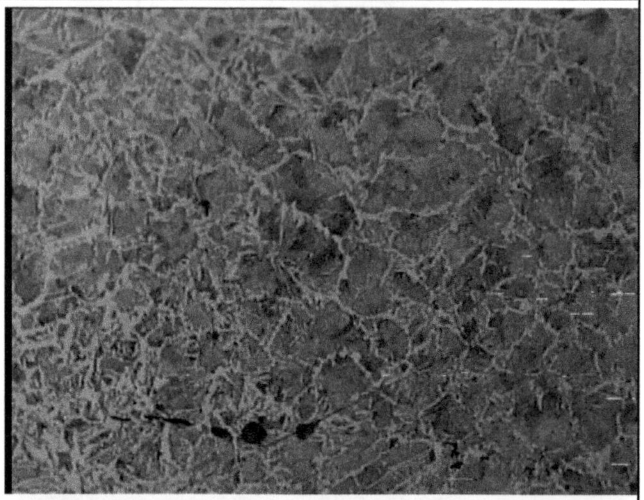

Spits

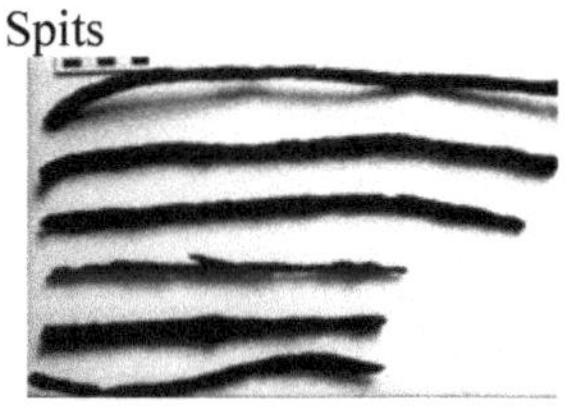

4th from the last Photomicrograph of MES 32 x66. Similar structure to MES 31, with slag inclusions along the welding lines moved to the direction of twisting.

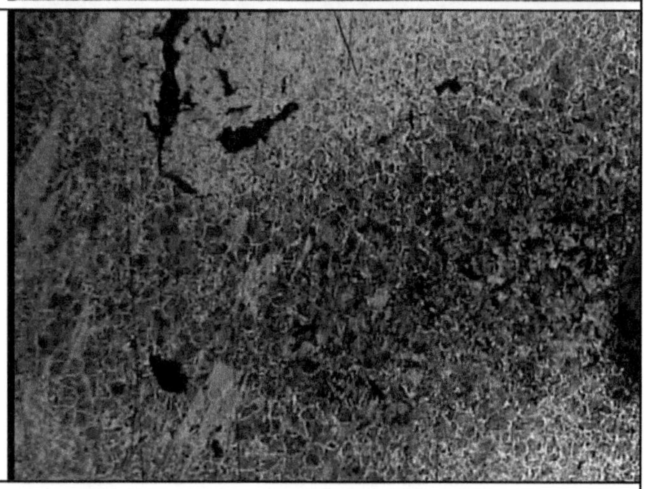

↓

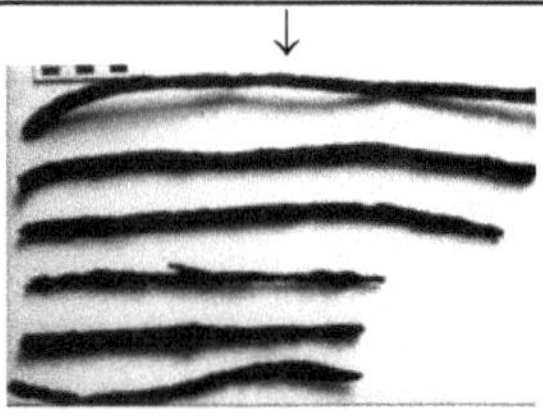

Photomicrograph of MES 33 x66. Spit mate by forging together and then twisting one medium and one low carbon sheets. Mg and Mn are present as impurities in the interstitial glass.

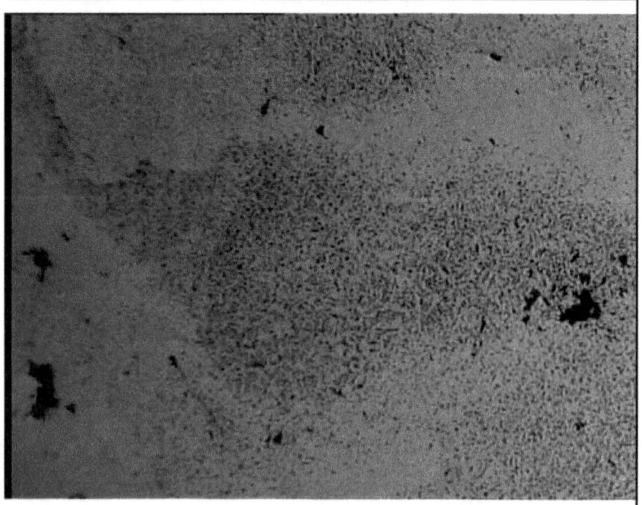

PLATE 7c. MESSEMVRIA-ZONE. ARTEFACTS OF WELDED STRUCTURE.

Curved knife ↓ 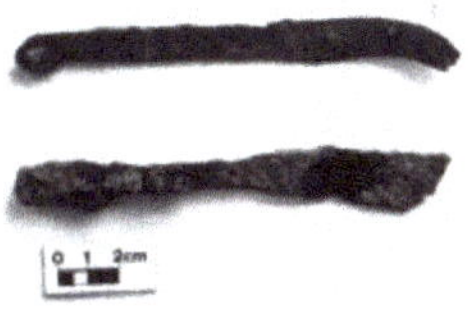 Photomicrograph of MES 65 x66 showing welding of high carbon steel to medium carbon steel. Slag stringers along the welding lines.	
Photomicrograph of MES 53 x264, edge. Single phase slag stringers probably from the forging and carburization of the bloom. The interstitial glass contents Mg (2.75%) and high Mn (4.58%) that points out to the smelt of Mn/Fe ores.	
Photomicrograph of MES 17 x66. Straight knife blade with obvious welding lines.	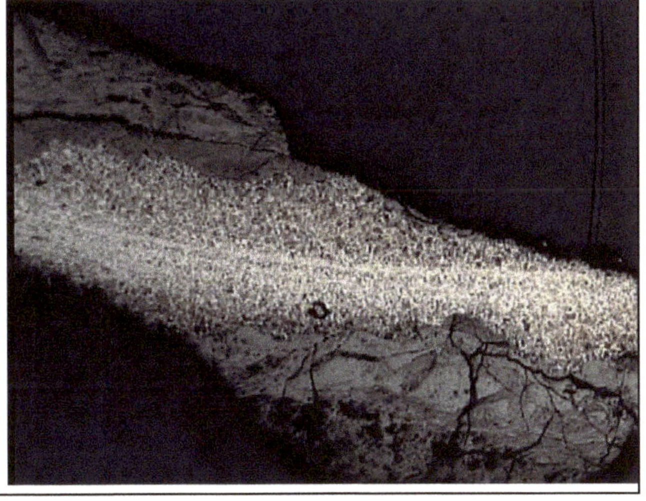

PLATE 8. MESSEMVRIA-ZONE. ARTEFACTS OF HIGH CARBON STEEL.

MES 9 x50 Straight knife blade made by on sheet of high carbon steel.	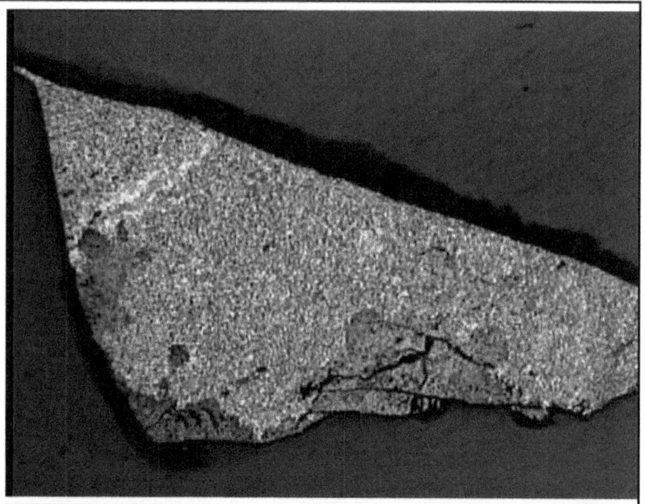
MES 9 x200. Spaerodised cementite and pearlite, caused by the heat working of a quenched structure, probably tempered martensite.	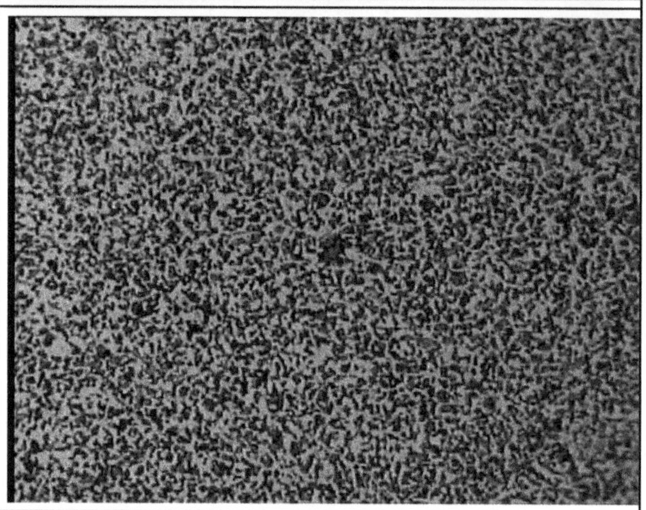
MES 17 x200 edge displaying eutectic steel structure with Pearlite and Ferrite in the grain boundaries.	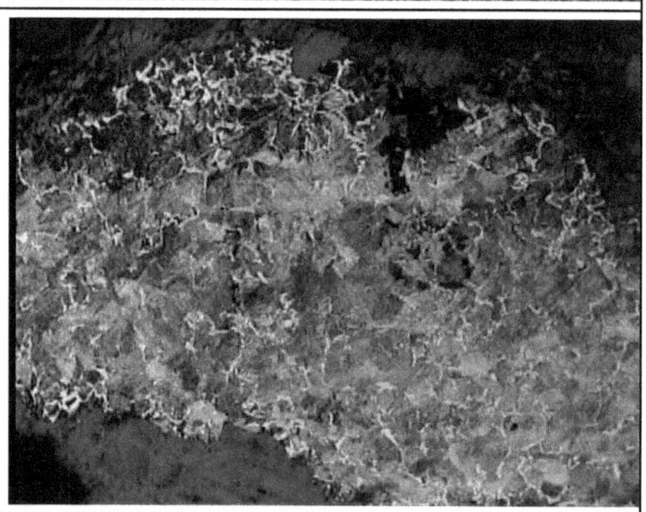

PLATE 9. AVDERA. SMELTING SLAG.

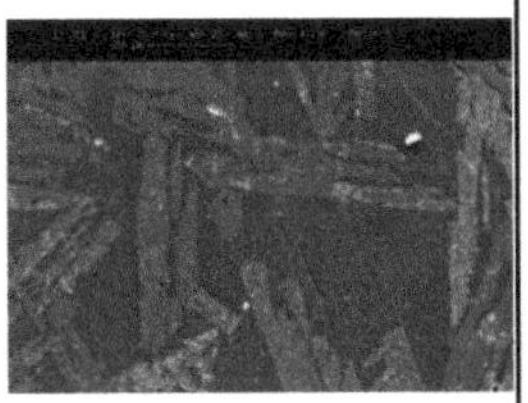

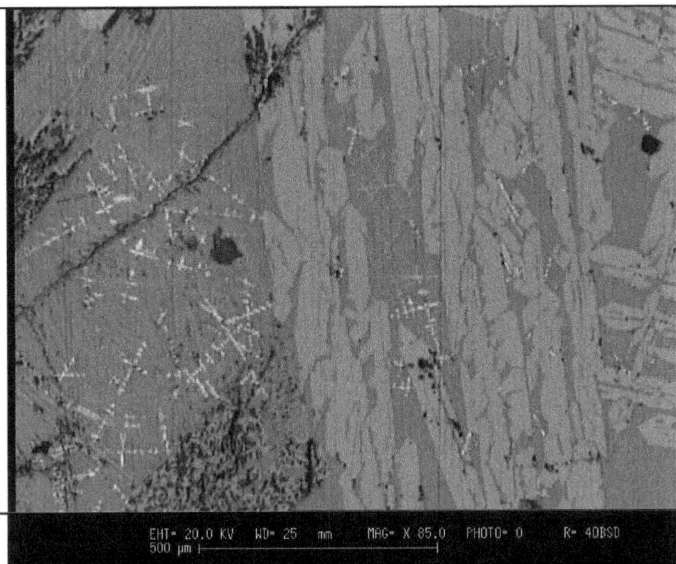

Photomicrograph of AVD 114 X132, showing skeletons of iron oxide (white) and needles of fayalite (light gray) in glassy matrix (dark gray).

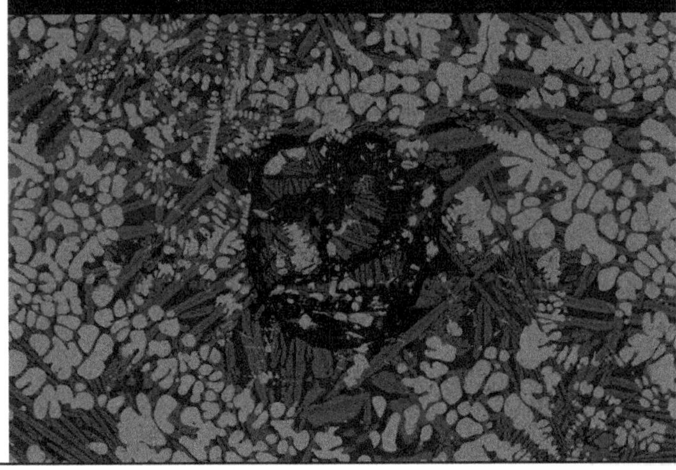

SEM photograph of AVD 123, showing iron oxide (white) and needles of fayalite (light gray) in glassy matrix (dark gray).

PLATE 10a AVDERA. SMITHING SLAG.

SEM photograph of AVD 116, showing globules of wustite (white) and needles of fayalite (light gray) in a glassy matrix (dark gray).	
Photomicrograph of AVD 122 X132, showing globules of wustite (white) and needles of fayalite (light gray) in a glassy matrix (dark gray).	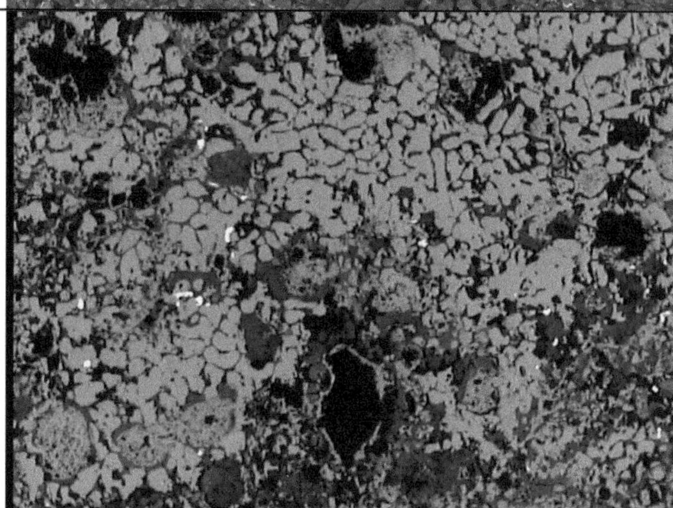
SEM photograph of AVD 129, showing globules of wustite (white) and needles of fayalite (light gray) in a glassy matrix (dark gray).	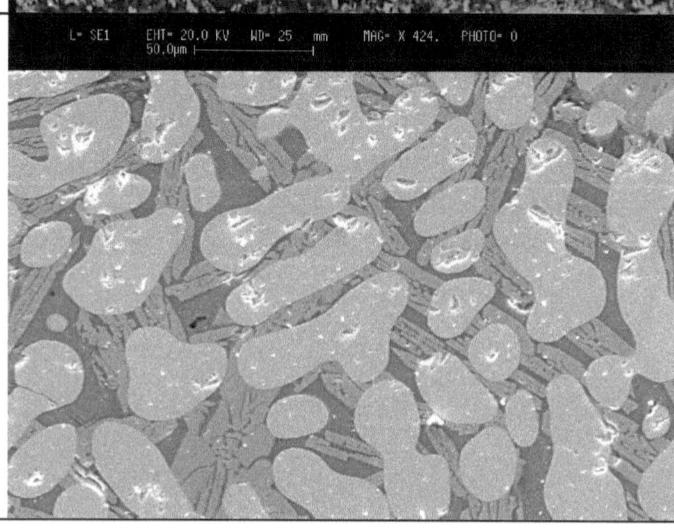

PLATE 10bAVDERA. SMITHING SLAG.

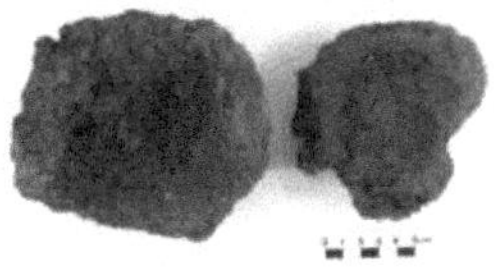

Photomicrograph of AVD108 X264 showing metallic iron (white) in a silicate phase (gray). Black areas are holes.

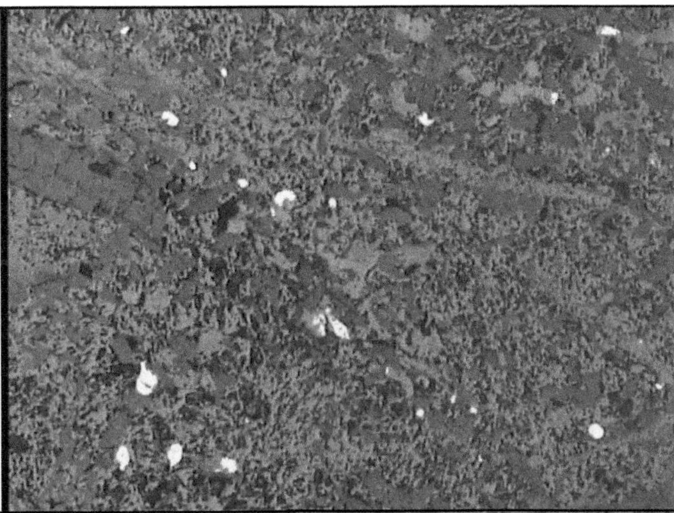

Photomicrograph of AVD 103 x528 showing metallic iron (white) in a silicate phase (gray). Black areas are holes.

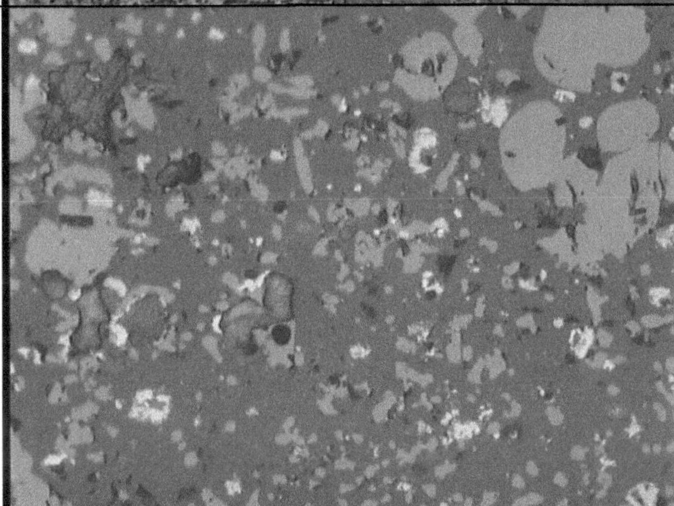

Photomicrograph of AVD 128 X132, showing globules of wustite (white) in a glassy matrix (gray). Black areas are holes.

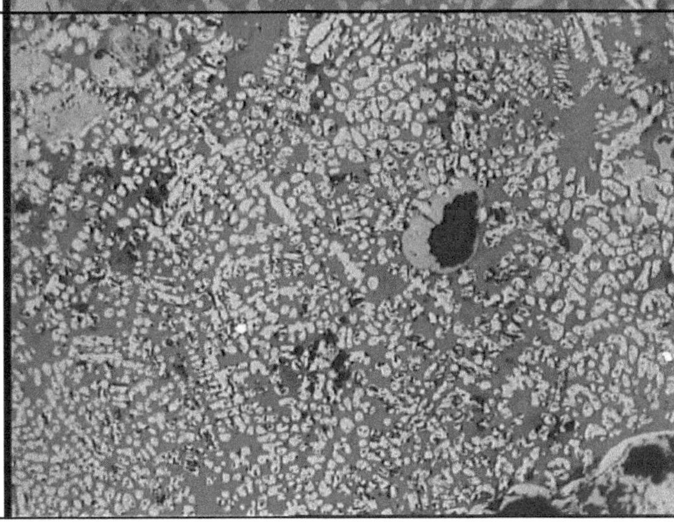

PLATE 11a. AVDERA. ARTEFACTS OF FERRITIC STRUCTURE.

AVD 1. Optical microscopy x66 showing large ferrite grains in body of the v shape sample. Ferrite grains are deformed and smaller on the edge, because of cold hammering to shape the Artefact.

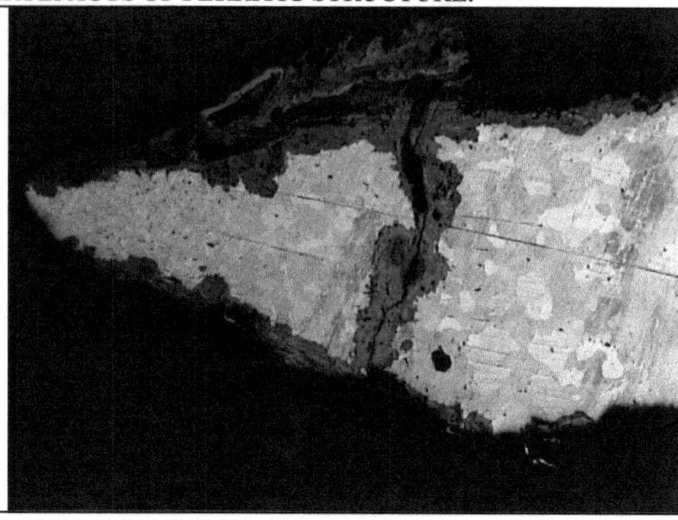

AVD 38 (top) AVD 33 (bottom)

Optical microscopy of curved knife AVD 37 x132. It is made by wrought iron (large ferrite in the middle) that has been deformed slightly on the edges due to cold hammering.

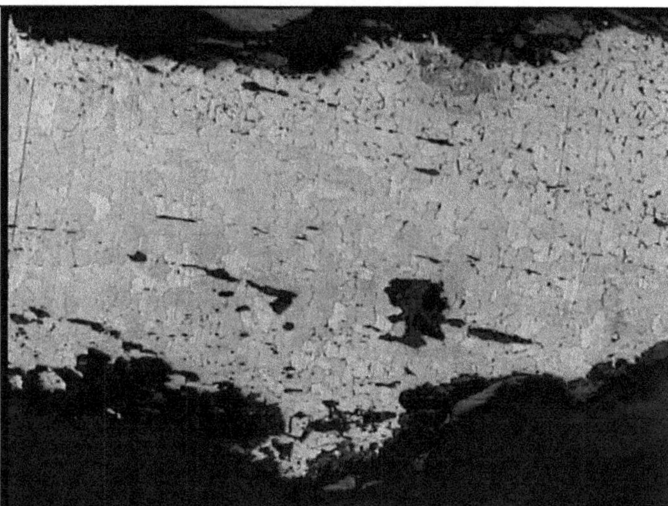

Spit AVD 43. Optical microscopy x132. Ferrite and some pearlite (low carbon steel) with slag inclusions with carburisation on the right edge.

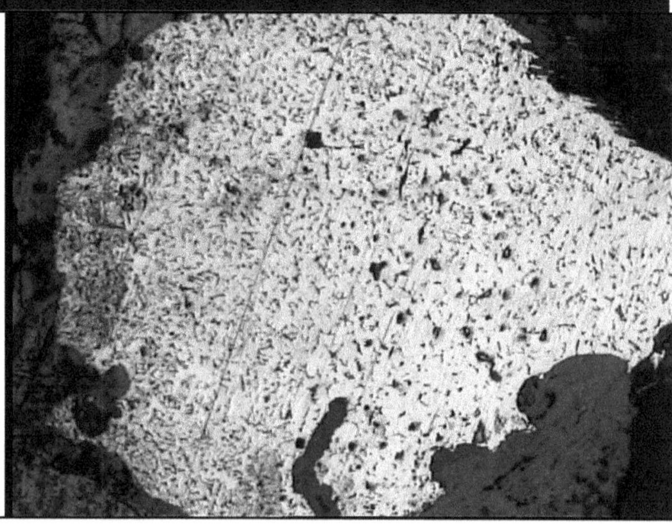

PLATE 11b. AVDERA. ARTEFACTS OF FERRITIC STRUCTURE.

AVD 23. Optical microscopy x132 showing uniform distribution of ferrite with some pearlite of hypo-eutectic composition.

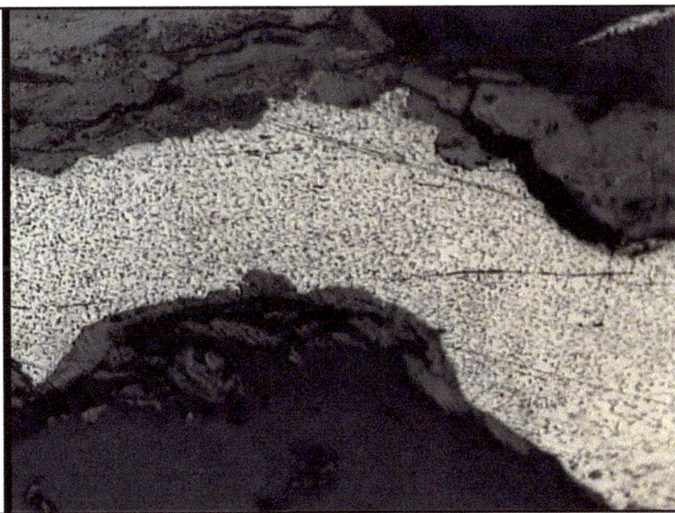

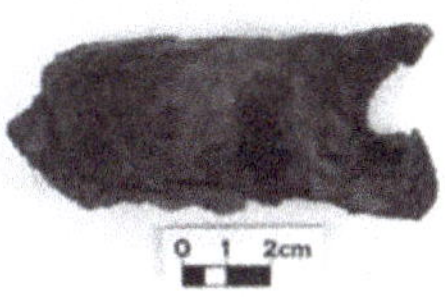

Optical microscopy of AVD 39 x 66. The v shaped section on the edge of the axe showed a ferrite structure through all the body with light deformation of ferrite on the edge due to cold hammering. Black areas are slag inclusions.

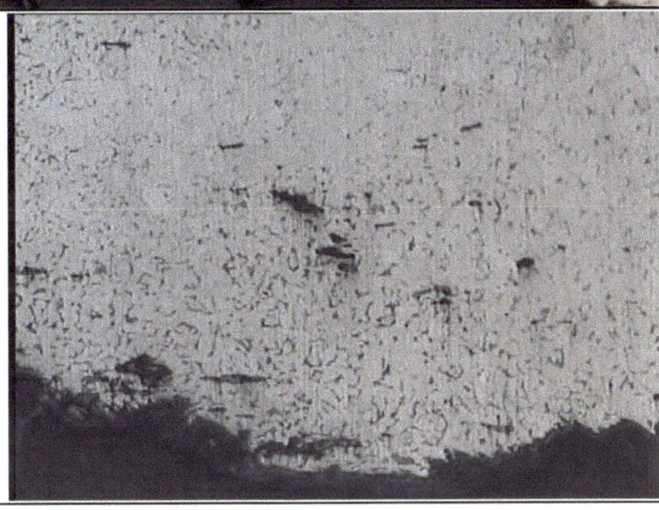

PLATE 12a.AVDERA. ARTEFACTS OF WELDED STRUCTURE.

Optical microscopy of straight knife blade AVD 38 x132, displaying uniform distribution of P through out the v section. Black lines are corrosion.

Straight knife blade Optical microscopy of AVD 36 x264, showing a carburised edge with pearlite (black) and bainite (white wings). The core displays a low carbon composition with ferrite and some pearlite in the grain boundaries. Slag inclusions indicate the welding lines.

PLATE 12b.AVDERA. ARTEFACTS OF WELDED STRUCTURE.

Curry comp AVD 5 Optical microscopy x132 displaying welding of a low carbon sheet (middle F and P) to wrought iron (F on the edges). Slag inclusion indicate the two zones of contact between the sheets.

Optical microscopy of AVD 26 X50, back showing welding lines between a wrought iron area (white on top) and low carbon area on the bottom (F and P).

PLATE 12c.AVDERA. ARTEFACTS OF WELDED STRUCTURE.

Straight knife AVD 53. Optical microscopy x66. The black area on the top is eutectic pearlite, the white area Ion the bottom is ferrite. Slag stringers along the welding lines. The white areas on pearlite matrix are probably bainite crystals formed during slow cooling after carburization.

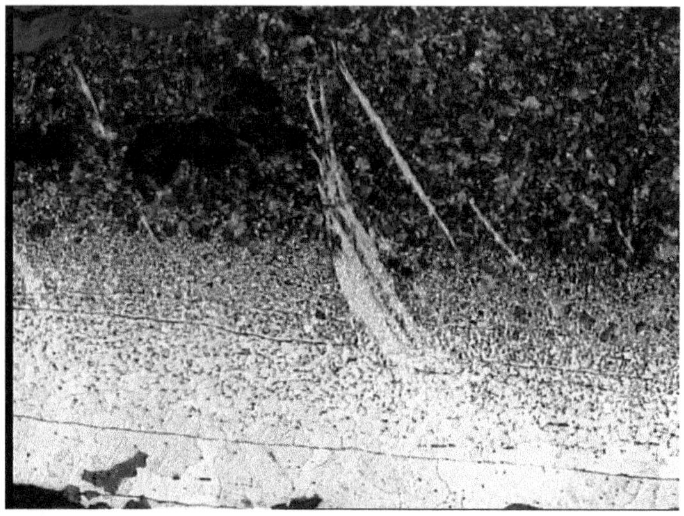

Axe AVD 30, similar to AVD 39. Optical microscopy of the v section x66. Gray is pearlite, white is spherodised cementite and black lines are slag inclusions along the welding lines.

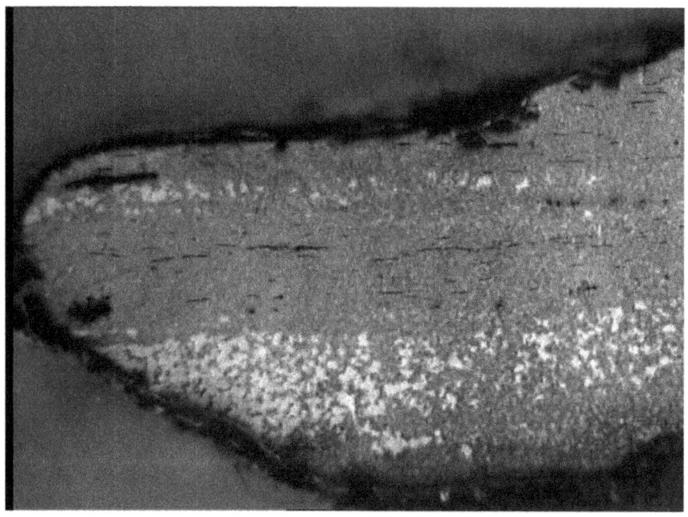

PLATE 13. AVDERA. ARTEFACTS OF HIGH CARBON STRUCTURE.

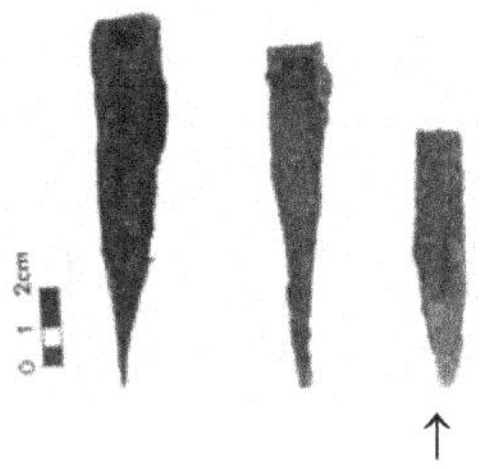

Photomicrograph of AVD 49 x66. Pearlite (gray) and spaerodised cementite (white). Black areas indicate corrosion.

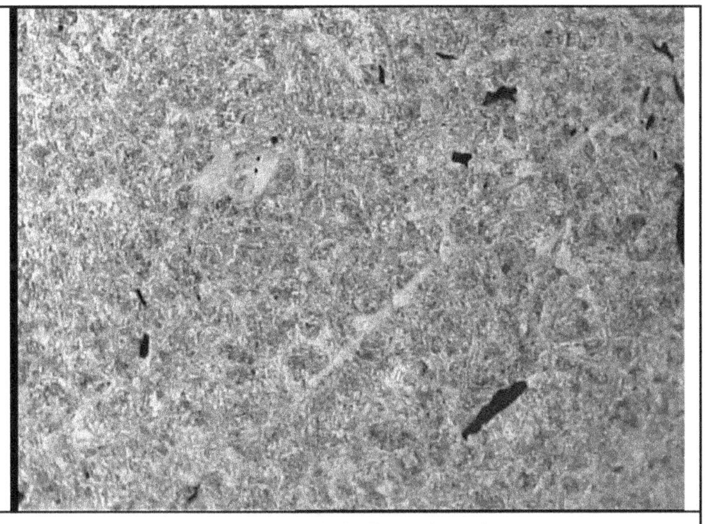

AVD 48.b
Photograph of iron two pointed pitchfork.

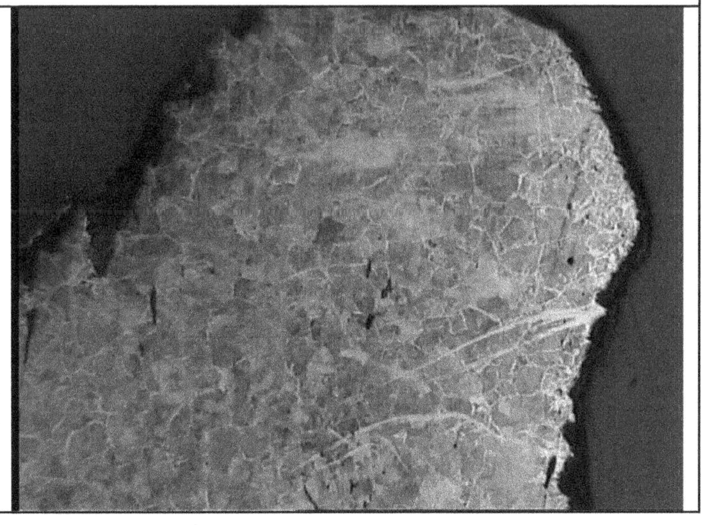

Optical microscopy of v shaped section from the edge of AVD 48 x66.Pearlite with cementite in the grain boundaries indicate the use of a high carbon steel for the fabrication of this edge. The white wing like formation of bainite suggest slow cooling.

PLATE 14. KALYVA-KASTRO. BLOOMS AND SMELTING SLAG.

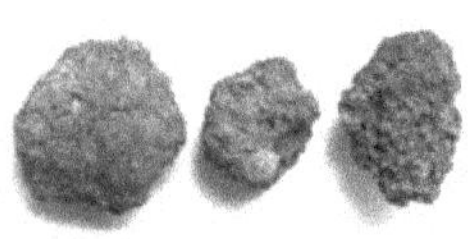 Optical microscopy of KL 75 x200, displaying round modules of bloomery iron in a silicate matrix. The light gray is weathered metallic iron (limonite). Metallic iron (white) is included in the modules on the right corner of the picture. Black areas are holes.	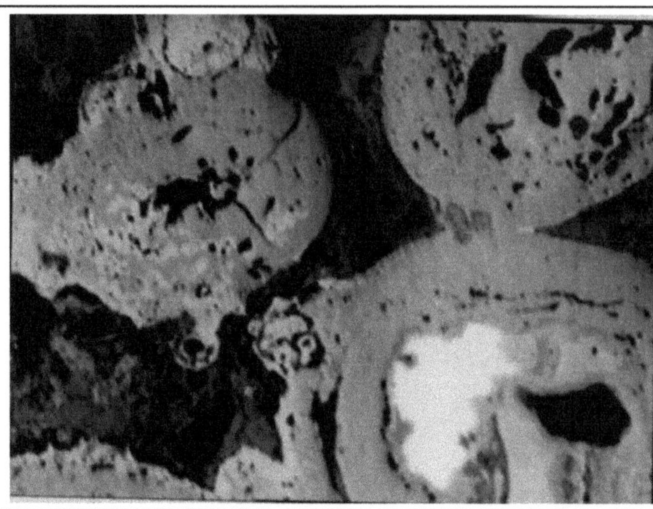
Optical microscopy of KL 87 x400. Globules of wustite (97.97% iron oxide) and Mg rich (1.50%) fayalite in interstitial glass rich in phosphorous (1.6%) .	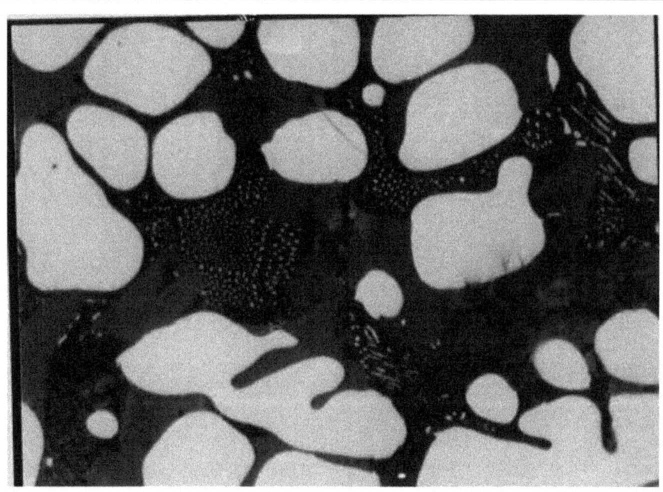

PLATE 15. KALYVA-KASTRO. SMITHING SLAG.

Optical microscopy of KL 73 x50 Highly porous slag with mainly wustite (97.80% FeO) in interstitial glass rich in K an Al. The metallic inclusions are 99.45% FeO.	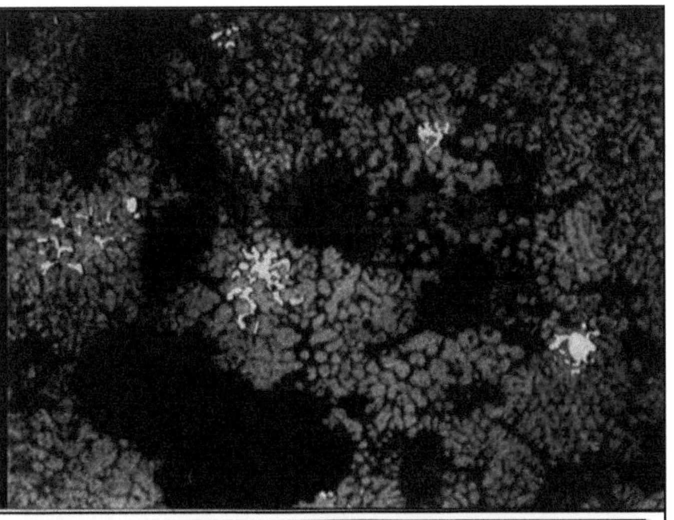
Optical microscopy of KL 102a x200, showing wustite rich in CaO, SiO2 and Al2O3, fine needles of Mg rich fayalite in interstitial glass rich in Mn. (6.04%).	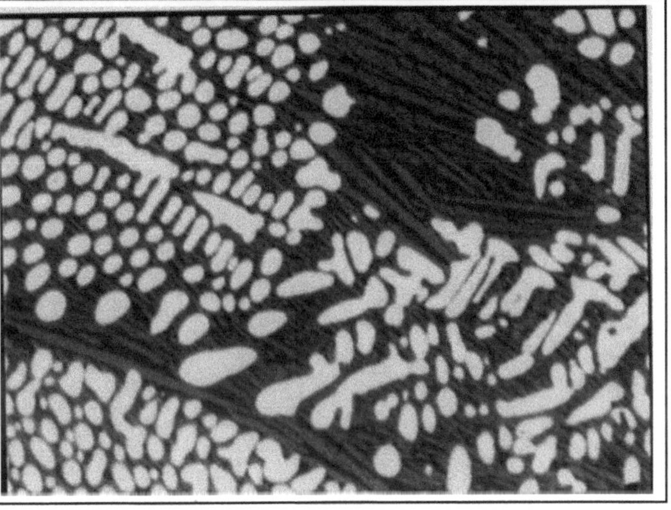

PLATE 16. KALYVA-KASTRO. FURNACE LINING.

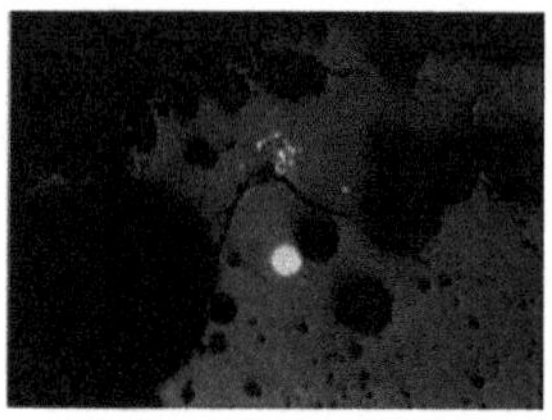

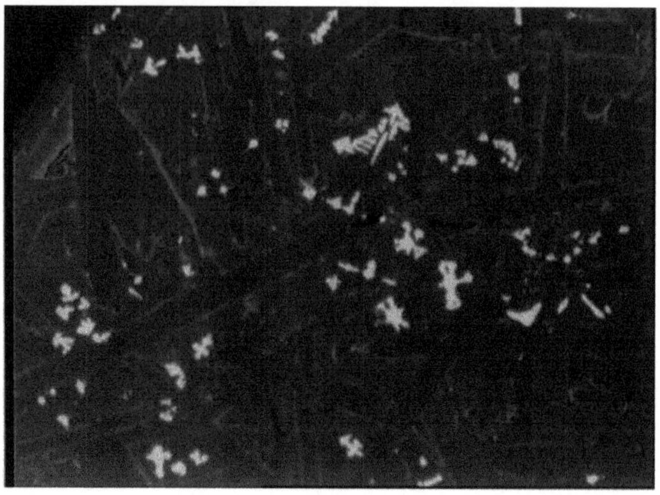

Optical microscopy photograph of KL 74 (above) showing a highly vitrified area with metallic prills.
Optical microscopy photograph of KL 74 x400 (right).

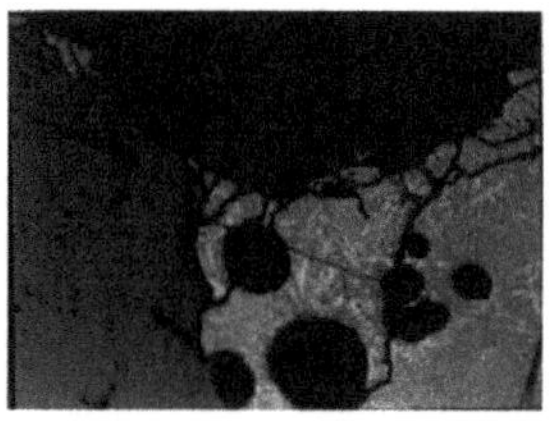

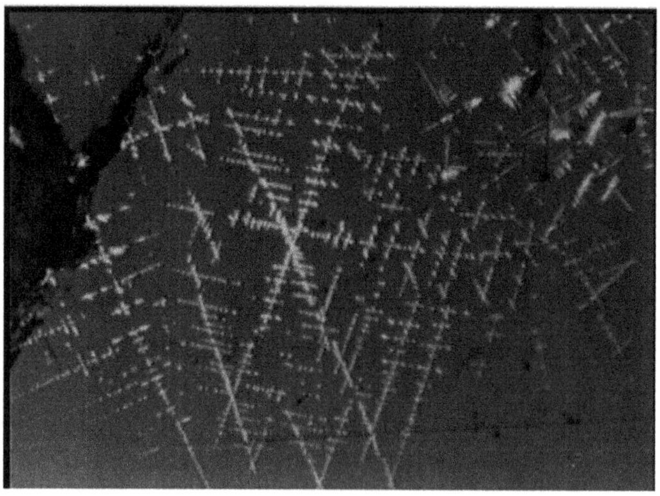

Optical microscopy of KL 85 x400. Thin dendrites of wustite rich in Al, Si, and C. W is also rich in phosphorus (1.07%).

PLATE 17. KALYVA-KASTRO. ARTEFACTS OF FERRITIC STRUCTURE.

Optical microscopy of scissors KL 59 x50 displaying deformed ferrite structure. The grains of ferrite and the slag inclusions are elongated to the direction of forging. The ferrite structure cannot explain the high hardness of the sample (Hv=339).
This might be due to a phosphorous content.

PLATE 18. KALYVA-KASTRO. ARTEFACTS WITH SURFACE CARBURISATION.

Optical microscopy of straight knife KL62 x50. The section shows a nice carburization of wrought iron on the right edge. Black holes are slag inclusions left behind from the bloom forging.

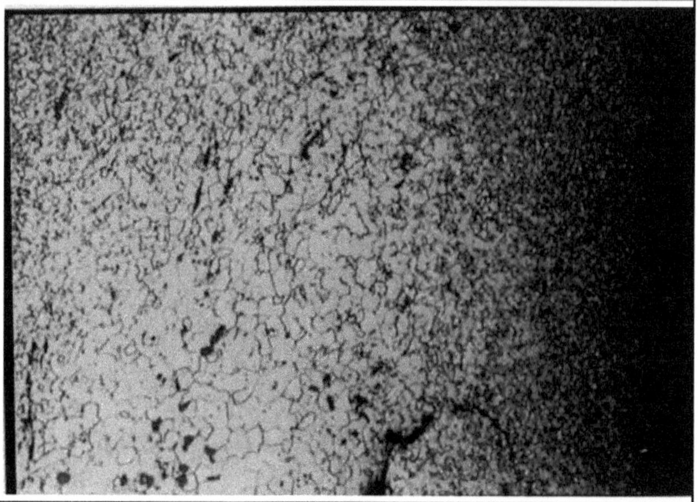

Optical microscopy of curved blade KL 65 x200 displaying a nice carburization of wrought iron on the left edge. The carbon is distributed evenly from the edge to the core. No slag inclusions or stringers are evident.
Hv= F 131

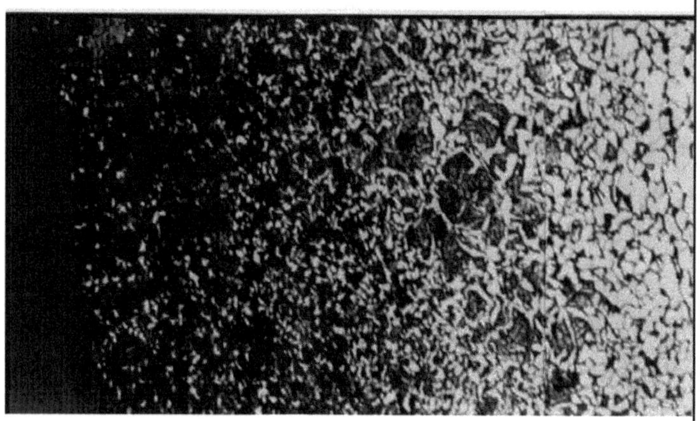

Optical microscopy of straight knife KL 20 x50 displaying carburization on both edges.
Hv edge=383
HvPearlite = 220

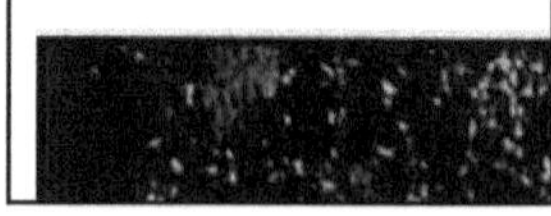

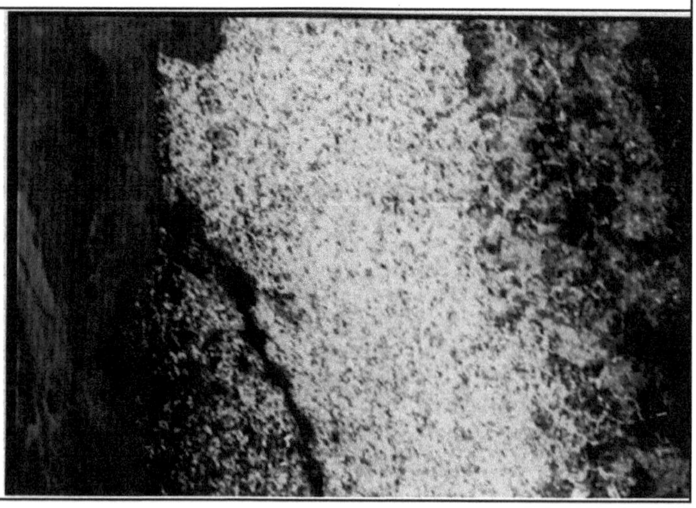

PLATE 19. KALYVA-KASTRO. ARTEFACTS OF WELDED STRUCTURE.

Optical microscopy of curved knife KL 12 x50. The v shaped section shows that this blade is made by welding of a low carbon steel (0.2%) to wrought iron. The long slag inclusion on the left indicates the zone of contact between the two sheets. Hv=123	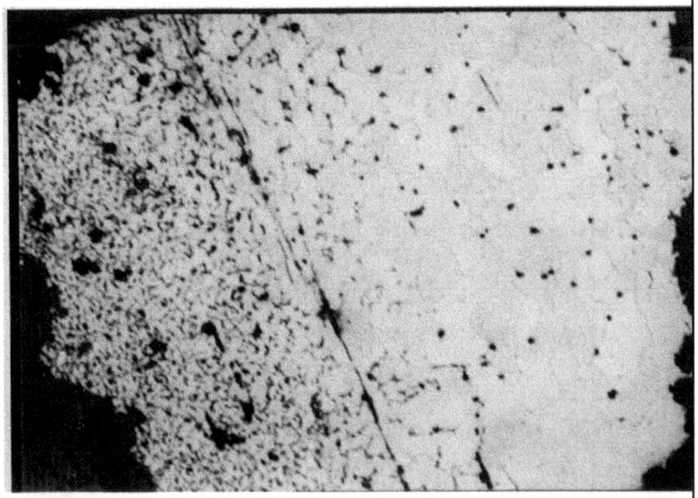
Optical microscopy of curved knife KL 4 x50. The blade is made by forge welding of low carbon steel at 0.2% (light gray) to wrought iron (white). The welding was not complete as the many elongated slag inclusions and the uneven distribution of carbon suggest. Hv=151	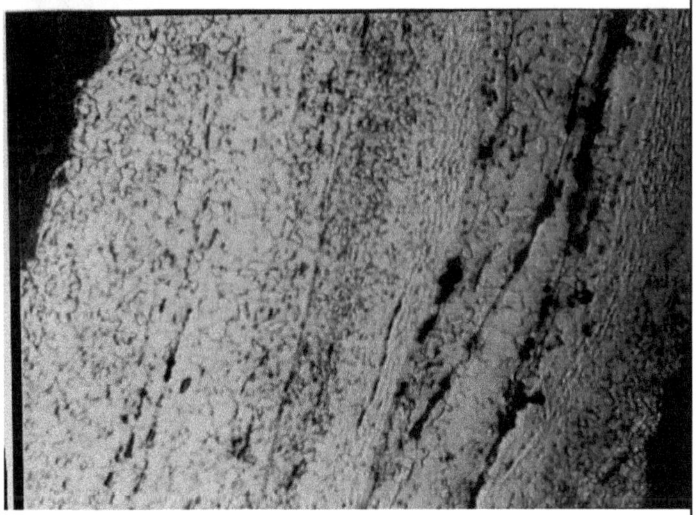

PLATE 20a. KALYVA-KASTRO. ARTEFACTS OF HIGH CARBON STEEL.

Optical microscopy of spearhead KL 60 x 50 displaying paerlitic structure with sperodised cementite.	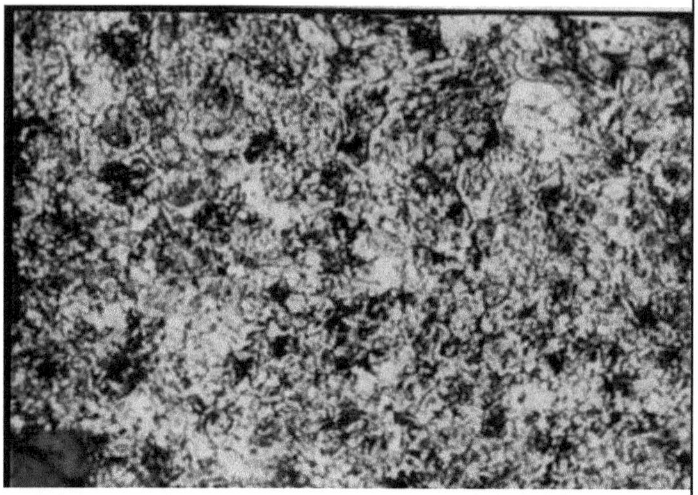
Optical microscopy of billhook KL 64 x400 displaying eutectic steel structure of lamellar pearlite (black). The white areas are crystals of ferrite. Carbon content should be very close to 0.8% and the hardness on pearlite is 227 Hv.	
Optical microscopy of axe KL 55x400 displaying tempered martensite structure resulting from quenching and heat treatment of high carbon steel. Hv=322. Also in sickle KL33, straight knives KL 46, 37, 39.	

PLATE 20a. KALYVA-KASTRO. ARTEFACTS OF HIGH CARBON STEEL.

Optical microscopy of curved blade KL 14 x50. The v shaped section displays uniform distribution of hyper-eutectic composition of steel with pearlite and cementite around the grain boundaries. The point of the v shaped section is de-carburised.

Hv on F+P=205
Core=F, 180

Similar is the straight blade kl41.

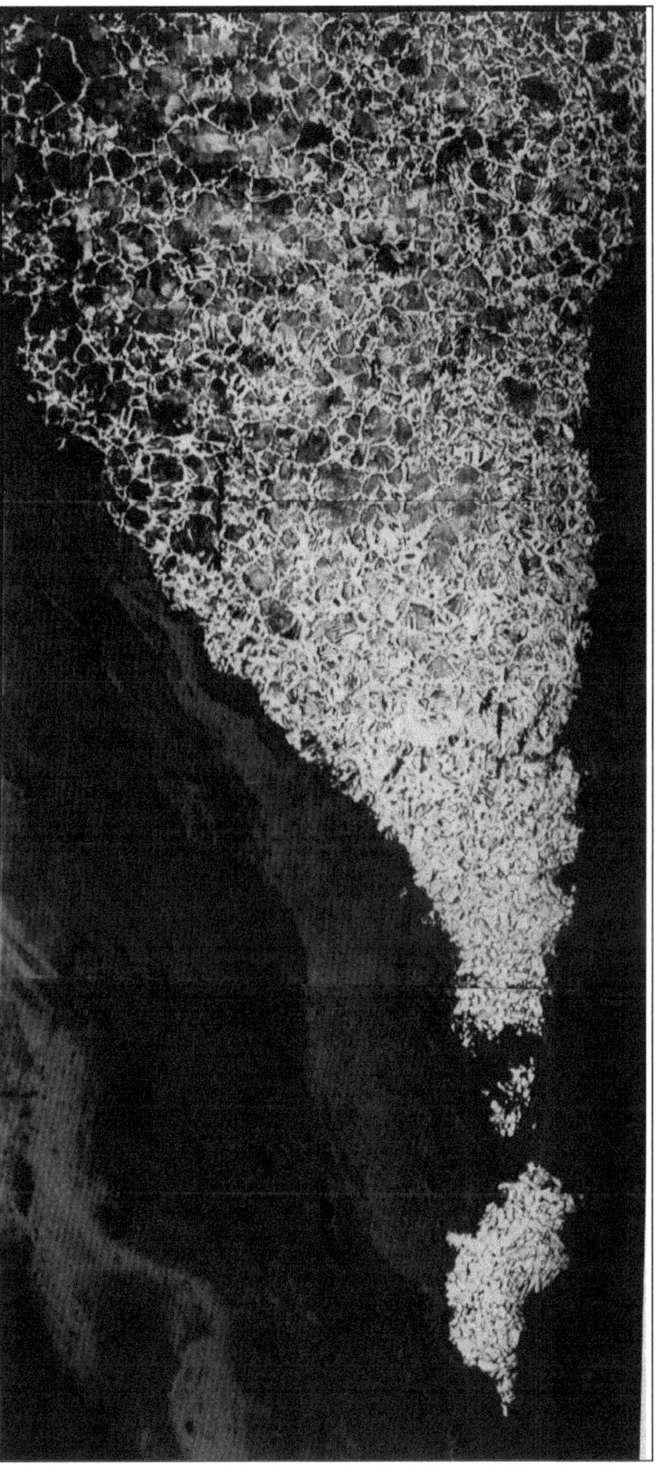

APPENDIX B
Arrowheads

Messemvria-Zone. Arrowhead MES 1

Photomicrograph of MES1 x66. Large ferrite grains with a lot of slag inclusions that are responsible for the corrosion.

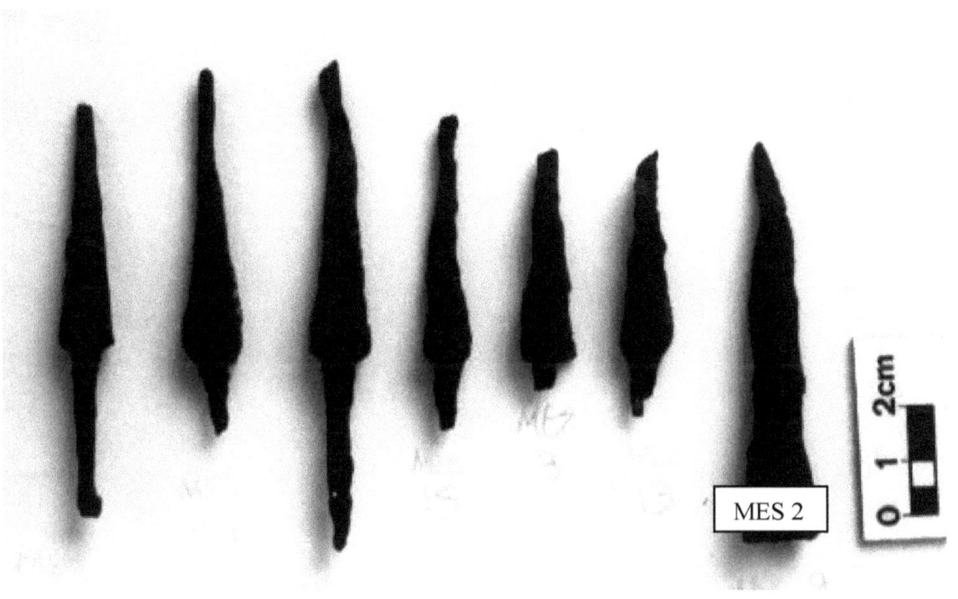

MES 2, Messemvria-Zone. Arrowhead

Photomicrograph of MES 2 x66. Ferrite with slag inclusions.

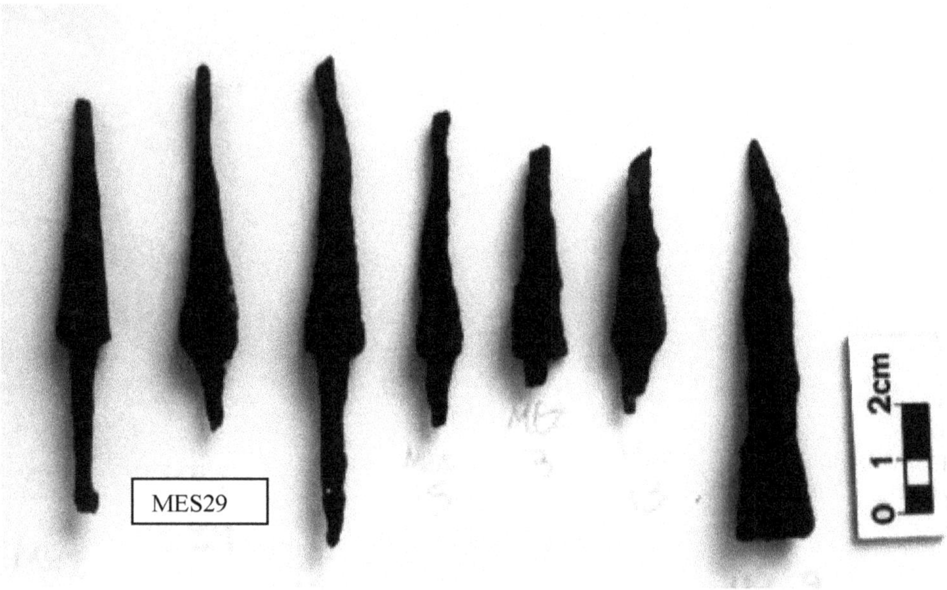

MES 29, Messemvria-Zone. Arrowhead

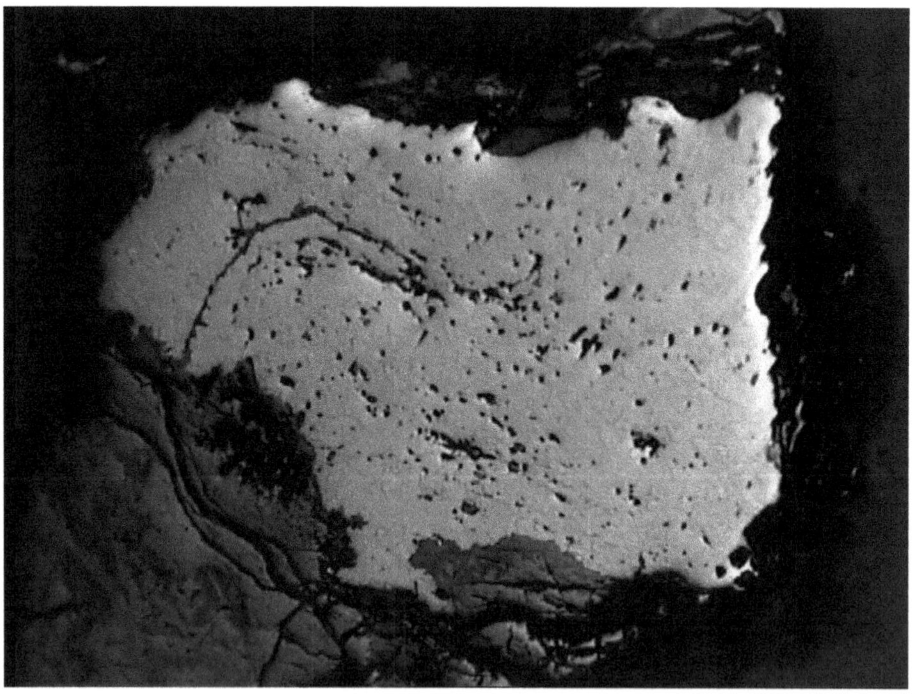

Photomicrograph of MES 29 x66. Ferrite with slag inclusions.

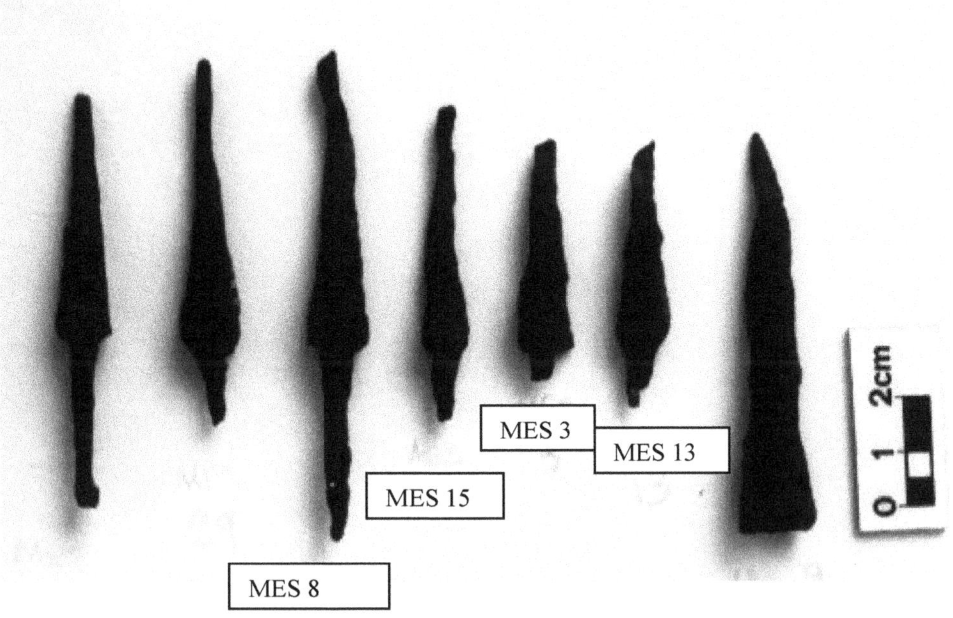

Messemvria-Zone. Arrowheads

Avdera, arrowheads. (MA 1079 on the left is similar to MES 2 and MA 971 on the right is similar to MES 1).

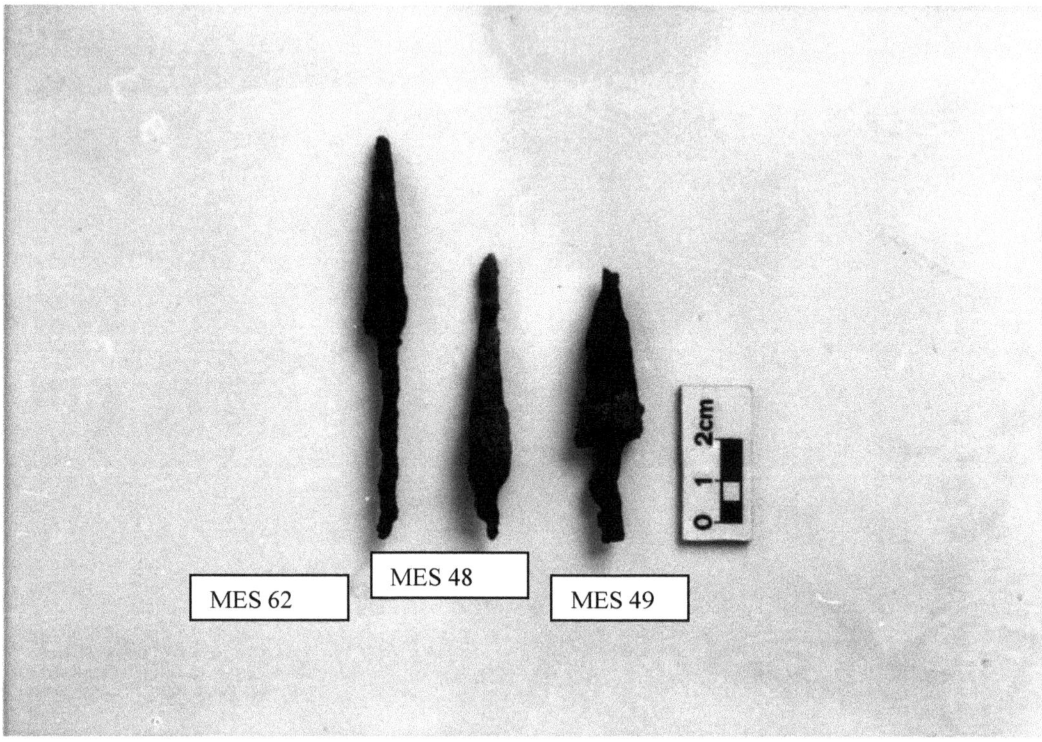

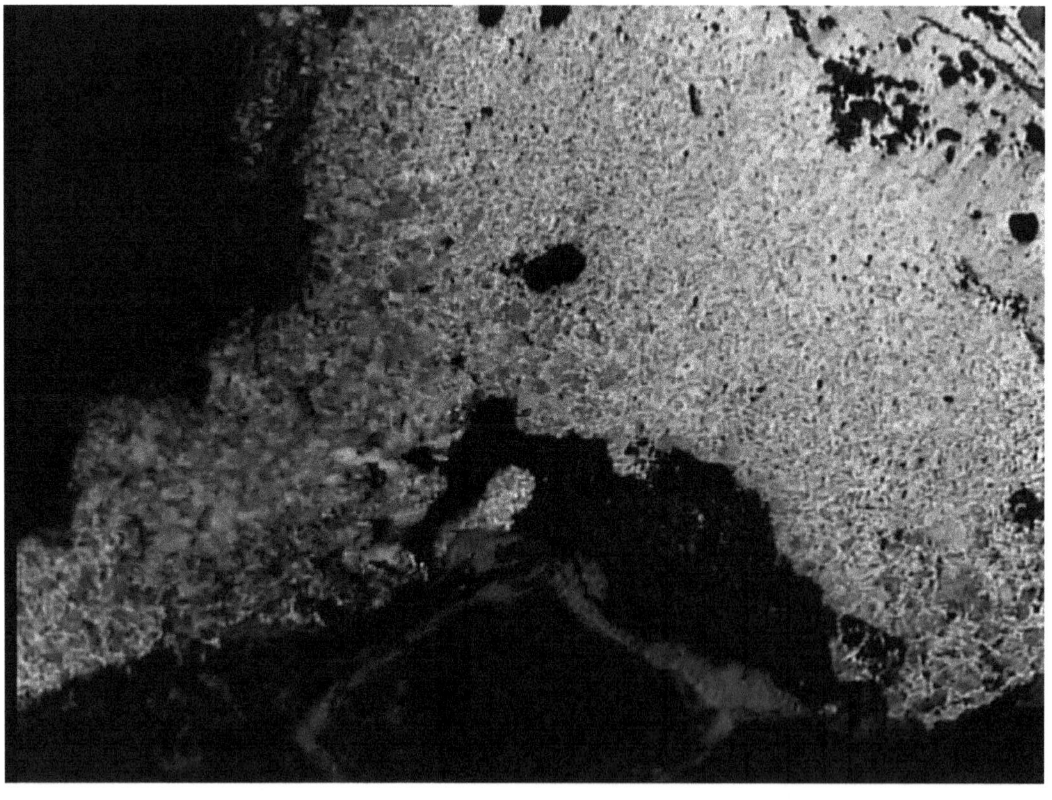

MES 62 x50. Low C steel with carburization along the edges. Slag inclusions contain two phases, wustite rich in Mn (1.29%) and interstitial glass rich in K (3.64%), Al (5.60%), Mn (3.22%) and P (2.80%).

Axes

AVD 39, Avdera.

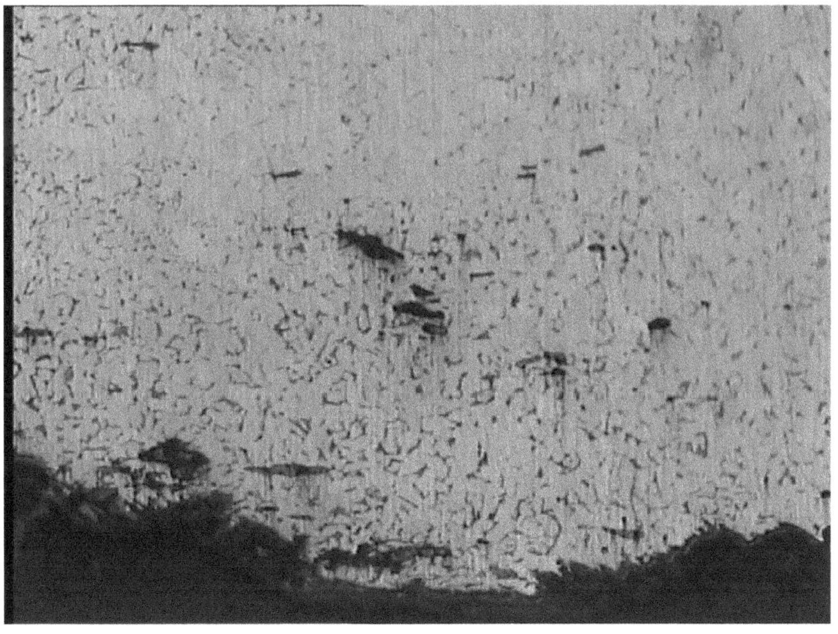

Optical microscopy of AVD 39 x 66. The v-shaped section on the edge of the axe showed a ferrite structure through all the body with light deformation of ferrite on the edge due to cold hammering. Black areas are slag inclusions.

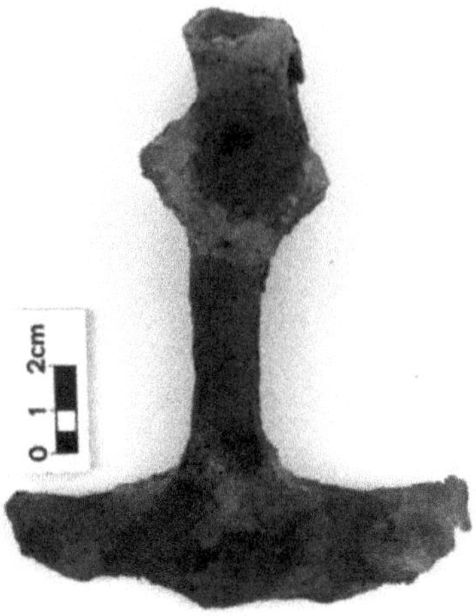

or

AVD 30, Avdera. AGK 766 AGK 2269

Other type of classical and helenistic Axes. Pleiner, 1969, Figures 11 and 10

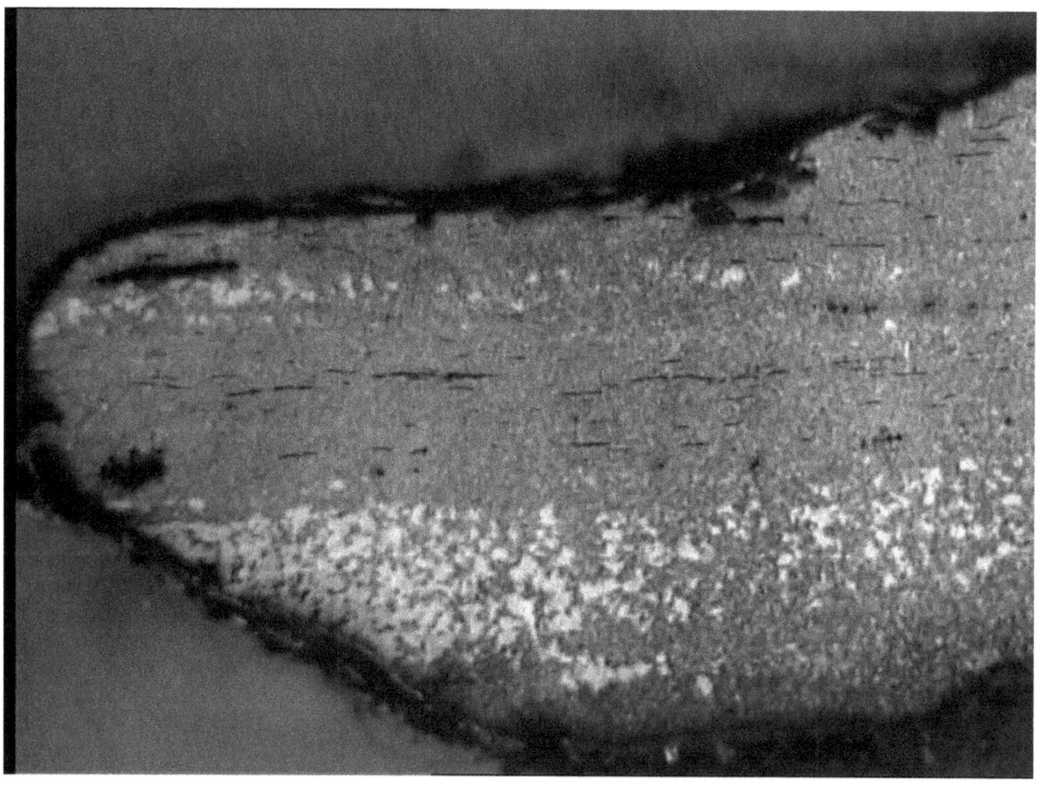

Axe AVD 30, similar to AVD 39. Optical microscopy of the v section x66. Gray is pearlite, white is spherodised cementite and black lines are slag inclusions along the welding lines.

KL 55, Kalyva – Kastro. This blade is wedge-shaped in section, its thickness decreasing from the narrow cutting edge to the broader haft end (which in our example is missing). In its Roman form it might be either single or double-edged, having a vertical cutting edge on the one arm and a horizontal on the other. Its Roman name was "*securis*" (White 1967, 60, Fig. 34).

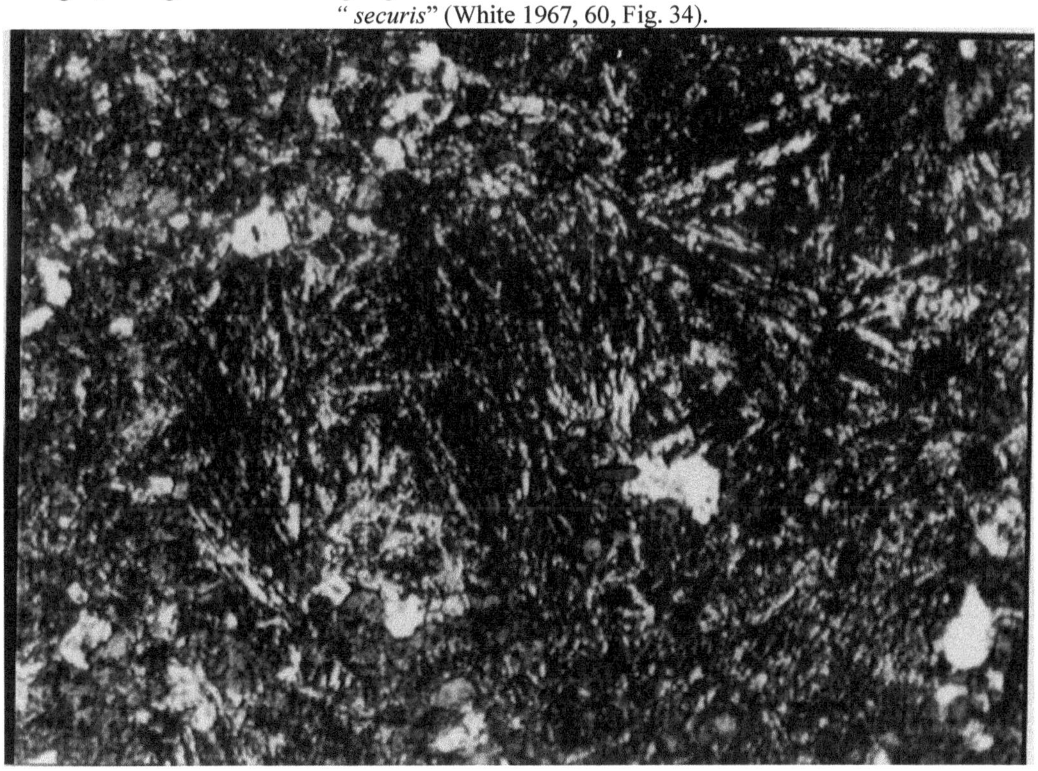

Optical microscopy of axe KL 55x400 displaying tempered martensite structure resulting from quenching and heat treatment of high carbon steel.
Hv=322. Also in sickle KL33, straight knives KL 46, 37, 39.

AVD 48. Two pointed axe (or pitch fork). The heavy two pointed axes, as literary and monumental evidence prove, were commonly used for breaking the soil in stony ground, especially where olives and vines were to be planted (White 1967, figs.26-27). Although such finds were also collected in Po valley in north Italy (White 1967, 50) this is the first time that such an implement is found in Roman Greece.

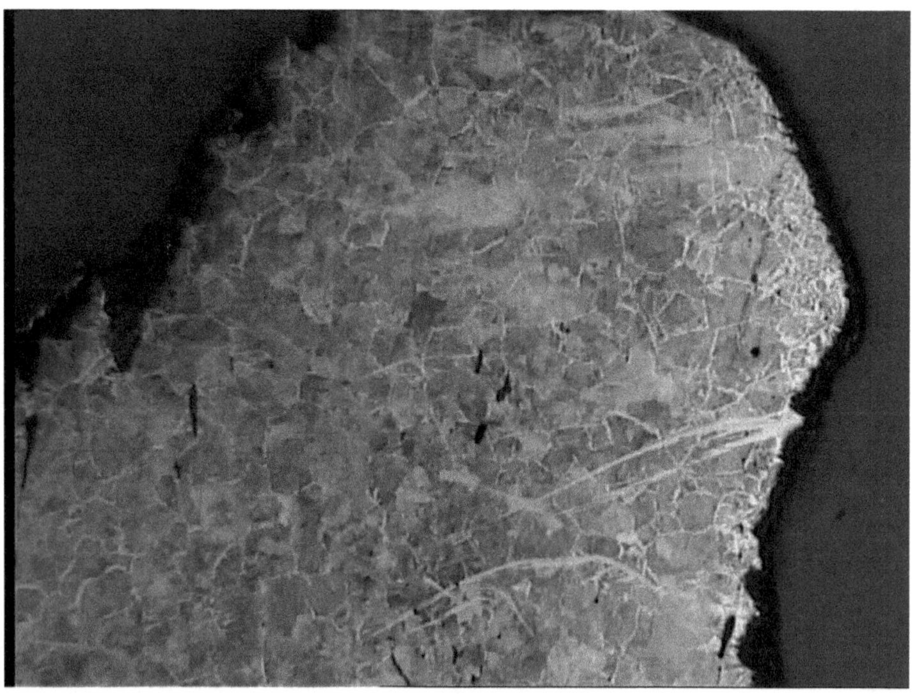

Optical microscopy of v shaped section from the edge of AVD 48 x66.Pearlite with cementite in the grain boundaries indicate the use of a high carbon steel for the fabrication of this edge. The white wing like formation of bainite suggest slow cooling.

This type is known as *ascia* or *rastrum* and it was used to prepare stony ground for cultivation (White 1967, 67, figs. 43-47).

Billhooks

KAL 64, billhook.

Optical microscopy of billhook KL 64 x400 displaying eutectic steel structure of lamellar pearlite (black). The white areas are crystals of ferrite. Carbon content should be very close to 0.8% and the hardness on pearlite is 227 Hv.

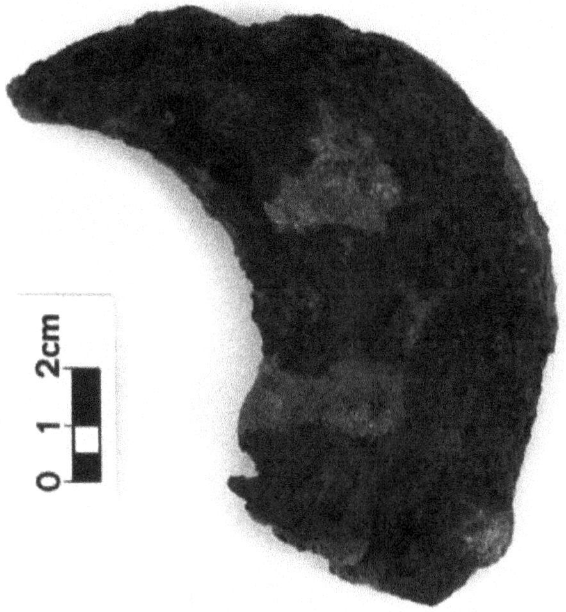

AVD 40, Avdera, billhook

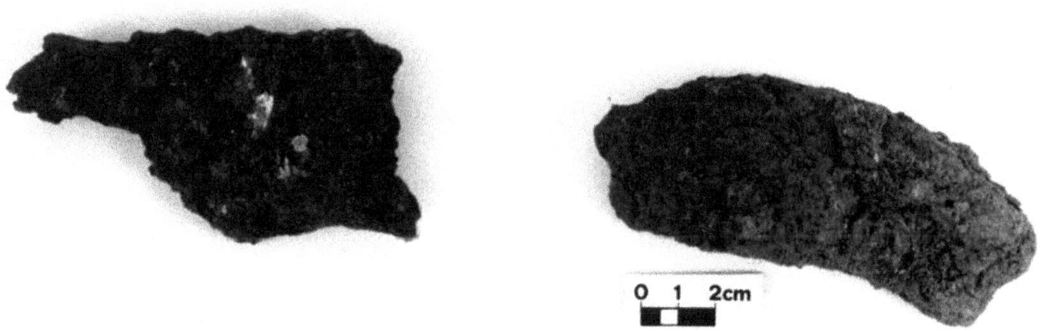

AVD 46, billhook

MES 53 and MES 67. Billhooks

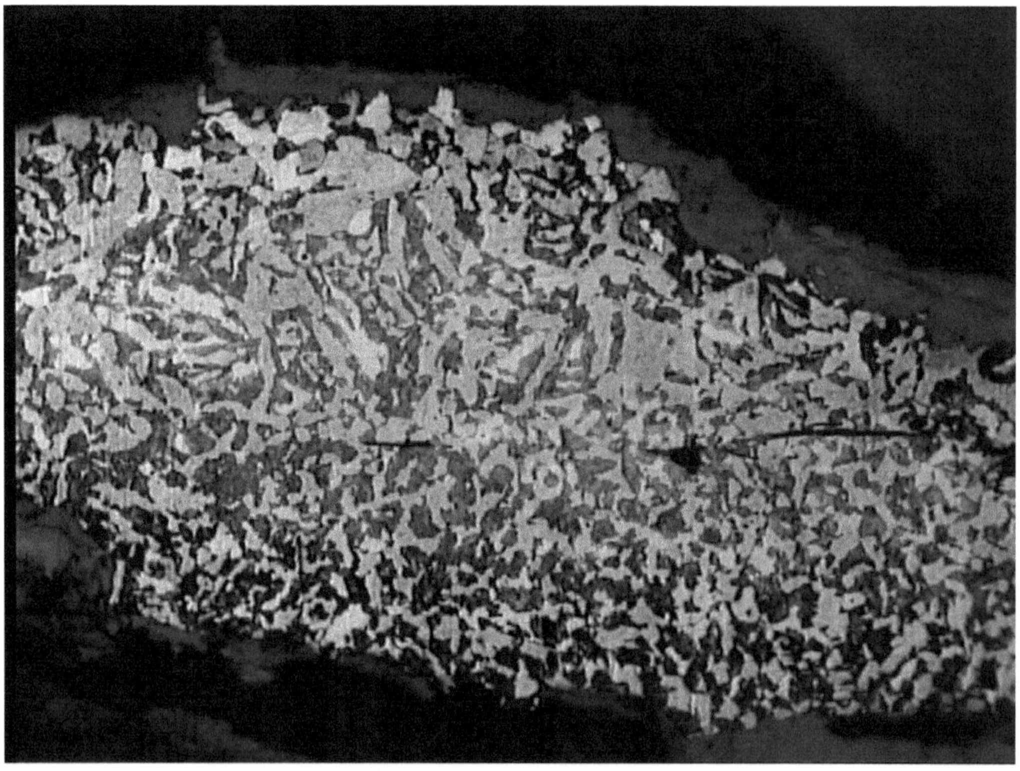

Photomicrograph of MES 53 x264, edge.
Single phase slag stringers probably from the forging and carburization of the bloom. The interstitial glass contents Mg (2.75%) and high Mn (4.58%) that points out to the smelt of Mn/Fe ores.

Chisels

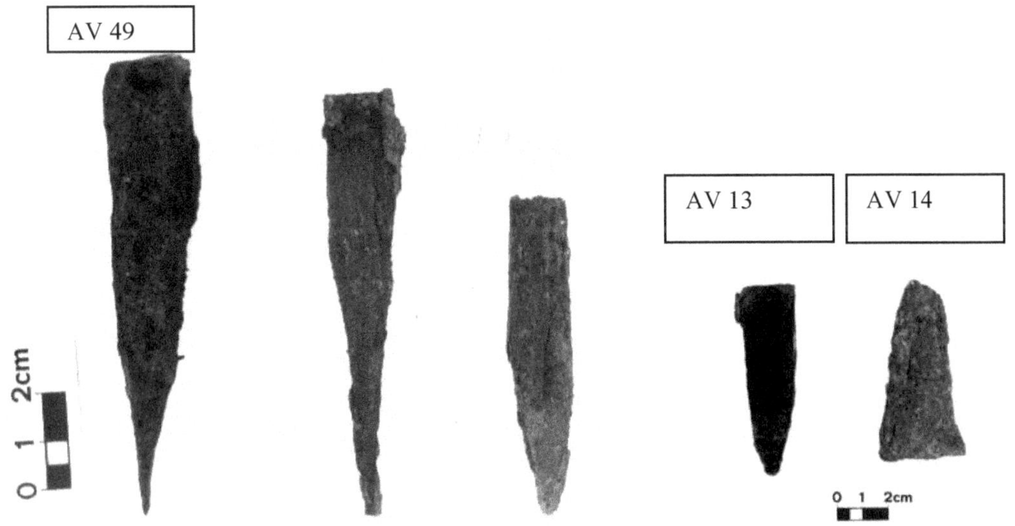

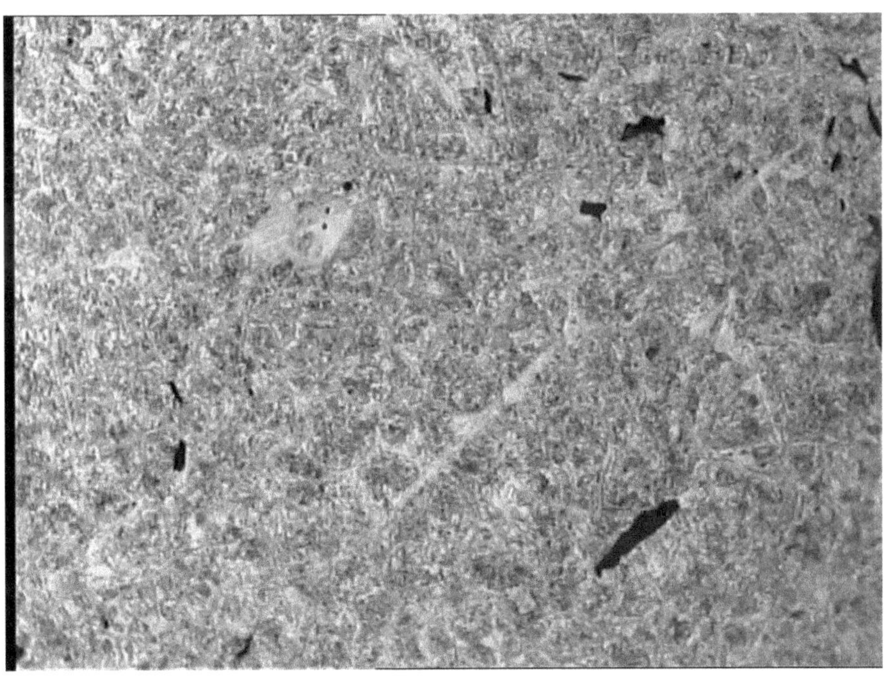

Photomicrograph of AVD 49 x66.
Pearlite (gray) and spaerodised cementite (white). Black areas indicate corrosion.

(for other types of classical and helenistic chisels. Pleiner, 1969, Figures 11 and 10)

Knives

Messemvria-Zone: curved knives

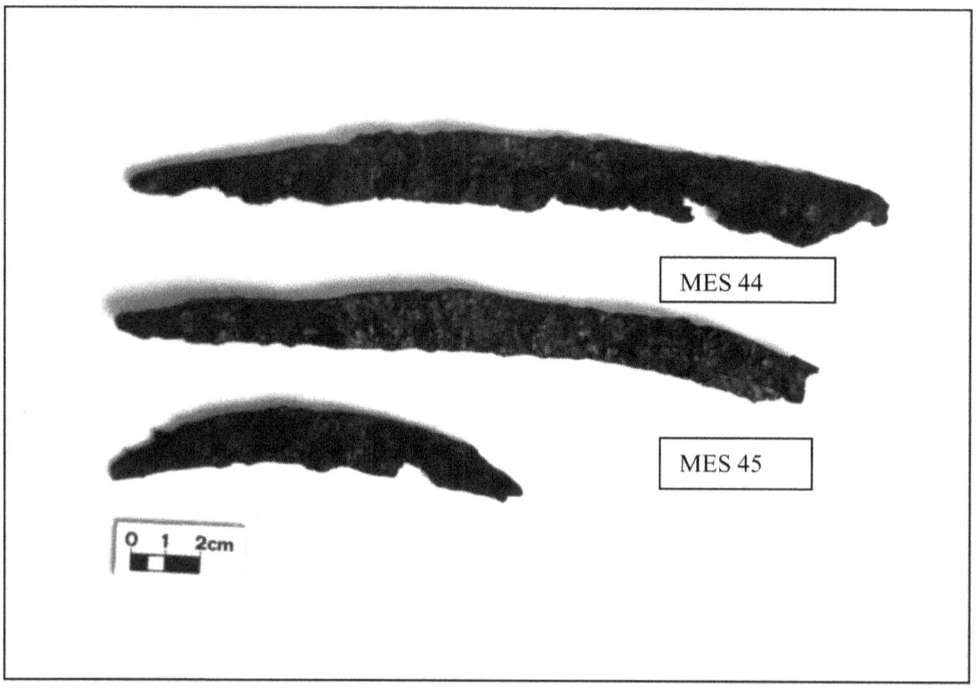

Photomicrograph of MES 65 x66 showing welding of high carbon steel to medium carbon steel. Slag stringers along the welding lines.

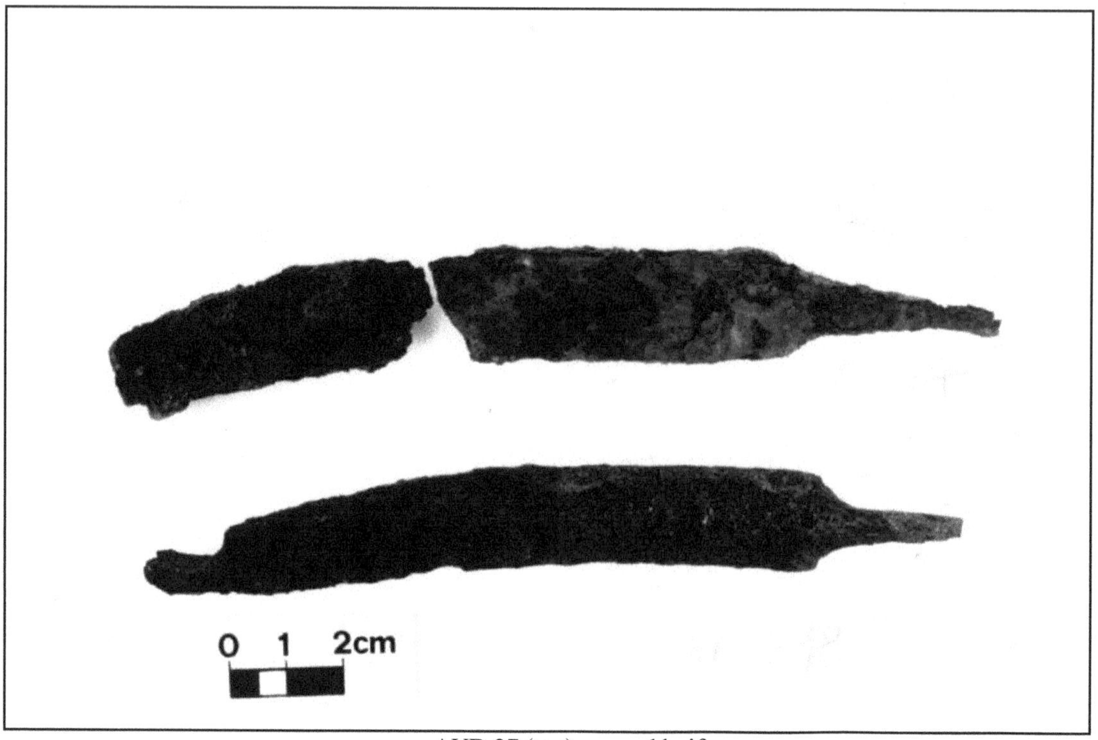

AVD 37 (top), curved knife

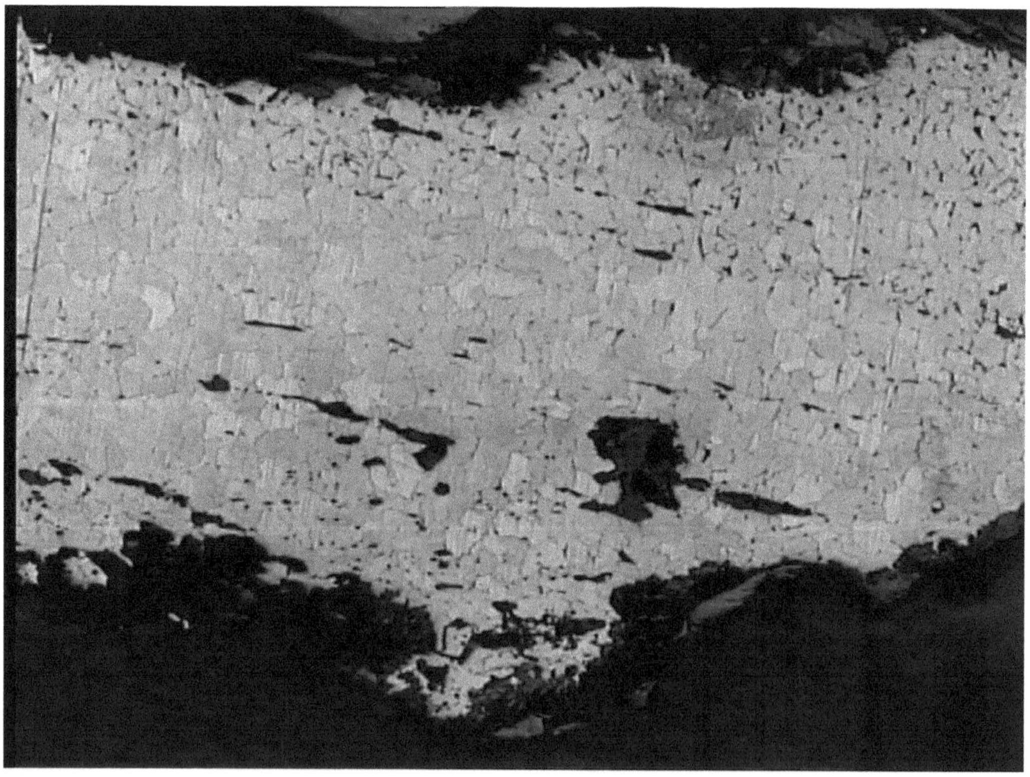

Optical microscopy of curved knife AVD 37 x132. It is made by wrought iron (large ferrite in the middle) that has been deformed slightly on the edges due to cold hammering.

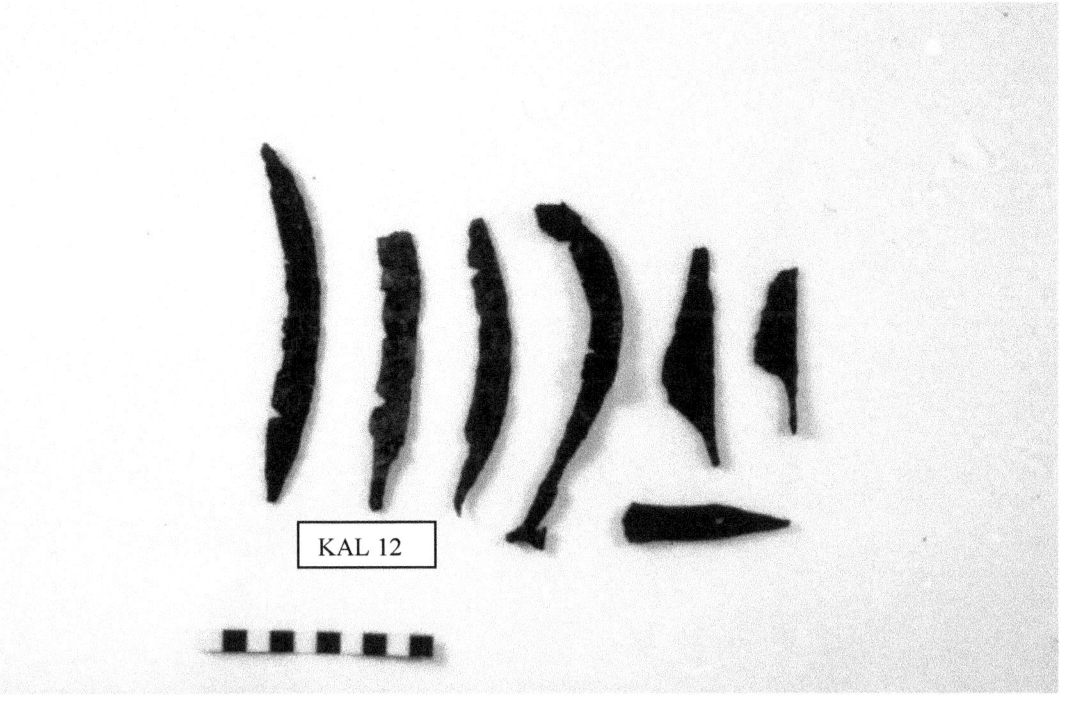

KAL 12, Curved Knife-

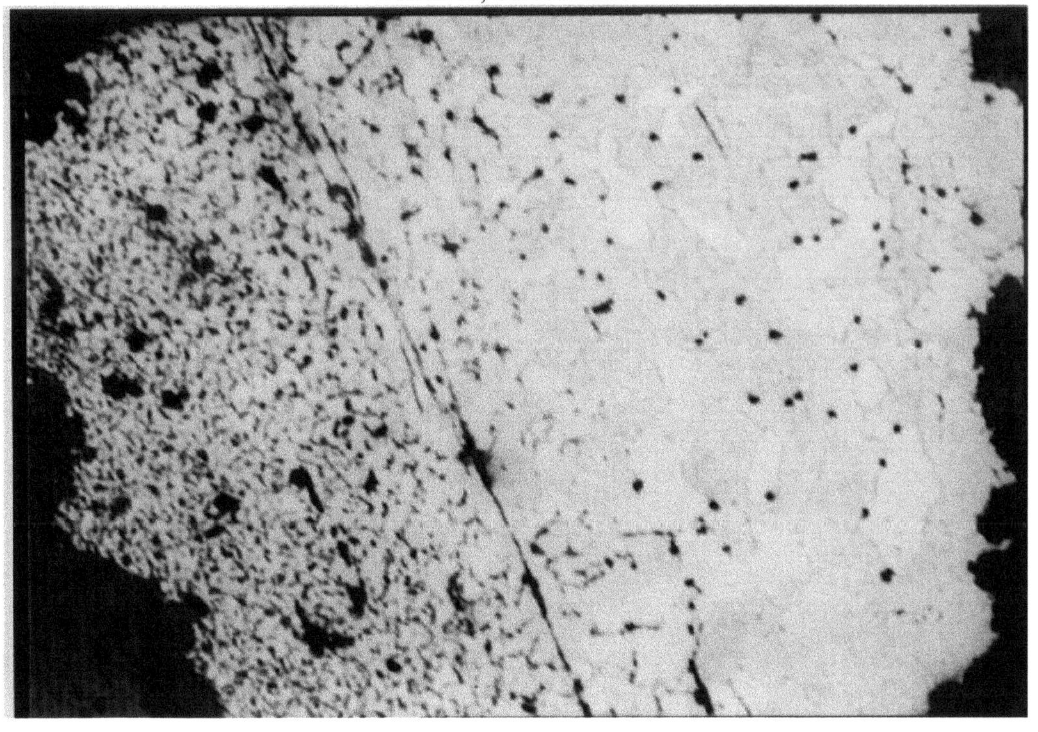

Optical microscopy of curved knife KL 12 x50. The v shaped section shows that this blade is made by welding of a low carbon steel (0.2%) to wrought iron. The long slag inclusion on the left indicates the zone of contact between the two sheets. Hv=123

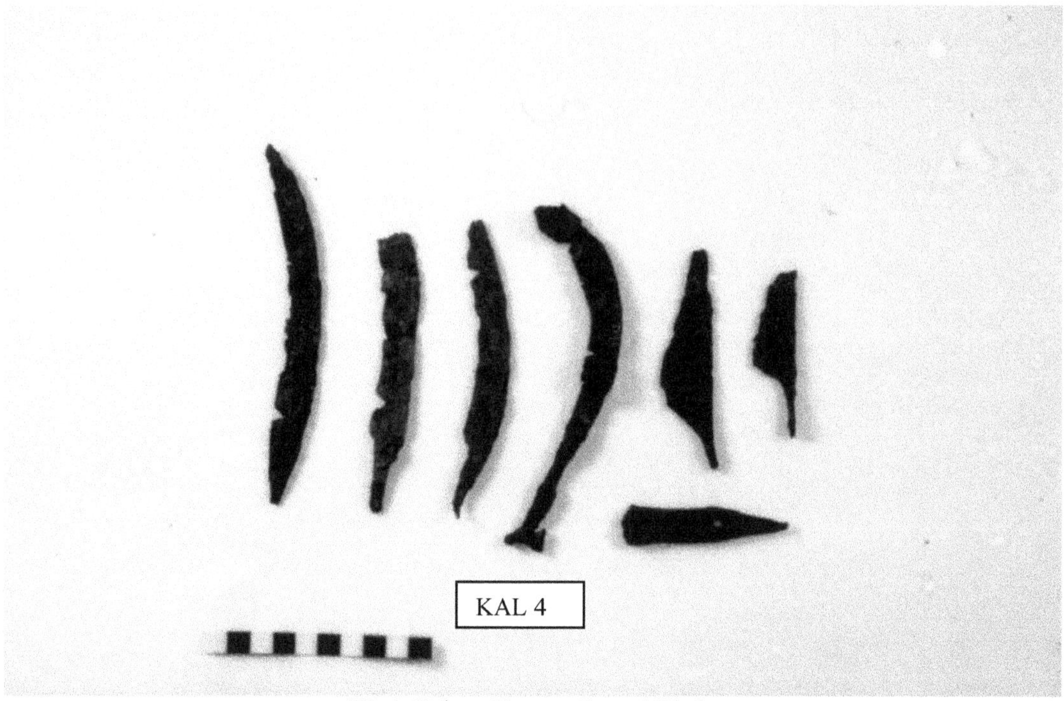

KL 4, Kalyva-Kastro, Curved Kinfe

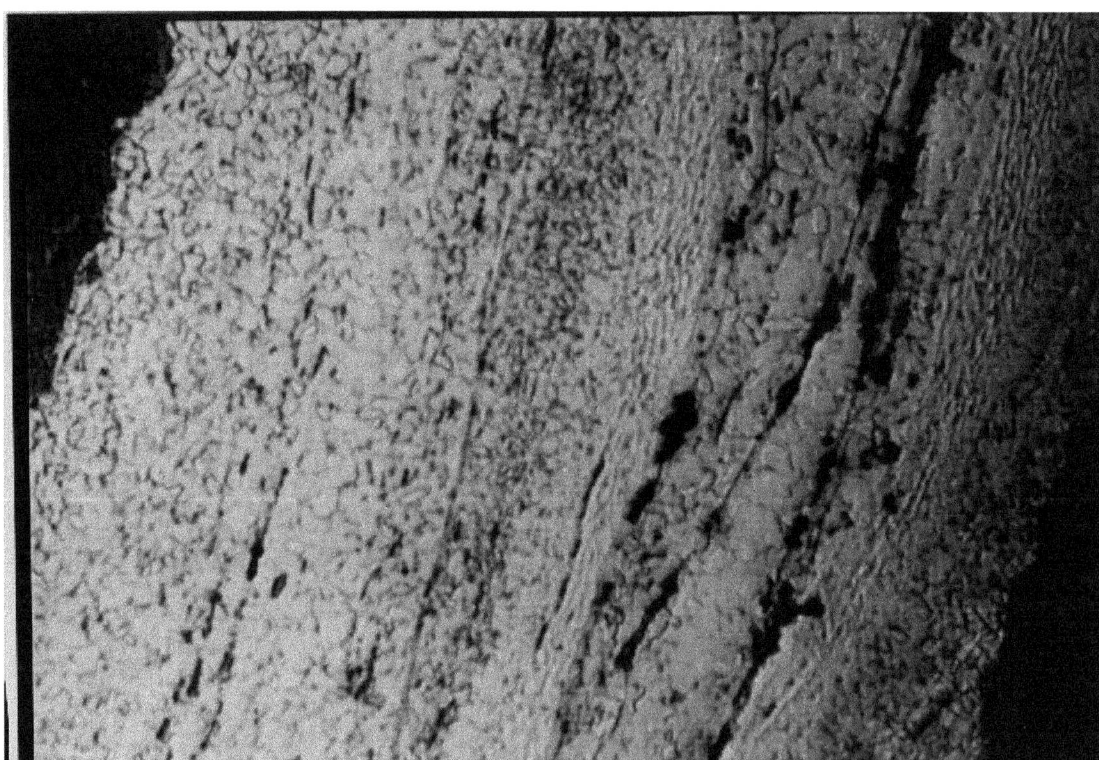

Optical microscopy of curved knife KL 4 x50. The blade is made by forge welding of low carbon steel at 0.2% (light gray) to wrought iron (white). The welding was not complete as the many elongated slag inclusions and the uneven distribution of carbon suggest. Hv=151

KL65

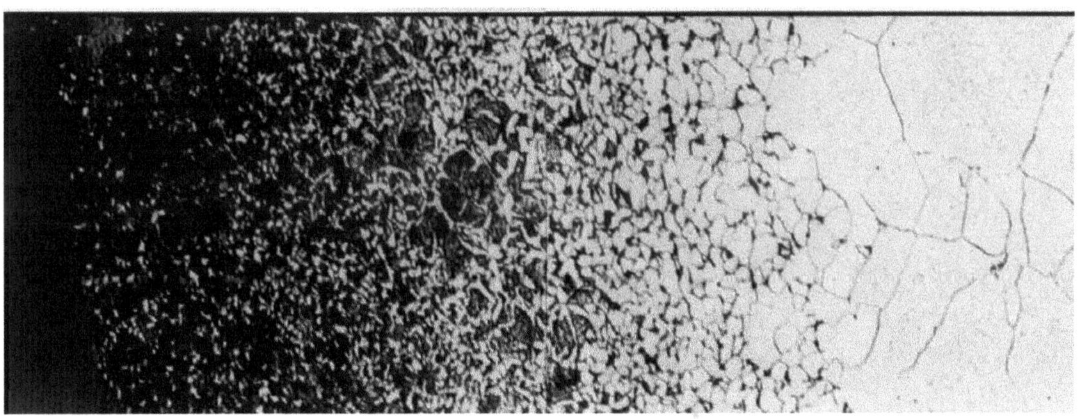

Optical microscopy of curved blade KL 65 x200 displaying a nice carburization of wrought iron on the left edge. The carbon is distributed evenly from the edge to the core. No slag inclusions or stringers are evident. Hv= F 131

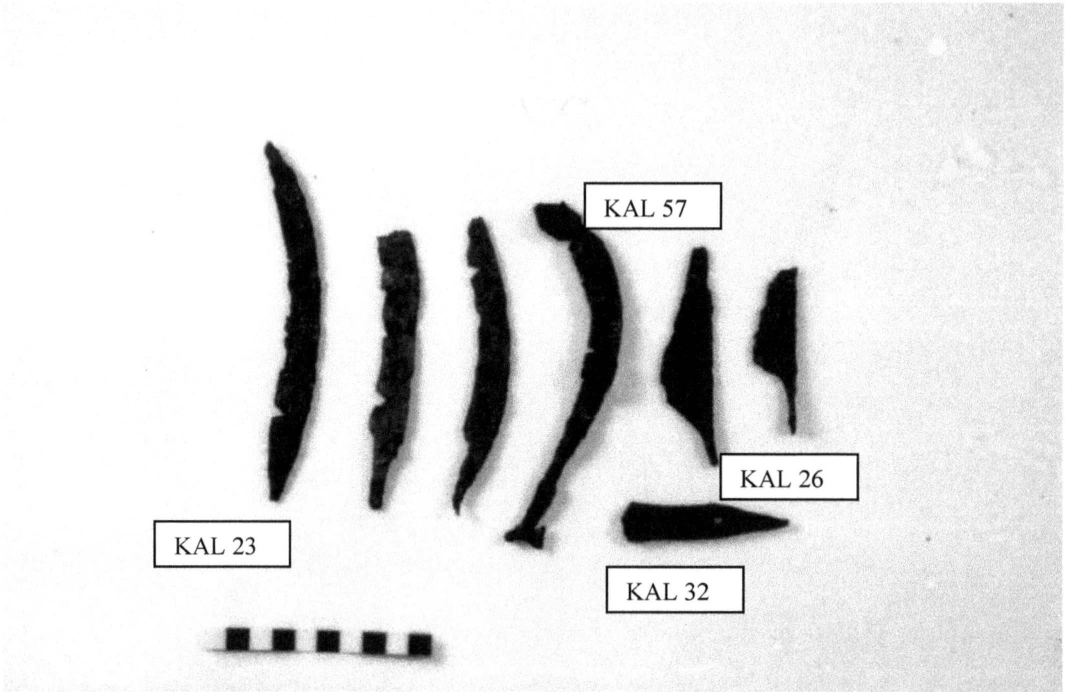

KAL 10, 1, 14, 98

KL 14

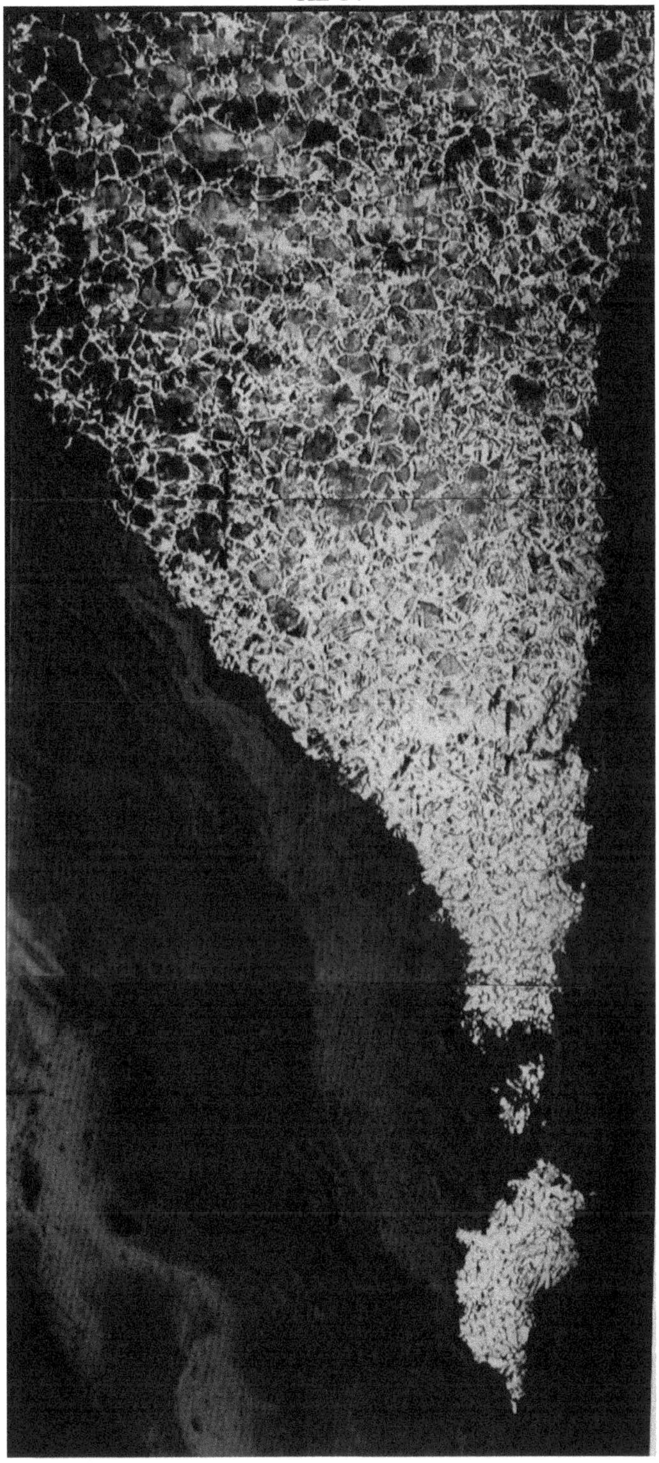

Optical microscopy of curved blade KL 14 x50. The v shaped section displays uniform distribution of hyper-eutectic composition of steel with pearlite and cementite around the grain boundaries. The point of the v shaped section is de-carburised. Hv on F+P=205 Core=F, 180
Similar is the straight blade kl41.

MES 17, Messemvria-Zone, Straight Knife

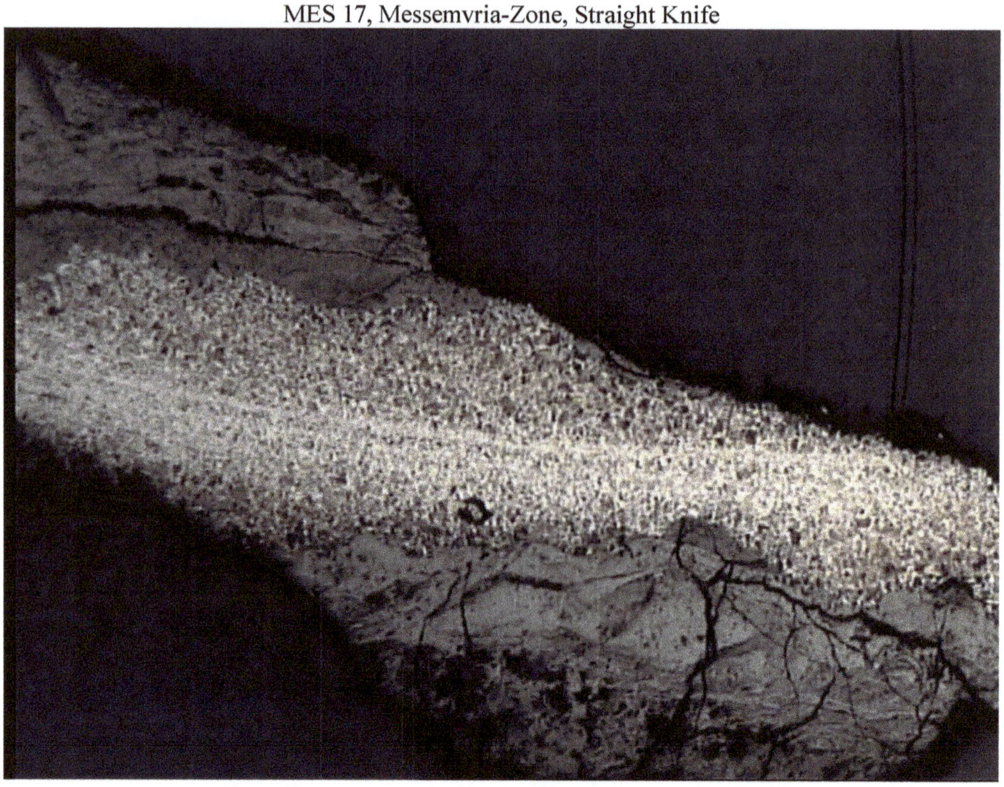

Photomicrograph of MES 17 x66. Straight knife blade with obvious welding lines.

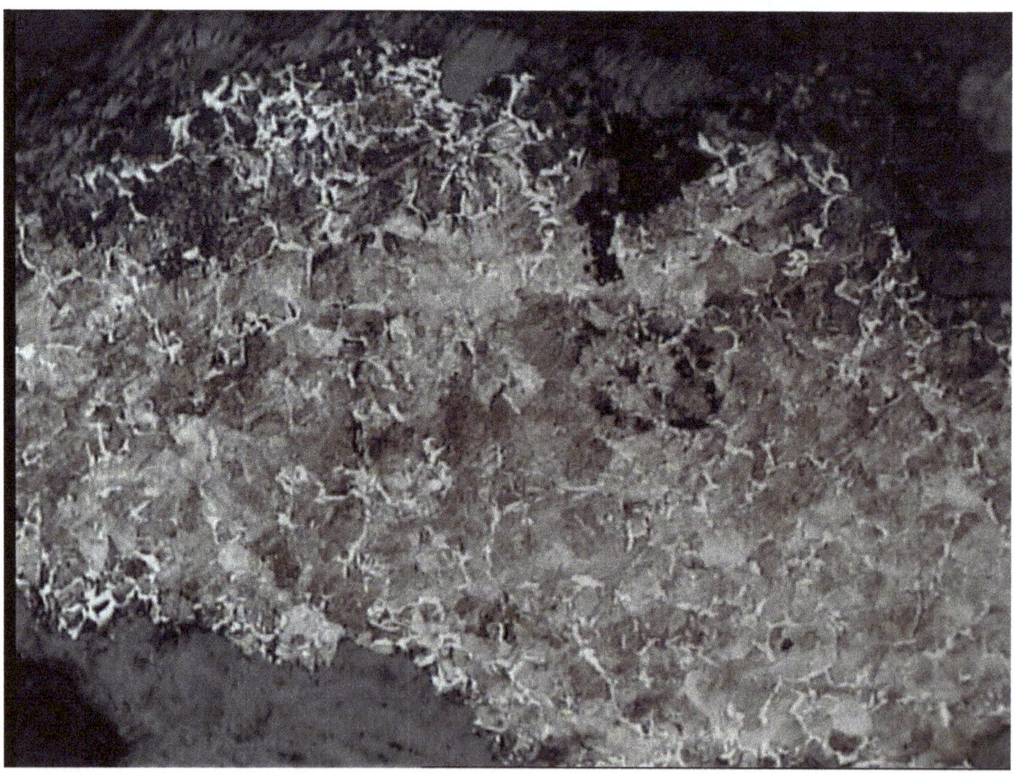

MES 17 x200 edge displaying eutectic steel structure with Pearlite and Ferrite in the grain boundaries.

MES 9 x50, Straight knife blade made by on sheet of high carbon steel.

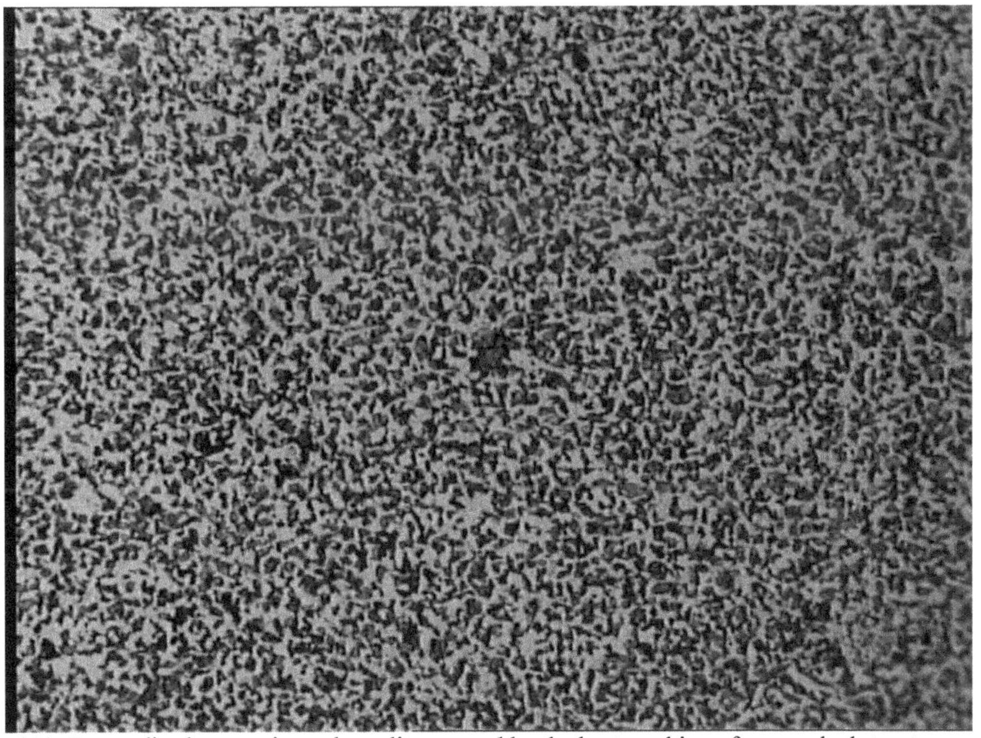

MES 9 x200, Spaerodised cementite and pearlite, caused by the heat working of a quenched structure, probably tempered martensite.

AVD 38, Avdera, Striaght Knife.

Optical microscopy of straight knife blade AVD 38 x132, displaying uniform distribution of P through out the v section. Black lines are corrosion.

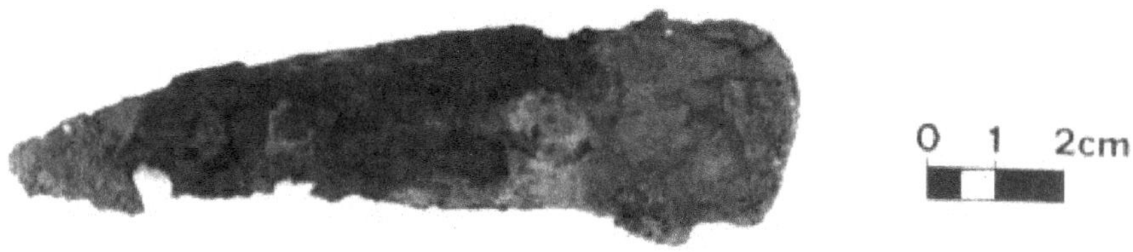

AVD 33

AVD 18

Knives from Avdera

Straight knife blade- Optical microscopy of AVD 36 x264, showing a carburised edge with pearlite (black) and bainite (white wings). The core displays a low carbon composition with ferrite and some pearlite in the grain boundaries. Slag inclusions indicate the welding lines.

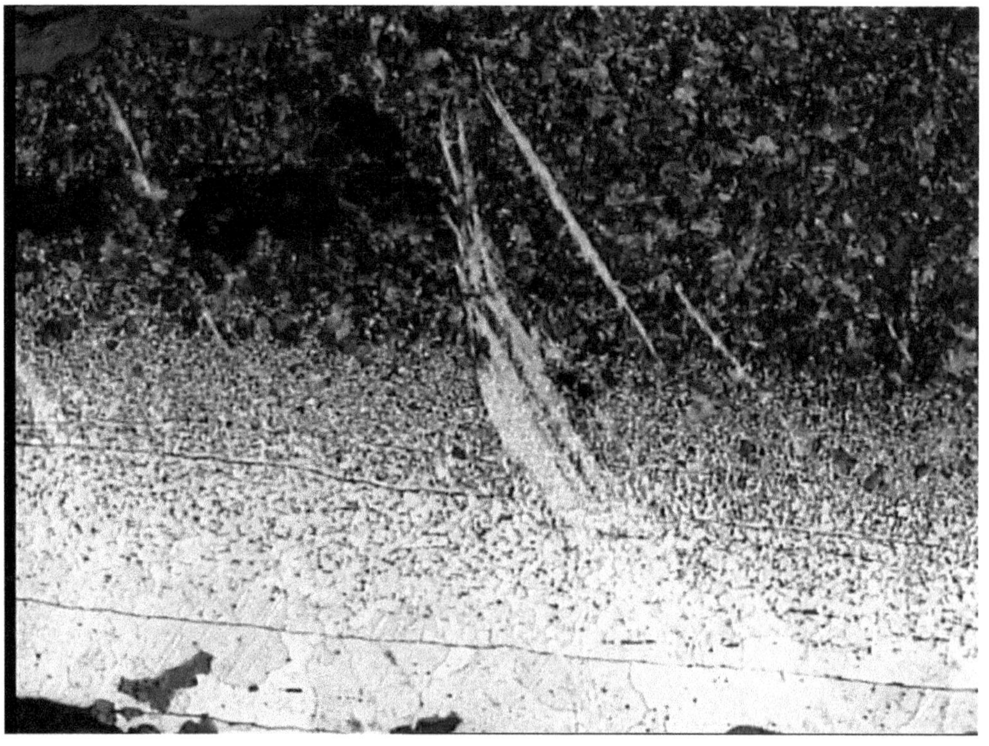

Straight knife AVD 53. Optical microscopy x66. The black area on the top is eutectic pearlite, the white area Ion the bottom is ferrite. Slag stringers along the welding lines. The white areas on pearlite matrix are probably bainite crystals formed during slow cooling after carburization.

(from left to right) KAL 49. 50, 16, 61, 20, 39, 48, 46, 62, 61

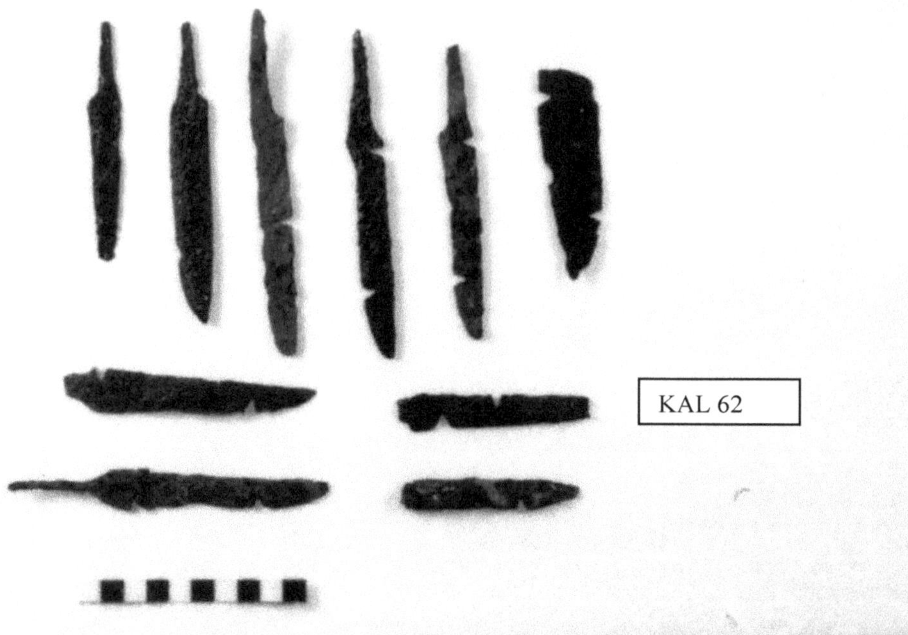

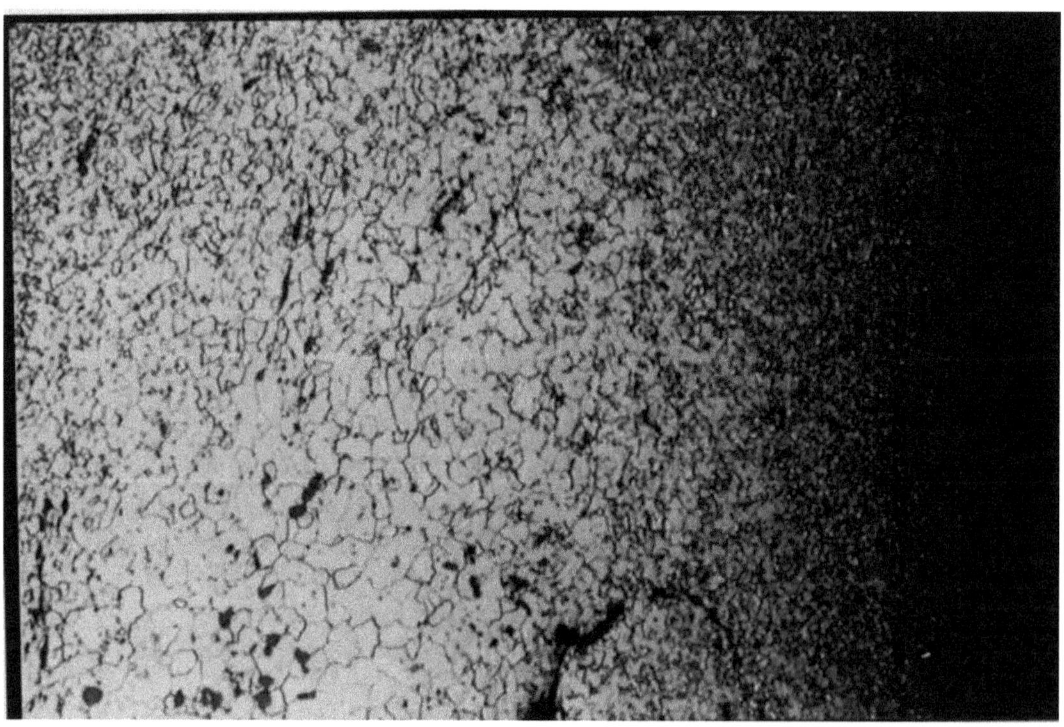

KL 62. Optical microscopy of straight knife KL62 x50. The section shows a nice carburization of wrought iron on the right edge. Black holes are slag inclusions left behind from the bloom forging.

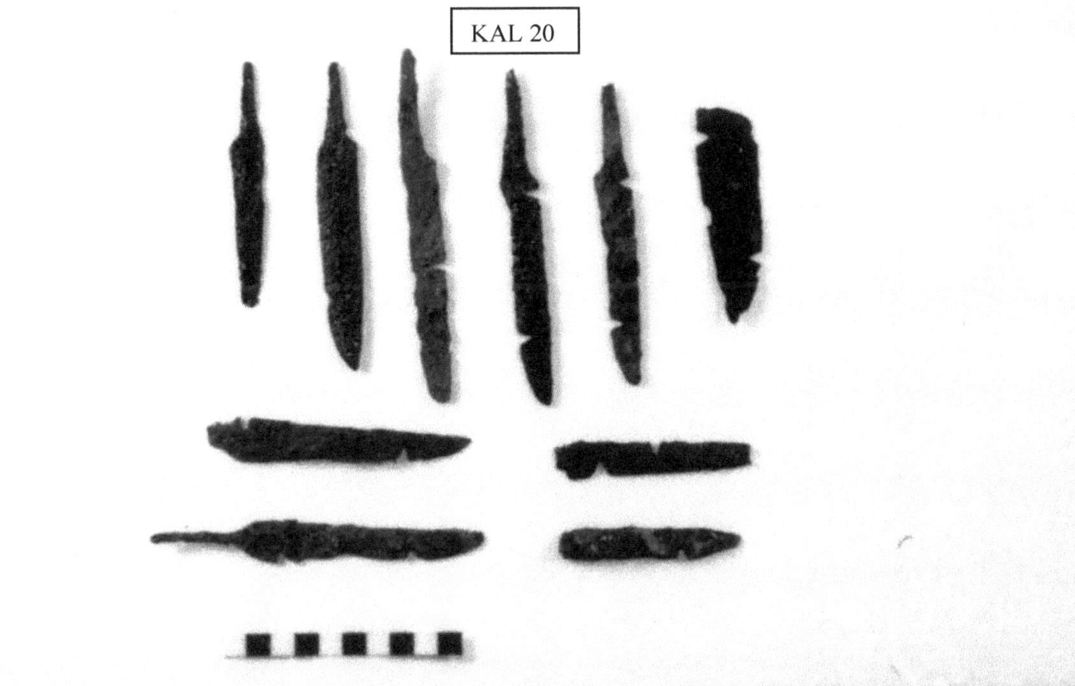

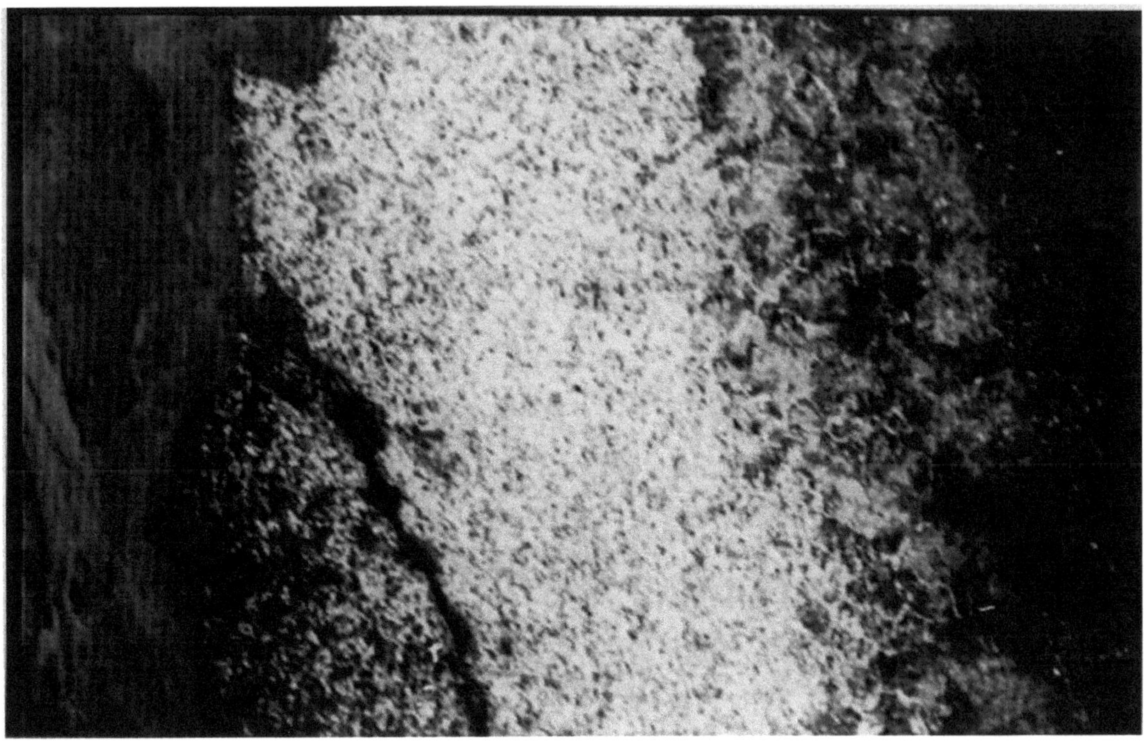

Optical microscopy of straight knife KL 20 x50 displaying carburization on both edges.
Hv edge=383
HvPearlite = 220

Saw

AVD 26

Cohlan 1959 relates the saw to metal cutting but no contextual info to say so for this one..

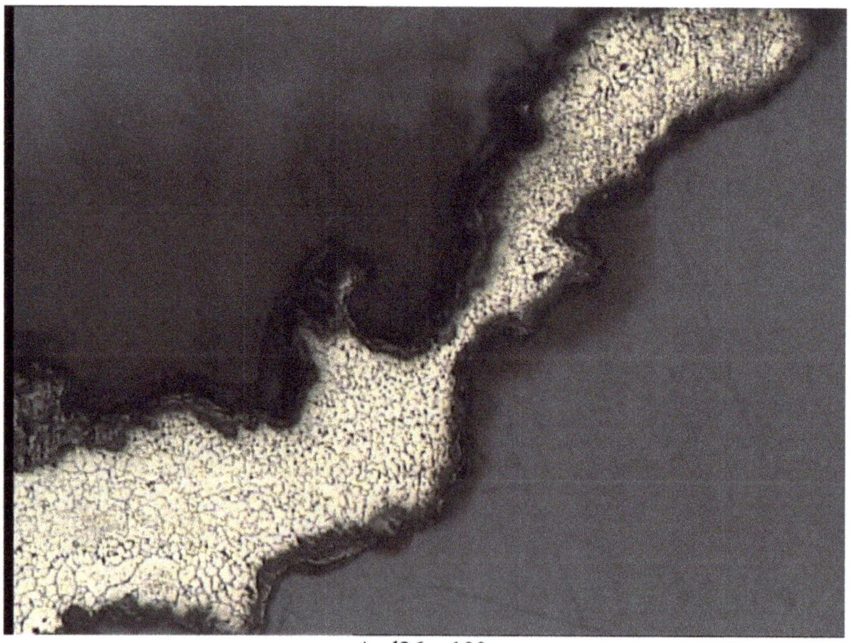

Avd26 x 100

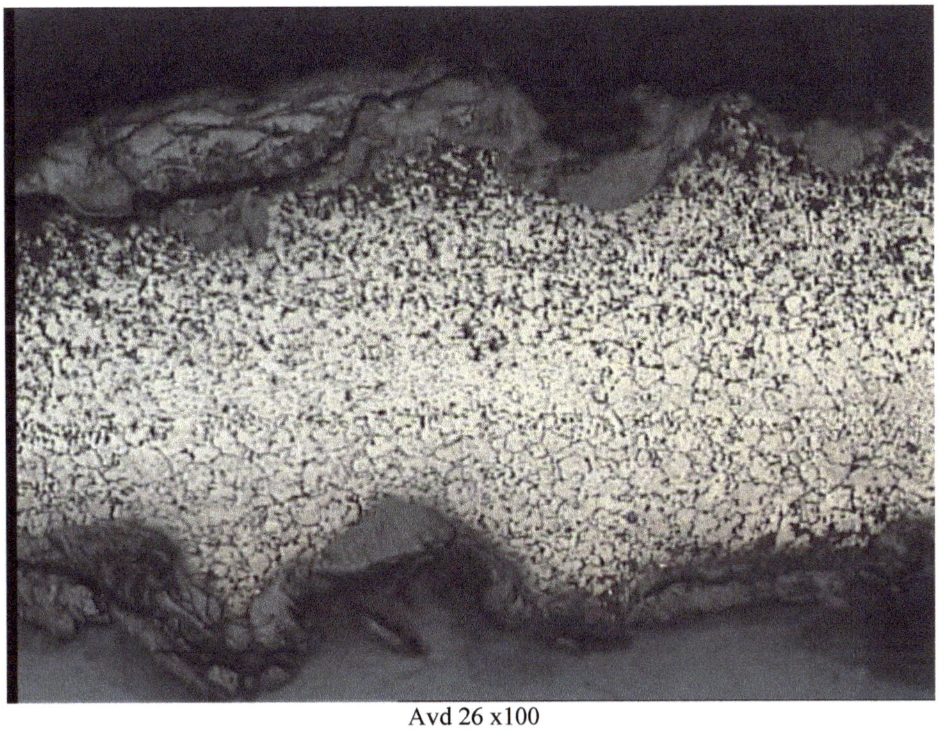

Avd 26 x100

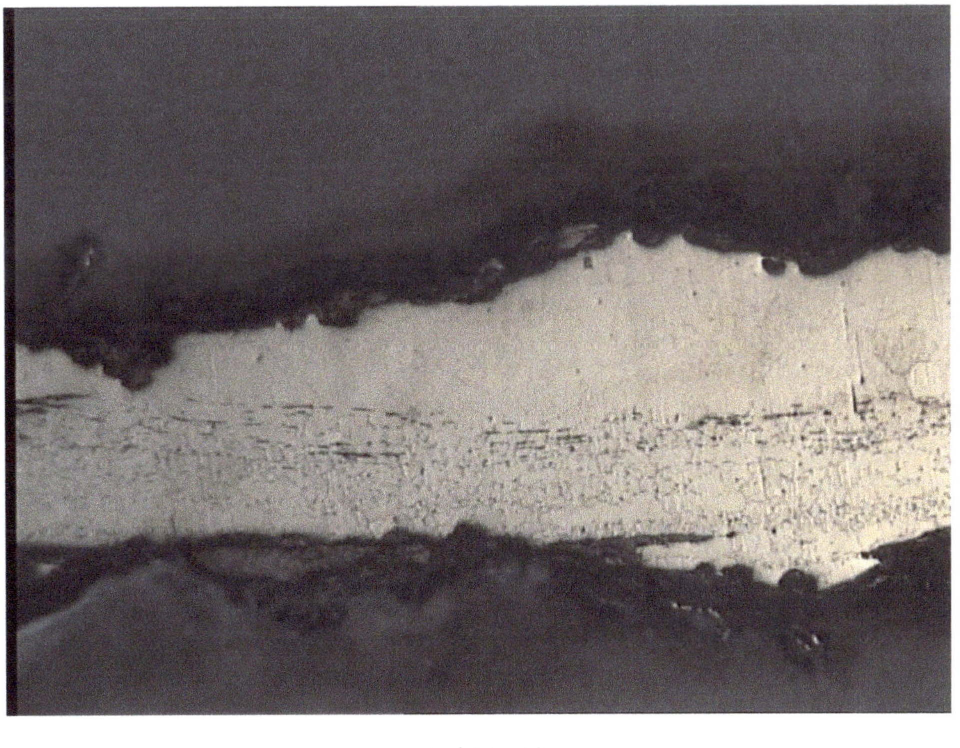

Avd 26 x 50

Sickles

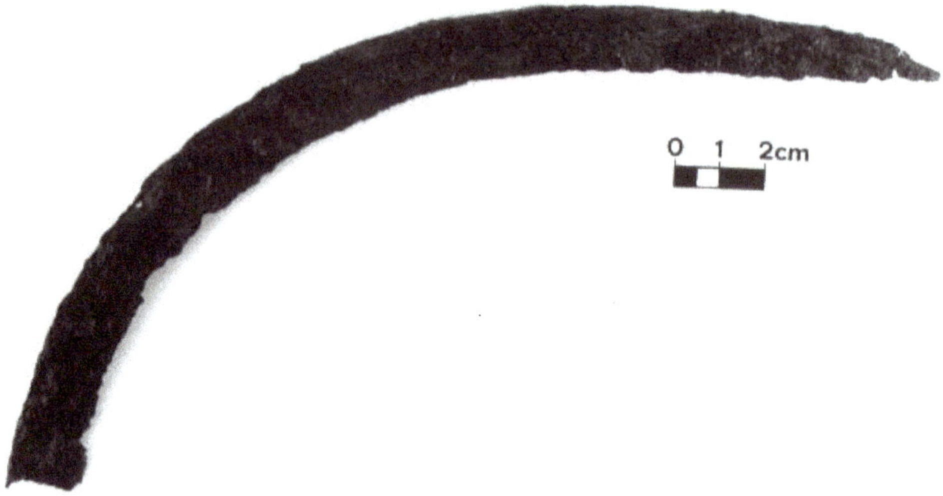

AVD 23

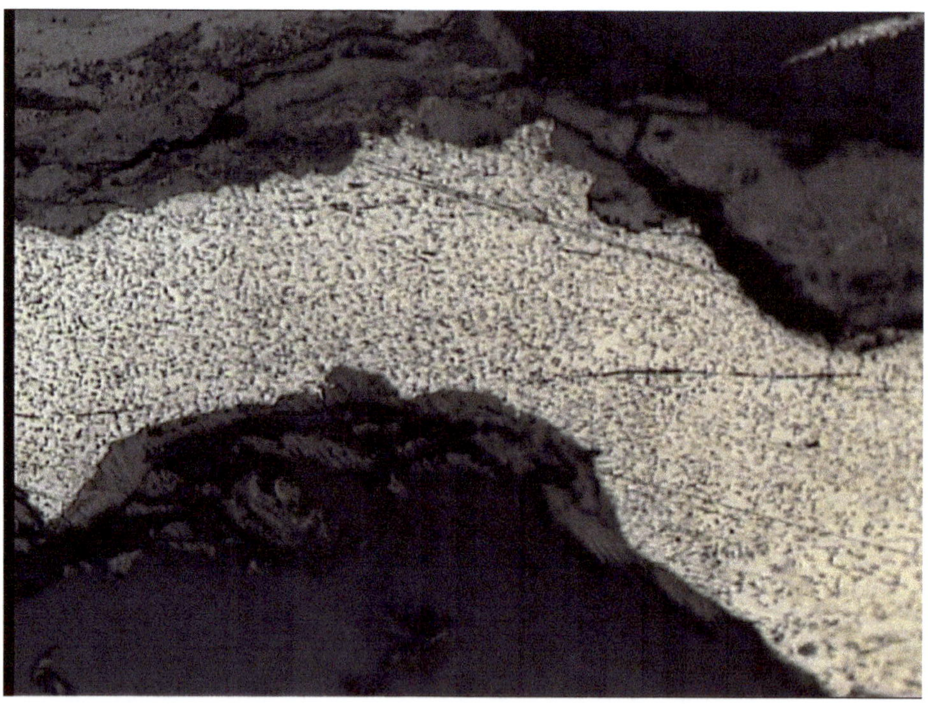

AVD 23. Optical microscopy x132 showing uniform distribution of ferrite with some pearlite of hypo-eutectic composition.

Shears AND Scissors (AVD 431`8 AND 4349 ALSO MES 49 AND 40)

Often used in operations such as sheep shearing where one hand is required to work the tools (White 1996, 119-120, figs. 95-96).

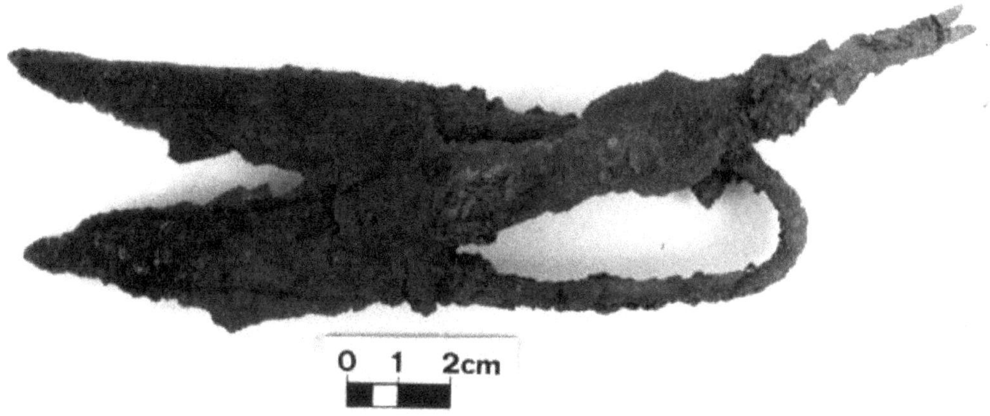

AVD 12

KL 59

Optical microscopy of scissors KL 59 x50 displaying deformed ferrite structure. The grains of ferrite and the slag inclusions are elongated to the direction of forging. The ferrite structure cannot explain the high hardness of the sample (Hv=339).
This might be due to a phosphorous content.

Spearheads

MES 23

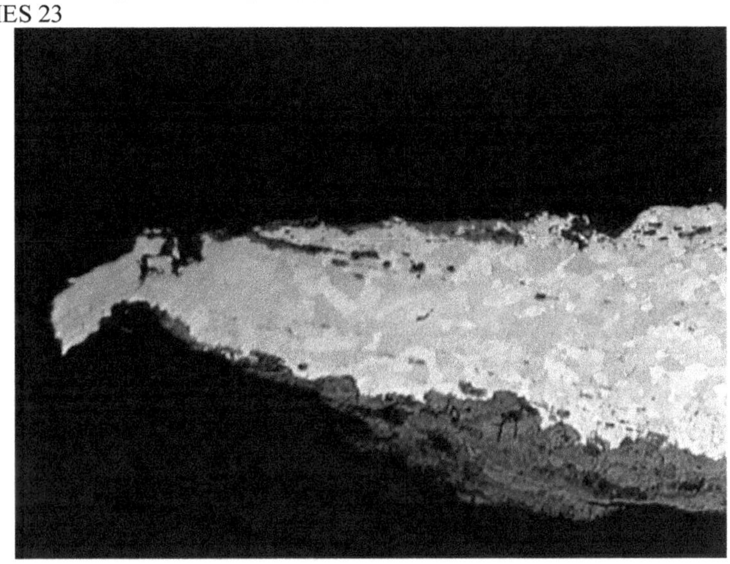

MES 23 x66, edge. Very soft and slightly deformed ferrite.

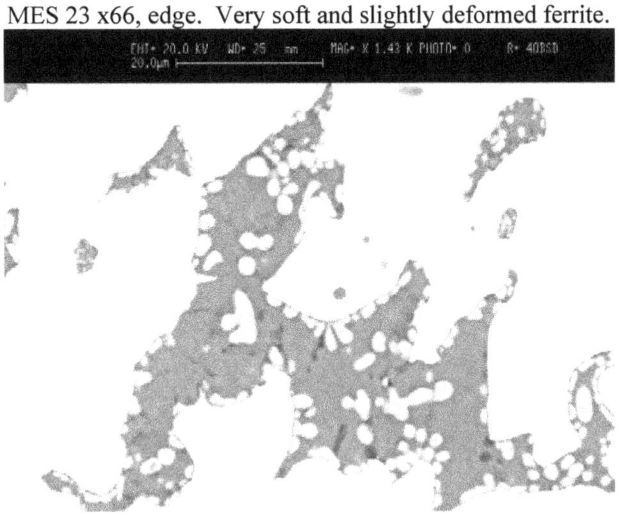

SEM photographs of slag inclusions of MES 23 showing Wustite, Fayalite and Interstitial glass. The absence of Mn points out to its not local origin.

MES 7

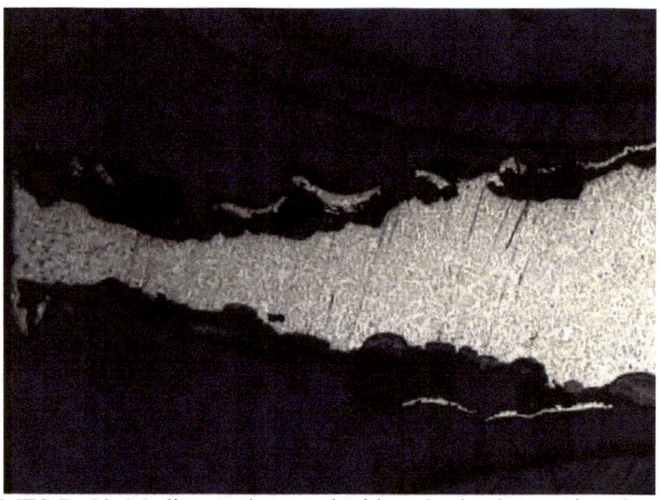

MES 7 x50. Medium carbon steel with carburization on the edge.

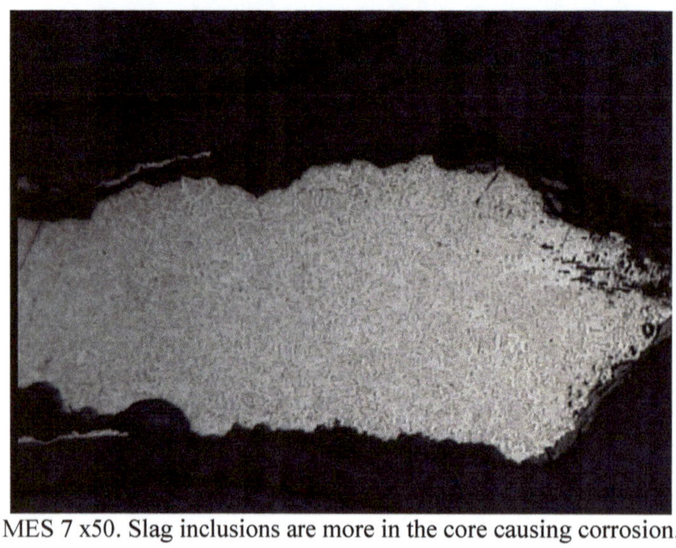

MES 7 x50. Slag inclusions are more in the core causing corrosion.

MES 25 (Exact one unknown)

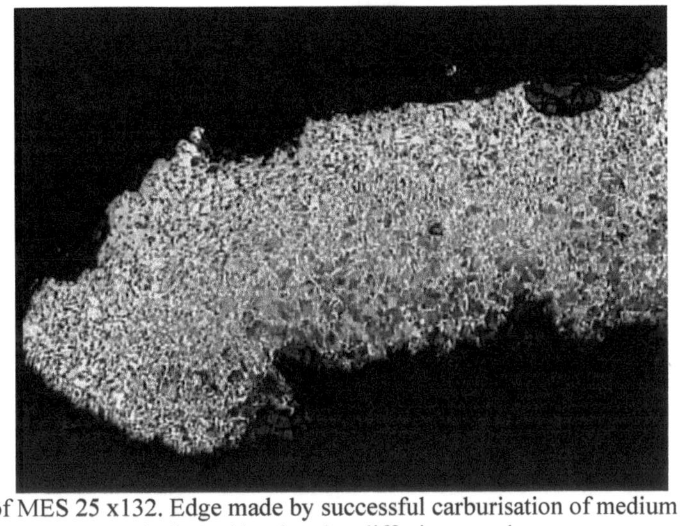

Photomicrograph of MES 25 x132. Edge made by successful carburisation of medium carbon steel as it is indicated by the nice diffusion o carbon
And the only few slag inclusions.

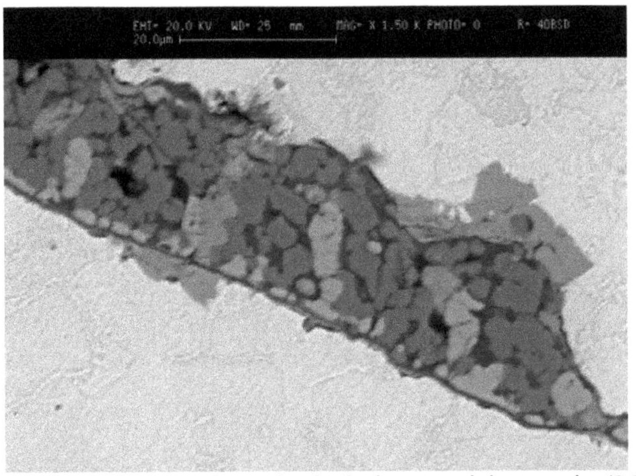

SEM photograph of MES 25 showing three phase slag inclusion containing wustite (96.51% FeO), fayalite and interstitial glass with no characteristic elements of local origin (note the absence of manganese oxide in the table below)

MES 10a (Exact one unknown)

Photomicrograph of MES 10a x66, edge. Edge made by welding steel with eutectic composition (P and F at 0.8% carbon) to low carbon steel.

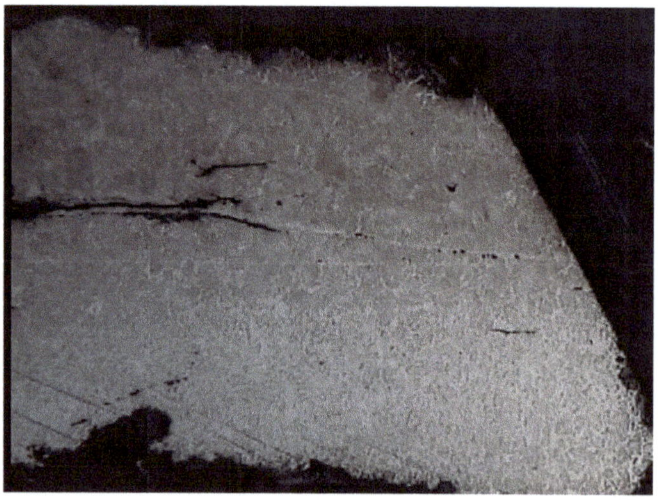

Photomicrograph of MES 10 x66, back.. Corrosion has been developed along the welding lines.

MES 22

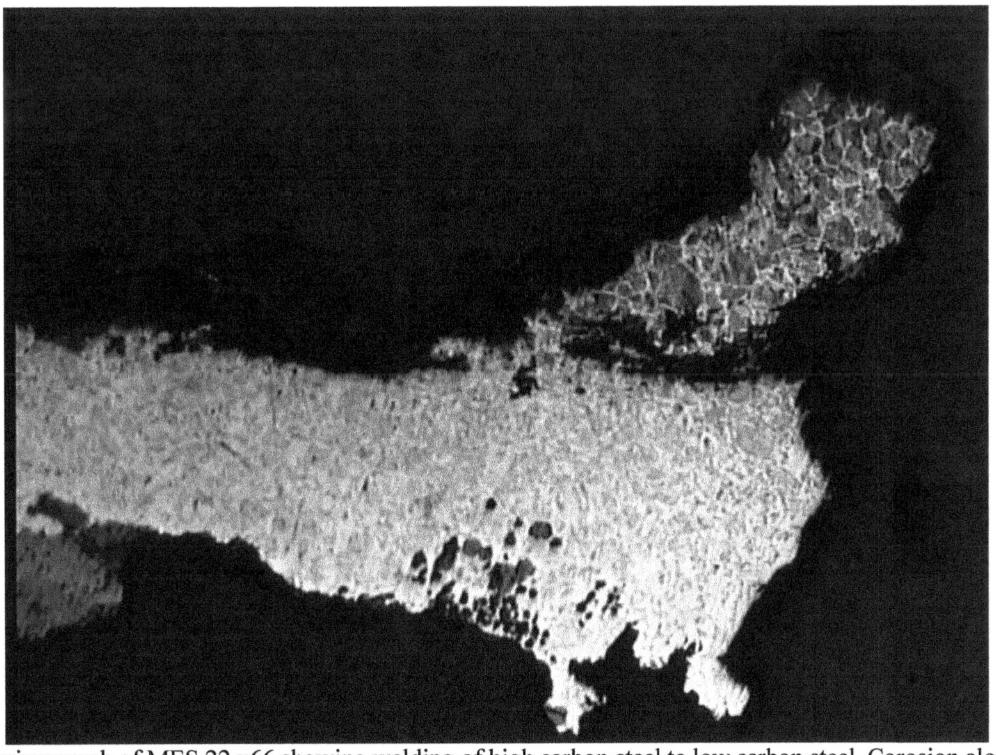

Photomicrograph of MES 22 x66 showing welding of high carbon steel to low carbon steel. Corosion along the welding lines.

2 spearheads (left and center from Avdera)

MES 55, 54, 51

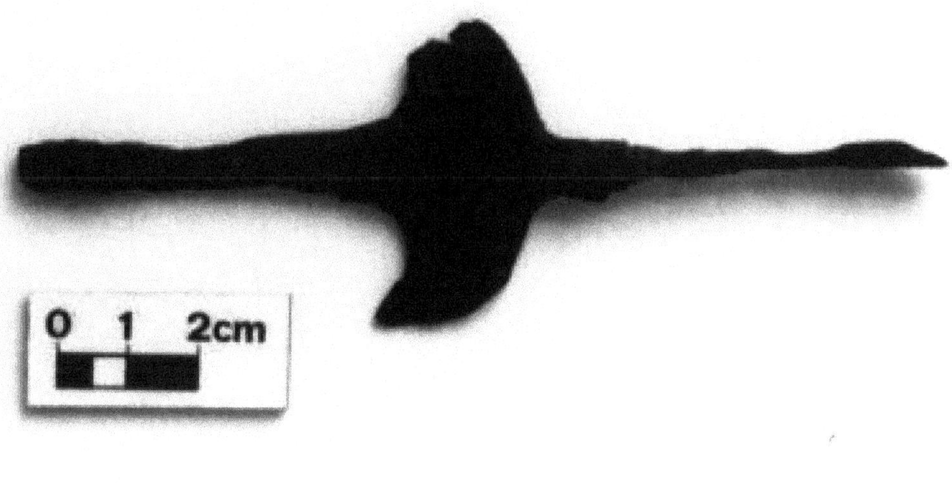

AVD 46

SPITS

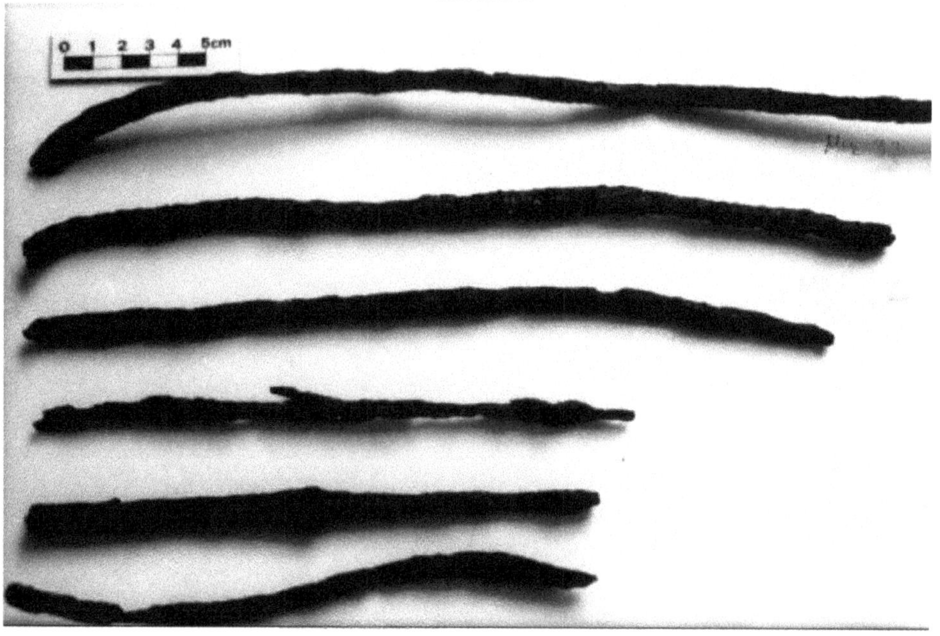

MES 31. (3rd from last)

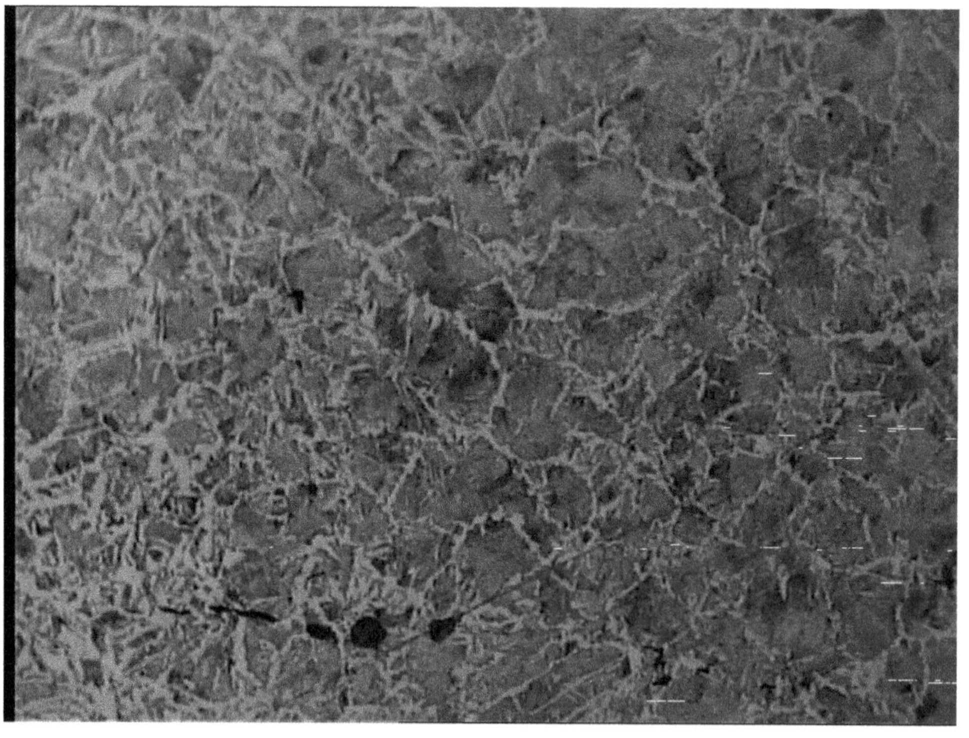

Photomicrograph of MES 31 x66. Spit made by forging together and then twisting one high carbon (in the photograph on the right) sheet with one medium carbon seat.

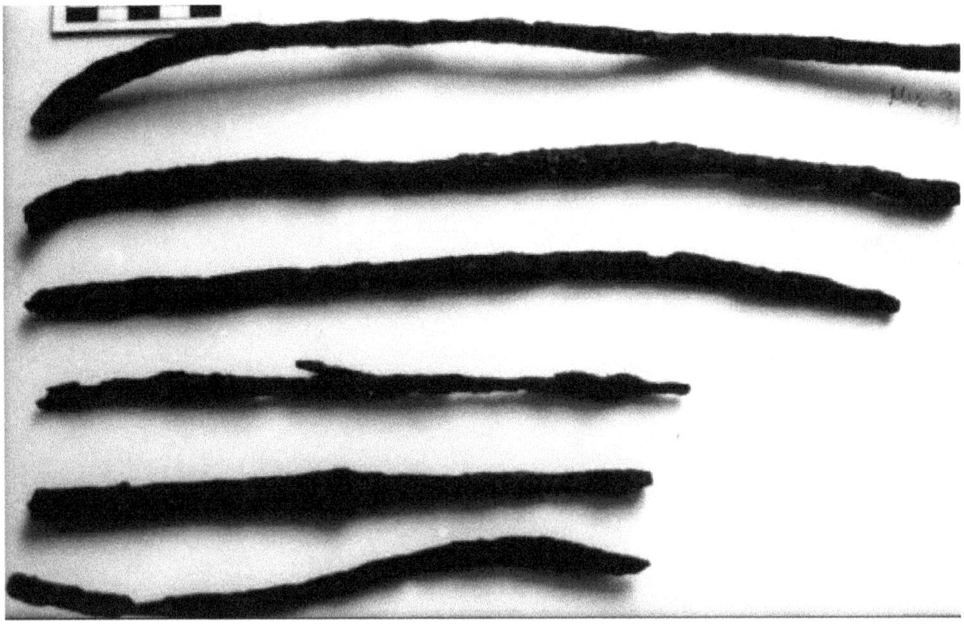

MES 32. (4th from last)

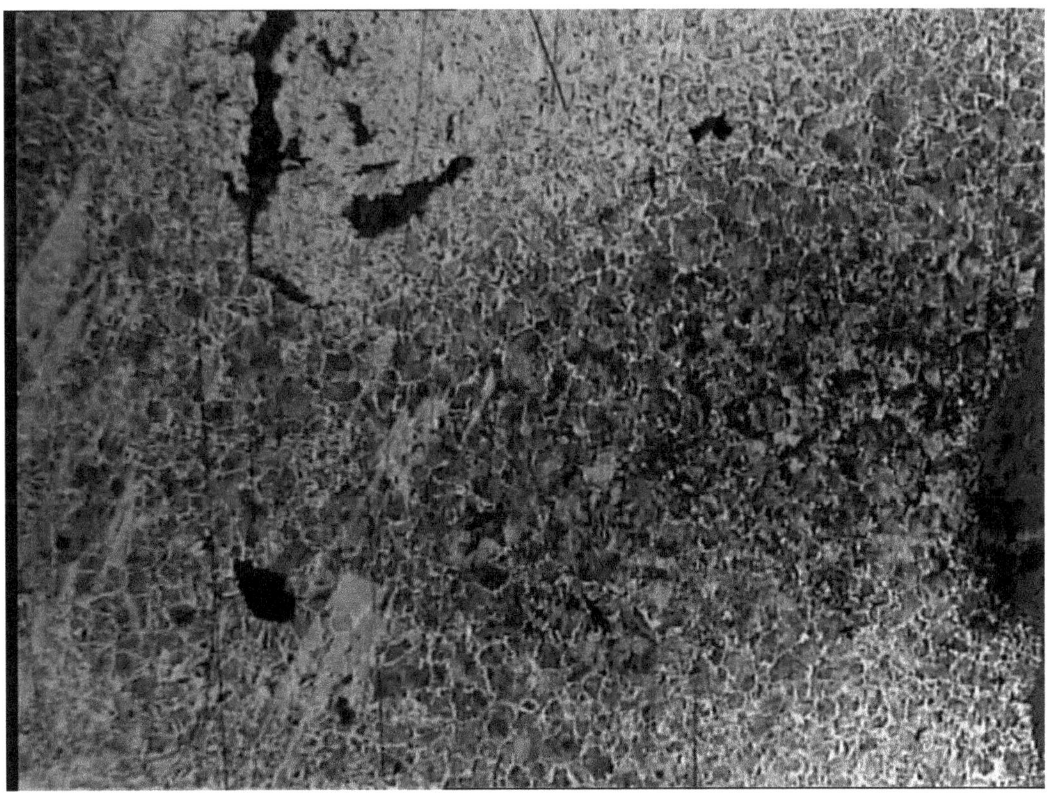

Photomicrograph of MES 32 x66. Similar structure to MES 31, with slag inclusions along the welding lines moved to the direction of twisting.

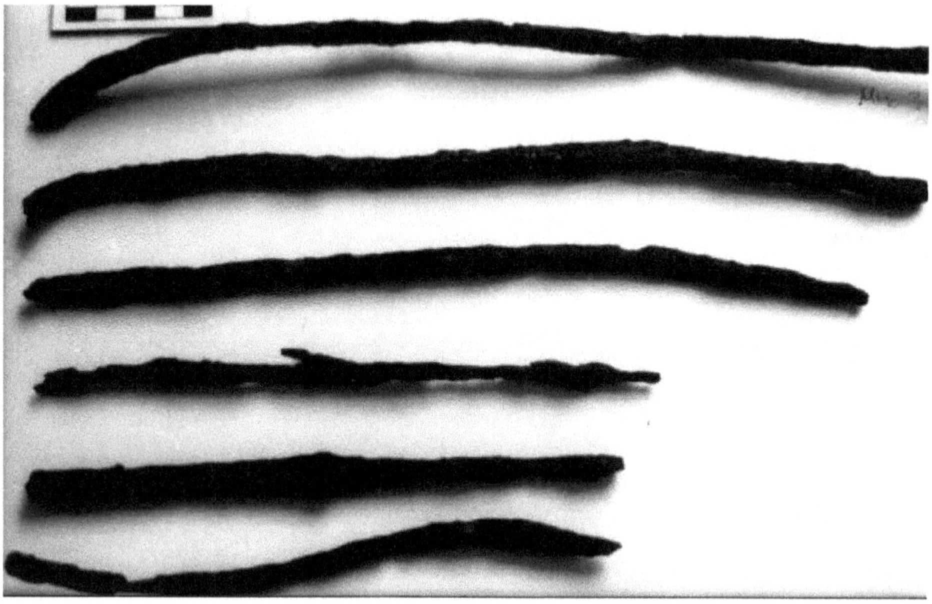

MES 33

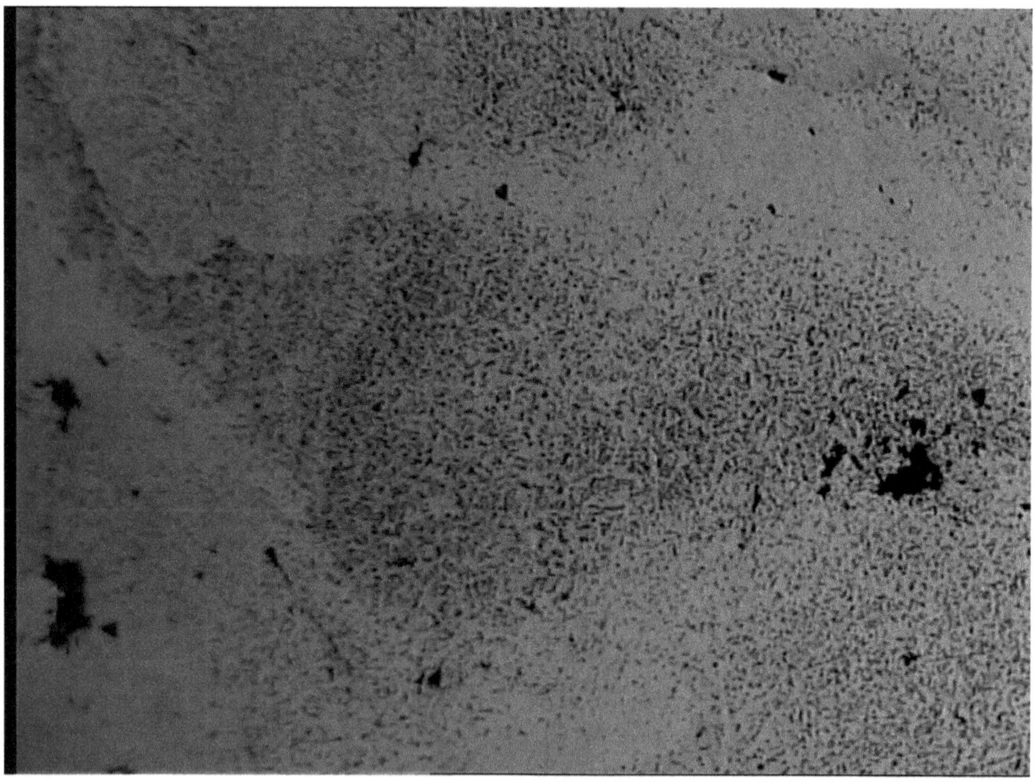

Photomicrograph of MES 33 x66. Spit mate by forging together and then twisting one medium and one low carbon sheets. Mg and Mn are present as impurities in the interstitial glass.

AVD 43 (one of the above)

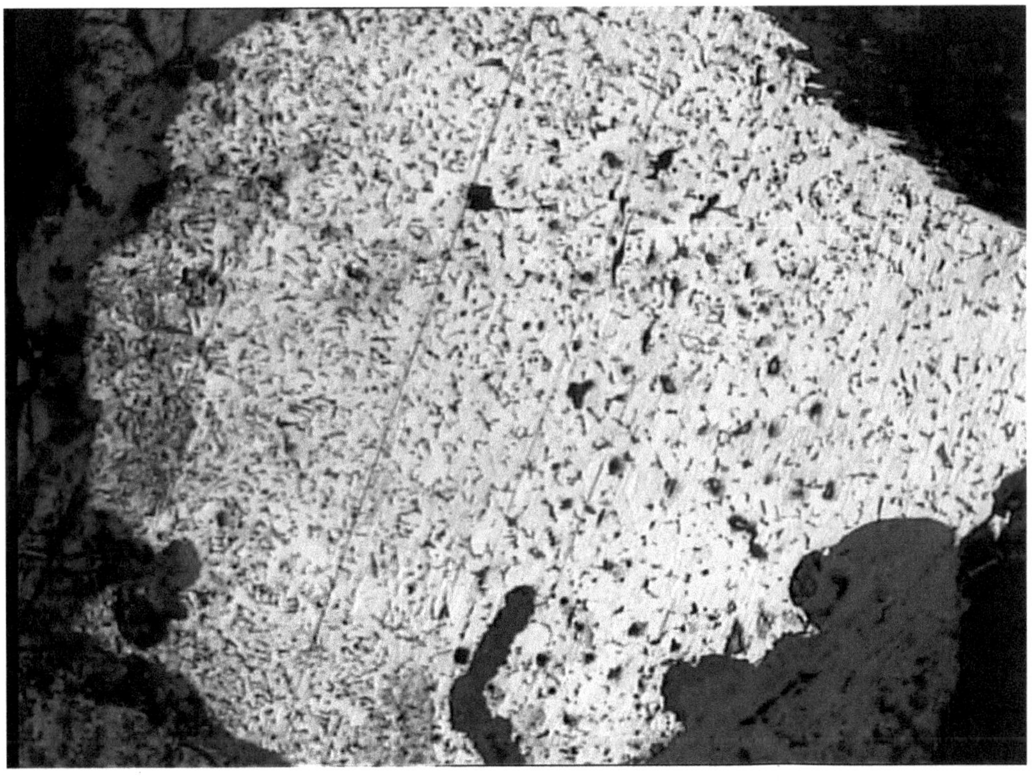

Spit AVD 43. Optical microscopy x132. Ferrite and some pearlite (low carbon steel) with slag inclusions with carburisation on the right edge.

Strigil

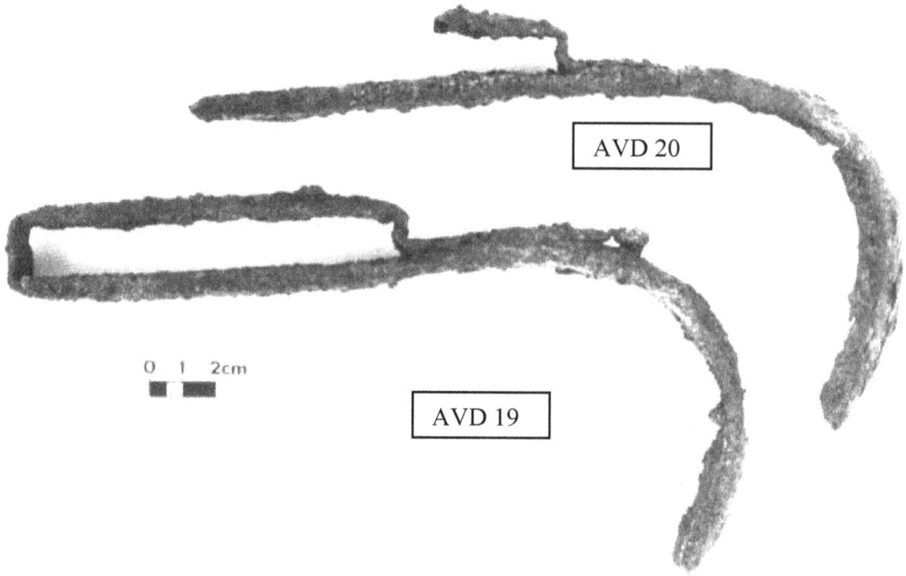

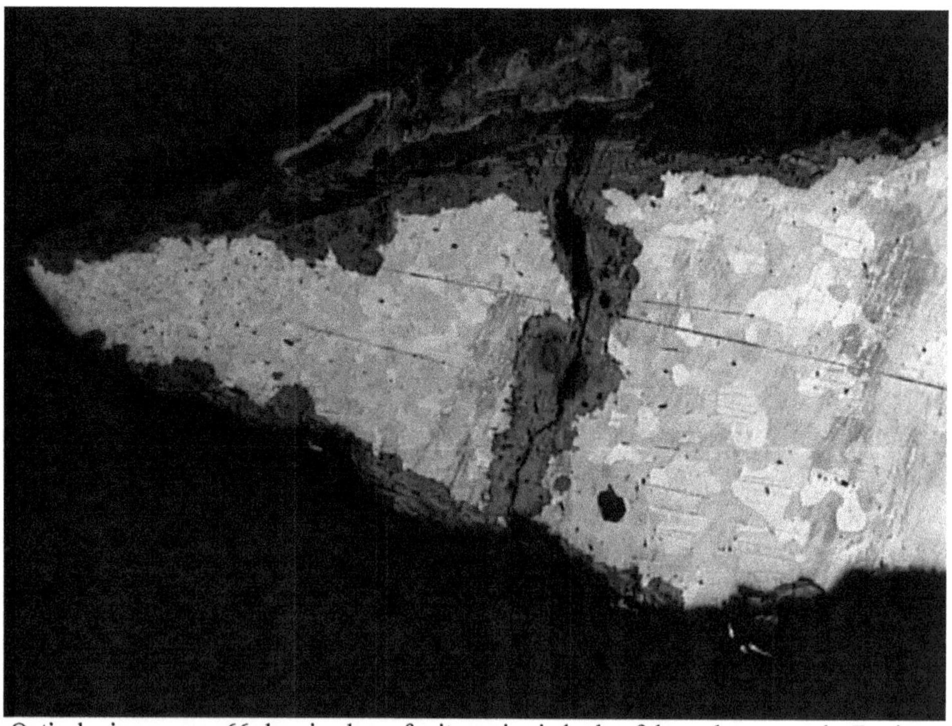

AVD 1. Optical microscopy x66 showing large ferrite grains in body of the v shape sample. Ferrite grains are deformed and smaller on the edge, because of cold hammering to shape the Artefact.

Fragmented strigil from Avdera

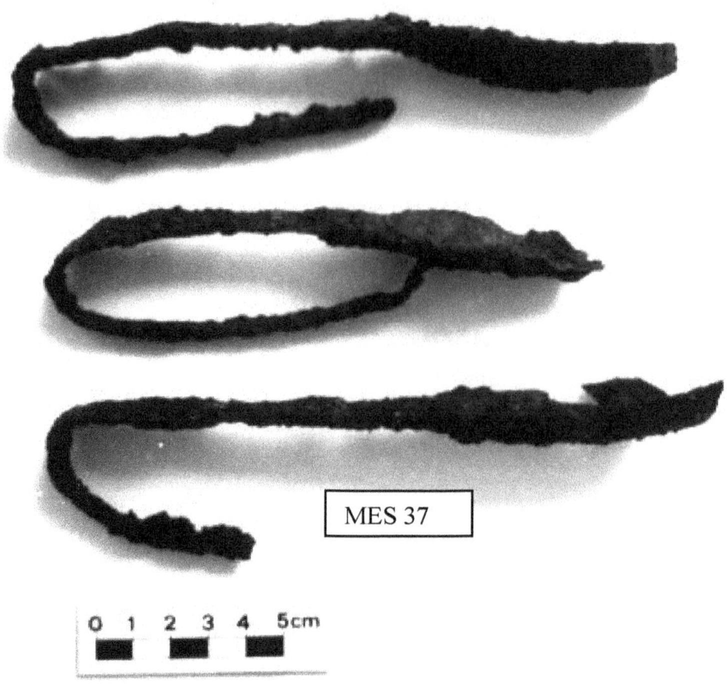

Strigil AVD 5 Optical microscopy x132 displaying welding of a low carbon sheet (middle F and P) to wrought iron (F on the edges). Slag inclusion indicate the two zones of contact between the sheets.

www.ingramcontent.com/pod-product-compliance
Lightning Source LLC
LaVergne TN
LVHW071206210926
848482LV00015B/735